Conservation of Ne[illegible]

Biological Resource Management in the Tropics

Edited by Michael J. Balick, Anthony B. Anderson, and Kent H. Redford

Anthony B. Anderson, ed., *Alternatives to Deforestation: Steps Toward Sustainable Use of the Amazon Rain Forest*

Anthony B. Anderson, Michael J. Balick, and Peter H. May, *The Subsidy from Nature: Palm Forests, Peasantry, and Development on an Amazon Frontier*

Michael J. Balick and Hans T. Beck, eds., *Useful Palms of the World: A Synoptic Bibliography*

Marianne Schmink and Charles H. Wood, *Contested Frontiers in Amazonia*

Conservation of Neotropical Forests

Working from Traditional Resource Use

Kent H. Redford and
Christine Padoch
EDITORS

Columbia University Press
NEW YORK

Columbia University Press
New York Oxford
Copyright © 1992 Columbia University Press
All rights reserved

Library of Congress Cataloging-in-Publication Data
Conservation of neotropical forests : working from traditional resource use/Kent H. Redford and Christine Padoch, editors
p. cm.
—(Biological resource management in the tropics)
Papers from a workshop requested and funded by the United States Office of the Man and the Biosphere Program and held in early 1989.
Includes bibliographical references and index.
ISBN 0-231-07602-9
0-231-07603-7 (pbk)
1. Man—Influence on nature—Latin America—Congresses.
2. Indians of South America—Economic conditions—Congresses.
3. Indians of Central America—Economic conditions—Congresses.
4. Rain forest ecology—Latin America—Congresses.
5. Rain forest conservation—Latin America—Congresses.
6. Forest management—Latin America—Congresses.
7. Traditional farming—Latin America—Congresses.
I. Redford, Kent Hubbard. II. Padoch, Christine.
GF514.C66 1992 92-4695

33.75′16′0913—dc20 CIP

Casebound editions of Columbia University Press books are Smyth-sewn and printed on permanent and durable acid-free paper.

Printed in the United States of America

c 10 9 8 7 6 5 4 3 2 1

p 10 9 8 7 6 5 4 3 2 1

Contents

Preface

Two of the greatest crises that humanity faces are the destruction of the tropical forests and the widespread poverty of the tens of millions of people who inhabit the rural areas of the globe. As many of these rural inhabitants live in, and make their livelihoods from, tropical forests, the fates of the forest and the rural poor are often inextricably linked.

A cross-disciplinary interest in this link between the forests and the people that inhabit them has recently emerged. Individuals interested in biological conservation have come to realize that the tropical forests are not "empty" and that conservation efforts must incorporate local human populations. Those interested in rural poverty and the rights of indigenous peoples have come to realize that, at least in the short term, their best allies are often the conservationists, who are interested in preserving vast areas of habitat. But cross-disciplinary cooperation, essential for any lasting success, has been hampered by lack of information on the interactions between forest peoples and the natural resources on which they rely. This book is designed to provide just such an information base for the lowland Neotropical forests. This volume, using the traditional division of rural inhabitants into "Indians," "*caboclos/ribereños*," and "colonists," examines only the first two groups, which may be regarded as traditional inhabitants of Neotropical forests whereas colonists are recent immigrants.

Conservation of Neotropical Forests: Working from Traditional Resource

Use contains contributions from a roster of the world's experts. It results from a workshop, Traditional Resource Use in Neotropical Forests, requested and funded by the United States office of the Man and the Biosphere Program, held in early 1989. The editors were asked to assemble a group of experts from both the natural and social sciences to present their work and review the state of the field. This charge, and this sponsor, shaped the form and content of the book.

Additional support for the workshop and preparation of this volume were provided by the University of Florida, the Program for Studies in Tropical Conservation, and the Tropical Conservation and Development Program. The original idea was brought to us by Marianne Schmink, who has steadfastly supported and encouraged us in all stages. Her intellectual and logistical support and advice were invaluable. Additional support was provided by Dr. Steven Sanderson and Dr. John Robinson, and an excellent cadre of graduate students: Connie Campbell, Chris Canaday, Jon Dain, Louie Forline, Amanda Jorgenson, Karen Kainer, Cynthia Lagueux, Marco Lascano, John Payne, Wendy Townsend, Ron Washman, and Sondra Wentzel, with particular thanks to Ann Edwards.

Contributors

ANTHONY B. ANDERSON
Banco Mundial, Sector Comercial Norte
Quadra 02, Lote A-Edificia
Corporate Financial Sector, Comjotus
303-304 Brasilia
Brazil

c/o The World Bank
1818 H Street NW
Washington, DC 20433

WILLIAM L. BALÉE
Department of Anthropology
Tulane University
New Orleans, LA 70118

F. WILLIAM BURLEY
P.O. Box 15
Powell Butte, OR 97753

JASON CLAY
Rights & Resources
Arlington, VA 22201

LILIANA C. CAMPOS DUDLEY
c/o Tanana Chiefs Conference, Inc.
122 First Ave. Suite 600
Fairbanks, Alaska 99701

DENNIS GLICK
Greater Yellowstone Coalition
P.O. Box 1874
Bozeman, MT 59715

ROBERT J. A. GOODLAND
Office of Environmental Affairs
Project Advisory Staff
World Bank
1818 H Street NW
Washington, DC 20433

SUSANNA B. HECHT
Dept. of Urban and Regional Planning
University of California, Los Angeles
Perloff Hall
405 Hilgard Ave.
Los Angeles, CA 90024

FLAVIO COELLO HINOJOSA
Metropolitan Touring
Av. Amazonas 239 y 18 de Sept.
P.O. Box 310, Suc. 12 de Oct.
Quito, Ecuador

MÁRIO HIRAOKA
Dept. of Geography
Millersville University
Millersville, PA 17551

EDVIGES MARTA IORIS
c/o Anthony Anderson
Banco Mundial, Sector Comercial Norte
Quadra 02, Lote A-Edificia
Corporate Financial Sector, Comjotus
303-304 Brasilia
Brazil

WIL DE JONG
Senior Scientist
Center for International Forestry Research
PO Box 6596 JKPWB, Jakarta 10065, Indonesia

HILLARD KAPLAN
Dept. of Anthropology
University of New Mexico
Albuquerque, NM 87131

BERT KLEIN
575 Wintergreen Avenue
Hamden, CT 06514

KATE KOPISCHKE
Dept. of Anthropology
University of New Mexico
Albuquerque, NM 87131

PETER H. MAY
Rua Senador Vergueiro, 232/1801
22230-001 Rio, RJ, Brazil

CAROLINA MURCIA
Wildlife Conservation Society
Colombia-Program
AA. 25527, Cali, Colombia

JAMES D. NATIONS
Vice President
Conservation International
1015 18th Street NW
Suite 1002
Washington, DC 20036

JORGE. E. OREJUELA G.
Chief, Environment and National Resource Are
Fundación para la Educación Superior
Calle 4a. No. 1-19
Aptdo. 5744
Cali, Colombia

CHRISTINE PADOCH
Institute of Economic Botany
New York Botanical Garden
Bronx, NY 10458-5126

DARRELL ADDISON POSEY
Oxford University
Oxford, OX1 3JA
United Kingdom

KENT H. REDFORD
Director for Biodiversity Analysis and Coordination
International Program
Wildlife Conservation Society
Bronx, NY 10460

JAN SALICK
Department of Botany
Porter Hall
Ohio University
Athens, OH 45601

MARIANNE SCHMINK
Center for Latin American Studies
319 Grinter Hall
University of Florida
Gainesville, FL 32611

ALLYN MACLEAN STEARMAN
Dept. of Sociology and Anthropology
University of Central Florida
Alafaya Trail
Orlando, FL 32816

MICHAEL WRIGHT
African Wildlife Foundation
1400 Sixteenth Street NW
Washington, DC 20036

Conservation of Neotropical Forests

ONE

Traditional Peoples and the Biosphere: Framing the Issues and Defining the Terms

MARIANNE SCHMINK, KENT H. REDFORD, AND CHRISTINE PADOCH

Over centuries traditional peoples have developed intimate knowledge of the tropical ecosystems on which they depend. Because of their cultural distinctiveness and the small scale of their economic systems—linked tenuously, if at all, to modern states and market systems—the intricacies of local resource-use systems in the tropics have been of interest primarily to anthropologists and social geographers. The stewardship of tropical resources has only recently become an issue of global concern. With a growing awareness of their potential to provide lessons for sustainable resource-use strategies, studies of traditional peoples have taken on a new importance.

During the 1970s and 1980s environmentalists and developers began to recognize the inextricable links between conservation and development. Pressure mounted for development agencies to reorient their perspectives to incorporate environmental conservation within their agendas, and conservation organizations explicitly affirmed their shift toward "development-oriented conservation" to promote sustained human use of resources. The World Conservation Strategy, published in 1980, and the Tropical Forestry Action Plan, published in 1987, document that change. There are now a number of projects in tropical countries (including several described in this volume) that are experimenting with ways to put these difficult goals into practice. One of the main challenges they face is understanding and working effectively with local peoples, a subject that is the major focus of this volume.

In the early 1980s the United States Man and the Biosphere Program (MAB) recognized the need to broaden the knowledge available about traditional forms of tropical resource use. This concern reflects MAB's goal of fostering sound interdisciplinary research on the relationships between people and their environment, in order to improve the scientific basis for policy formulation and problem-solving intervention. United States MAB began as the U.S. partner of the international MAB program directed by UNESCO, and has continued to participate informally since the United States withdrew from UNESCO (United Nations Educational, Scientific and Cultural Organization). MAB promotes an integrated, interdisciplinary, problem-focused (rather than discipline-focused) research approach to problems arising from the interactions between human activities and natural systems. It aims to provide a bridge between basic science and technological applications. Within this general philosophy, both international and U.S. MAB programs have placed considerable emphasis on understanding the complexity of interactions between local peoples and tropical forests (Hadley and Schreckinberg 1989).

United States MAB is run by a chairman and a national committee representing the scientific community and the various U.S. agencies that provide the program's financial support. The scientific direction of the U.S. MAB program is provided by several directorates, small committees whose members have interdisciplinary expertise and interest in specific ecosystems. During the 1980s, the MAB-1 Directorate for Tropical and Sub-Tropical Forests developed a series of activities with two goals: to expand understanding of traditional forest peoples and their resource use and to explore ways of implementing participatory conservation and development projects in the tropics. The strategy included a review and dissemination of MAB-supported tropical research results to date; a commissioned state-of-the-art review paper; a symposium at a national meeting to discuss MAB-sponsored tropical research; a workshop bringing together researchers and practitioners to discuss resource use by tropical peoples; and the publication of this volume of papers from that workshop. Through these activities the directorate sought to provide focus and intellectual leadership for research and action related to traditional forest-management issues.

In 1987 the MAB-1 Directorate brought together and published a summary of the results of a dozen projects carried out in tropical areas (Lugo et al. 1987). The directorate took the opportunity to reflect on how well this impressive research effort attended to MAB's overall program goals. This led to a reconsideration of two concerns that have persistently plagued the efforts of MAB and other organizations: how to more effectively integrate insights from the social and natural sciences, and how to link research more directly to the needs of policy and implementation.

The directorate found that tropical projects financed by the U.S. MAB program varied in their approaches to the human-biological interface. They ranged from pure biology, unrelated to human interventions, to those studies that thoroughly integrated the analysis of interacting social and natural systems. Intermediate approaches took as their point of departure the impact of human disturbance, but did not analyze the causes or patterns of the human behavior itself.

Studies that epitomized the MAB approach were those that focused on the interaction of social and biophysical factors in influencing localized, specific resource-use patterns. Such studies required a sensitivity to both social and natural principles and attention to the dynamics of their interactions. They were carried out by interdisciplinary teams, or by individuals with unusually broad training. Despite the potential value of their findings, these case studies tended to generate site-specific information whose broader significance remained unclear and whose applicability to policies and projects was addressed only in general terms.

The link between research and implementation was another general concern of the directorate's discussions. MAB had funded some biological studies directly related to environmental policy or specific management interventions, but no equivalent social studies. The directorate's review highlighted the need to broaden the scope of MAB to address key sociological and political questions related to resource management, based on a well-conceived scientific framework.

Next the MAB-1 Directorate decided to use its limited resources to commission state-of-the-art review papers on selected topics of pioneering importance. One of the priority issues was traditional management of tropical forests. Anthropologist Jason Clay was contracted to review the existing literature on indigenous tropical forest management, summarize the dispersed findings, analyze the results and their relevance to pressing management questions, and point to directions for future work. After wide circulation and discussion, the review paper with its extensive bibliography was published as a monograph (Clay 1988).

The workshop that produced the papers included in the present volume was the next step in the directorate's strategy to strengthen the knowledge base on traditional use of forest resources. U.S. MAB and the University of Florida (UF) provided financial support for the workshop, organized by Kent Redford (UF) and Christine Padoch (New York Botanical Garden), with assistance from Marianne Schmink (UF). Both Padoch and Schmink are anthropologists and were members of the MAB-1 Directorate. The workshop was cosponsored by the MAB-13 Directorate for Human Settlements, whose member Susan Fletcher helped to expand the workshop to address her directorate's interest in management of protected areas.

The workshop's goals were to bring together experienced researchers to 1) evaluate the strengths and weaknesses of research to date; 2) examine recent efforts by scientists and managers attempting to integrate local populations into management and decision-making processes concerning natural resource use and conservation; 3) identify gaps in existing disciplinary and interdisciplinary work and promising new directions for research (appendix 1.1); and 4) define a number of basic terms including "traditional," "management," and "traditional management" (see below). It further was designed to promote a dialogue between researchers and practitioners regarding priorities for future research and action. The results of these discussions are presented in the remainder of this volume.

The challenge of understanding the practices of traditional people, and of applying those insights to benefit conservation and development efforts, has continued to mount in the years since MAB-1 began its program of work in this area. Despite increased rhetoric, attempts at reform, and a growing number of actual projects in the tropics, many obstacles remain to the effective incorporation of traditional knowledge in the design and implementation of conservation and development efforts. This is particularly true in the inclusion of local peoples in MAB reserves (Chicchón 1991).

One part of this problem stems from the inadequacies of the concepts we use. "Man in the Biosphere" is a case in point. The "biosphere" concept has the merit of emphasizing global linkages but directs attention towards macrolevel concerns and processes at the expense of small-scale, local adaptations that are now of increasing interest. The use of the generic *Man* to refer to *people* reinforces an individual, static viewpoint in which such complexities as the division of labor by sex and social standing are likely to be obscured. The conservation community is far behind development practitioners in addressing gender differences, and in analyzing and incorporating the essential role women play in conservation and resource management.

There are also problems with our conceptualization of traditional tropical peoples and their practices. Although there is no ready substitute for it, the term *traditional* is misleading. The term derives from a world view that places what is "modern" at the other end of a continuum of progressive change from all that is traditional (and, by comparison, backward and unchanging). The optimism of postwar technological development reinforced the notion that, with modernization, everyone around the globe would become more alike as they increasingly subscribed to universal principles of "rational" decision making, leaving behind the outmoded and conservative customs of the past. Despite these problems, use of the term ͵ pervasive. Recognizing the importance of defining terms before discussing them, the participants in the workshop were asked to define *tradition/traditional*, as well as *manage/management* and *traditional management*.

The term *tradition* is defined by The American Heritage Dictionary as "a mode of thought or behavior followed by a people continuously from generation to generation; a cultural custom or usage" (Morris 1970). One of the problems with the definition is the degree to which traditional activities incorporate change: are Kayapó Indians who hunt with flashlights traditional hunters? Another is the suggestion that traditional groups are homogeneous in their behavior. Three definitions of *traditional* were proposed by the workshop participants: 1) historically constituted practices that respond to the local environment, to internal cultural demands and values, and to external forces; 2) practices which do not involve the use of technologies based on fossil fuels; and 3) "low-input, low-output" systems.

The second term, *manage* as in *management*, is defined by American Heritage as "to direct or control the use of; handle, wield, or use" (Morris 1970). To many people, *management* is held to be tne application of universally "rational" principles and is therefore both conscious and correct, by definition—but can there be "bad management"? And must management be conscious at the level of the individual performing the act, requiring intent and action? Or can it be the accumulated result of actions by generations? The choice between these two alternatives—an extremely important choice which determines how to interpret the sustainability of actions by individual humans involved in resource harvesting—generated much discussion amongst participants. There are clearly many distinct "rationalities" that may logically induce governments, social groups, or individuals to harvest resources in unsustainable ways (Schmink 1987). Is this, then, "management"? According to whose criteria should such judgments be made? How can management plans attend to competing interests in an equitable way without comprising conservation objectives? These are key issues in negotiating the trade-offs between market demands, political interests, the rights of native peoples, and the fate of the natural environment they inhabit.

Finally, the participants grappled with the term *traditional management*. Depending on the choice of definitions for the term *manage*, two alternatives were suggested: 1) a set of customary practices designed to modify an ecosystem with specific goals in mind; and 2) human intervention and or manipulation of the environment utilizing ideas or customs perceived as historically associated with group identity.

In the past, traditional management or "sustainable" resource use among many Neotropical indigenous peoples likely was due to low population densities and the lack of a technology that allowed them to destroy a resource. The important question now is not which traditional practices, as practiced in the past, are sustainable, but rather which conditions cause people to conserve their resources, and which conditions favor destruction, or overexploitation, of local resources.

Part of the recent interest in understanding the practices of "traditional" peoples stems from the recognition that the logic of many sustenance systems is not wholly governed by the market calculus we associate with developed societies. Traditional societies often operate according to different concepts of production, space, time, ownership, and wealth. Community goals and institutions may be stronger than the short-term considerations involved in individual market exchanges. To the extent that their survival, and that of their children, depends on the future of the resources in that same site, local communities may have a built-in incentive to use natural resources sustainably. The strength of these community-level incentives varies from community to community based on factors whose operation is not at all well understood.

"Traditional" people are often viewed as conservative people who avoid the risks entailed in innovation. By extension, they are seen as backward and unchanging. This misconception underlies the willingness of some conservationists to accept the presence of traditional peoples in protected areas, assuming that they will not change their time-honored, carefully honed customs and practices that are in harmony with the local environment. This assumption that traditional peoples will remain "ecological noble savages" (Redford 1990) is not only incorrect but potentially disastrous to the people themselves.

The studies in this volume, and many others published during the last decade, illustrate the incredible diversity of traditional resource-use patterns and the ability of traditional peoples to incorporate new methods, tools, and species into these systems. Astoundingly complex and diverse resource-use systems—and patterns of daily, seasonal, generational, and historical change—reflect the active adaptations of traditional peoples to their changing circumstances. The survival of communities over generations in their habitats is evidence of their success, especially when compared to other contemporary land uses in the Amazon. A diverse array of resource-use strategies is probably part of the key to this success. Technological shifts occur in response to interactions between internal and external demographic, socioeconomic, and political factors including migration, market penetration, sociopolitical battles, and resource depletion. Some of these strategies are undoubtedly more sustainable than others, yet very little is known about the dynamics of tropical resource use by local populations.

The pressures of rapid change on most traditional populations can be expected to keep increasing. Most of these pressures emanate from forces in the regional, national, and international arenas. To be sustainable, conservation and development strategies must recognize this constant change, even among the most remote forest peoples. The overwhelming diversity of situations makes it imperative to develop effective local strategies. The ideas and practices of traditional peoples are based on historical processes of adaptation to local settings, using principles and strategies whose effectiveness can be measured and tested. We need not assume that traditional peoples are perfect resource managers in order to recognize their important role as partners in the effort to improve the conservation of tropical forests.

There is an unfortunate tendency to see this partnership as a mandate to tap or appropriate indigenous knowledge. Because of its profound cultural and social roots, ethnobiological information is much more than simply another commodity. Traditional concepts, analysis, and experimentation can suggest new directions for research and policy, especially in cases when local people are active participants in resource management projects and, for example, can themselves be the experimenters. Traditional peoples can teach us a new appreciation for diversity in ways of seeing a changing world and our own place in it.

Recommendations from Participants

The following is a summary of the recommendations made by participants in the MAB workshop.

Topics for Research

1. Existing research consists primarily of site-specific case studies of traditional resource management systems. More case studies are needed in order to build towards more general models of ecosystems and extractive economies. Suggestions for types of case studies needed include:

- documentation and testing of indigenous experimentation to improve ecological and economic productivity of resource management systems.
- independent assessments of specific development interventions, protected or managed areas, and technology transfer.

2. More research is needed on the existing and potential uses of tropical forest flora and fauna (wildlife, plants, lesser-known hardwoods) for subsistence and to provide local and export markets. Such research should focus on: marketing systems; microeconomics and functioning of local extractive systems; and biological parameters of sustainability.

3. Studies are needed to analyze the impact of commercial fishing,

dams, deforestation, and mining on local populations and aquatic ecosystems.

4. Much greater research emphasis must be placed on the institutional, political, and socioeconomic context of local resource-use systems. Specific topics for needed research include:

- the impact of fiscal and regional policies on land use and decision-making by different populations.
- historical analysis of property regimes (not just land rights), including statutory and constitutional law in regard to property and access to resources.
- forms of conflict resolution and mediation of access rights.

Research Methods

1. Interdisciplinary or multidisciplinary team approaches need to be encouraged.

2. Applied research with direct (not just potential) applicability to specific problems of sustainable land use should be the priority. This requires an interdisciplinary approach, linked to research and extension.

3. Participatory or advocacy research methods should be developed, which directly involve local populations that are the object of research and which can provide benefits to those populations in the form of information, training, protection from the potential harmful effects of technology transfer, and compensation for the commercial value of their knowledge.

4. Research strategies should include better institutional linkages between local, national and international NGOs, public agencies, and research and extension organizations. Such an approach will build local institutional capability and stimulate interdisciplinary training.

Specific Recommendations for MAB

1. Change MAB's name to People and the Biosphere (PAB).

2. Evaluate the success of the biosphere reserve concept in practice. Encourage synthetic summaries and the development of useful models.

3. Forge stronger links with MAB programs in Latin American countries, including involvement in research activities.

4. Improve dissemination of results of MAB-sponsored research.

LITERATURE CITED

Chicchón, A. 1991. Can indigenous people and conservationists be allies? *TCD Newsletter* 23:1–5.

Clay, J. W. 1988. *Indigenous Peoples and Tropical Forests: Models of Land Use and Management from Latin America.* Cultural Survival Report 27. Cambridge, Mass.: Cultural Survival.

Hadley, M. and K. Schreckinberg. 1989. *Contributing To Sustained Resource Use in the Humid and Sub-Humid Tropics: Some Research Approaches and Insights.* Paris: UNESCO, MAB Digest 3.

Lugo, A., J. Ewel, S. Hecht, P. Murphy, C. Padoch, M. Schmink, and D. Stone. 1987. *People and the Tropical Forest.* Washington, D.C.: U.S. State Department, Man and the Biosphere Program.

Morris, W., ed. 1970. *The American Heritage Dictionary of the English Language.* New York: Houghton Mifflin.

Redford, K. H. 1990. The ecologically noble savage. *Cultural Survival Quarterly* 15(1):46–48.

Schmink, M. 1987. The "rationality" of tropical forest destruction. In J. C. Figueroa Colón, F. H. Wadsworth, and S. Branham, eds., *Management of the Forests of Tropical America: Prospects and Technologies,* 11–30. Rio Piedras, Puerto Rico: U.S.D.A. Forest Service, Institute of Tropical Forestry, Southern Forest Experiment Station.

PART I

Indigenous Peoples

Introduction

KENT H. REDFORD AND
CHRISTINE PADOCH

The resource management practices of indigenous tribal groups of South America have long been of interest to anthropologists and cultural geographers. Until recently, however, development planners and conservation experts paid little attention to the complexities of these traditional patterns. Dismissed as archetypical "primitive" practices that development projects sought to eliminate and condemned largely as the "slash-and-burn" agriculture considered anathema by conservationists, tribal resource use was examined in detail by few.

Popularly held views of traditional forest management practices have changed in recent years. The more benign, although equally false, belief that forest foragers and agriculturists are unfailing conservationists living in complete equilibrium and harmony with their environments is now commonly held by urban people concerned with tropical forest conservation. But even many professional foresters and development experts have come to realize that traditional folk and their misunderstood and maligned technologies may be among the most important allies and tools modern problem solvers have.

Whether regarded as the answer or the problem, however, rapidly changing traditional resource management practices remain little understood. The five essays in part I all explore the complexities of resource management by indigenous tribal groups in Neotropical lowland forests. They take the reader far beyond simple generalizations about equilibrium and perfect harmony. The first paper, by one of the recent pioneers of research into traditional resource management, Darrell A. Posey, lays out many of the intriguing problems and questions that the later papers discuss in depth. Using anecdotes from his own extensive field experience, Posey defines and examines some of the assumptions and misunderstandings that make ethnobiological research always challenging and make interdisciplinary collaborations often taxing. He notes the need for researchers and others to rid their thinking of "archaic dichotomies" between what is "natural" and what is "cultural" and urges both natural and social scientists to adopt a historical perspective in studying people and landscapes in the Amazon. Posey also addresses several questions that were central to the conference on which this volume is based. Among these are definitions of resource management and how management can be said to be occurring, even if the actors are not consciously "managing."

Posey goes on to suggest an agenda for future research, pointing out areas of traditional expertise that have yet to be examined or described in any detail. He concludes with a brief but provocative discussion of the uses of ethnobotanical research. He suggests that an important challenge to those interested in indigenous resource management is to ensure that indigenous people are recognized as intellectual authors and participants in the planning and development of products and programs based on their knowledge. Furthermore, it is imperative that international agreements be made that guarantee indigenous people just compensation for their intellectual property.

One of the points brought up by Posey is documented and discussed in the next essay, by William Balée. Reporting on the resource use practices of three foraging groups, the Guajá, Hetá, and Aché, Balée argues convincingly that these peoples, far from being foragers frozen in time, have undergone a process of "agricultural regression." While they practice no active cultivation now, their languages carry the echoes of an agricultural past, and their resource-use practices depend on the ecological remains of past farming. Using an impressive combination of research tools ranging from linguistics to ecology, Balée shows that both Amazonian peoples and their environments have complex histories. He documents the existence and use of "artifactual landscapes" such as old fallows and anthropogenic palm groves by present-day foragers. Stressing the inadequacy of evolutionary typologies that dominated anthropological thinking in the past, Balée argues for the use of "historical ecology" in further research.

This blurring of false dichotomies and inadequate typologies, suggested by Posey and documented by Balée, is among the interesting recent developments in studies of traditional resource management. The subjects of the next essay, the Awa of the montane forests of Colombia and Ecuador, are agriculturists, although their resource use defies classification into one simple category. Their subsistence system is highly complex and includes a variant of shifting cultivation known as slash-and-mulch agriculture, as well as diverse agroforestry fields, the raising of animals, hunting, fishing, gathering, and other activities. After outlining Awa production systems and the group's impressive knowledge of fauna and flora, author Jorge E. Orejuela G. goes on to discuss the place of the Awa in regional society and as integral participants in the Binational Awa Biosphere Reserve. He sees the goal of the reserve as not merely the protection of the area's wildlands, but also the improvement of the lot of the Awa and their nontribal neighbors.

The last two papers in this section return to the discussion of foraging technologies among South American groups, focusing again on diversity and change. Hillard Kaplan and Kate Kopischke describe the variety and flexibility of traditional subsistence practices, presenting detailed data on the Aché of Paraguay and two communities of Machiguenga in Peru. They document the variety of adjustments traditional peoples make to new technologies and to new environmental challenges. The authors then go on to comment, on the basis of their extensive and detailed field observations, on research and policy implications for conservation and development among "highly traditional" people. They argue that indigenous resource use practices are never static and that any attempt to preserve native cultures "must be tempered by a respect for native peoples to make their own choices and respond to opportunities as they see them."

Kaplan and Kopischke end their article with a commentary on the extent to which native peoples actively manage their environments to ensure continued resource availability. Their conclusion, that conservationist patterns are highly variable, again supports the emphasis on diversity and the importance of history found throughout this section.

The last essay is a discussion by Allyn Stearman of hunting by three Bolivian indigenous groups—the Sirionó, Chimane, and Yuquí—and their relations with nontribal neighbors. Basing her analysis on extensive field research among each group, Stearman points out that specific environmental and historical situations result in dramatically different indigene-settler interactions. Stearman's conclusions emphasize diversity, complexity, and the importance of site- and situation-specific understanding, echoing the findings of the other contributors to this section. She demonstrates most effectively that "the presence of populations contiguous to indigenous groups does not inherently constitute a threat to foraging activities." Among her provocative conclusions are that ranching, when practiced in appropriate tropical environments, need not be harmful to the environment or to native populations.

TWO

Interpreting and Applying the "Reality" of Indigenous Concepts: What Is Necessary to Learn from the Natives?

DARRELL ADDISON POSEY

Several scientists in the MAB workshop have studied the role of indigenous peoples in the modification, molding, and conservation of environmental landscapes. They have convinced even their most skeptical colleagues that traditional knowledge is a rich source of information about the environment. But some nagging problems remain regarding the extent to which native peoples are "conscious" of their management activities, and indeed, if they have any "real" concepts of ecological conservation at all. These questions are at the center of the most fundamental problems that have driven the evolution of anthropological theory and practice since the inception of cross-cultural investigation: how to separate scientific interpretation from the "reality" of those societies under study.

The structural and functional theorists that dominated anthropology for many decades of this century established the "right" of researchers to "flesh out" structures and activities not "seen" by the native and even to interpret behavior in ways not agreed upon, or in some cases acceptable to, the local society. Thus marriage could be interpreted as an economic union between two kin groups rather than the love match espoused by the newlyweds. And schools could be shown to function to keep young adults off the job market rather than quench their thirsts for knowledge, as argued by jobless teachers. All societies have their myths, and anthropologists work to demythologize human culture, exposing the structural elements that function to preserve societies.

Emic and Etic Distinctions

In an attempt to separate cultural interpretations by the investigator from explanations by the native, anthropologists and ethnobiologists adopted the linguistic distinction between *emic* and *etic*. *Emic* interpretations reflect cognitive and linguistic categories of the natives, whereas *etic* interpretations are those that have been developed by the researcher for purposes of analyses.

It would be nice if such distinctions clearly existed. After twelve years of studies of the Kayapó Indians and their environmental management activities in southern Pará state, Brazil, however, I have learned and relearned the dialogue between researcher and informant all too often obscures these categories. This is only natural, since individuals from two different cultures inevitably think and speak with different cognitive "realities." For mutual interpretation to occur, sharing of realities must also occur.

Sharing Realities

I have written extensively about Kayapó planting along trailsides, in old fields, in "forest islands," at old village sites, and other locations (Posey 1985). The concept of "planting" obviously implies a certain degree of consciousness on the part of indigenous peoples. But some Kapayó do not agree with my use of the word *planting* to describe their activities.

Recently one of the young chiefs, who can read Portuguese, commented on an article I had published in a Brazilian journal about Kayapó planting. "We don't really do that," he told me.

I responded in a manner considered most unbefitting of an ethnobiologist. "I'm terribly sorry," I said, "but you're wrong. You do do that!"

I continued to defend my position: "I have seen many nondomesticated plants being taken out of the ground by your people and then planted in their backyards and along the trailsides. I've seen Beptopoop take orchids and epiphytes from the forest and tie them on trees near his medicinal garden. And I've watched many times Uté and

Kwyra-ka take tubers and seedlings from way over there and plant them along the trails near the village."

"Yes," said the chief, "but these are 'natural' plants. They grow in the forest anyway. They are not planted; they are natural." He then continued in his critique: "We do not plant piquia (*Caryocar villosum*) trees, as you said we do. This is not the kind of plant that we plant."

"But," I argued, "I've seen the people throw piquia seeds into the holes and I've also seen them step on those seeds (to push them into the soil)." My defense continued: "I've had people tell me: 'If you step on the seeds, then they'll grow. If you don't, then they don't come up.' "

"Yes," he somewhat relented, "but that is not planting—that is something else completely different."

For me, these were all obvious examples of conscious planting. But not for my Kayapó friend. He insisted that only plants that could not grow without the help of humans are planted; all other species are "natural."

We then proceeded to spend most of the evening discussing why it was that I thought that the Kayapó did something that he contended they did not do. We finally came to the source of the problem: our differing concepts of *domesticated* and *natural*.

My Kayapó friend was using concepts that were more restrictive than mine. To him, "domesticated" plants are those species that would not exist if Indians did not plant them. Knowledge about the care and propagation of such domesticated plants are handed down through generations. All other plants that can survive alone in the forest or savanna are "natural," and, consequently, not "planted"—even though their seeds, tubers, or cuttings might be intentionally spread great distances from site to site and reflect millenia of genetic selection by the Kayapó.

Thus most of the fruit trees and medicinal species that I have described as planted by the Kayapó belong, to my Kayapó friend, to the "natural" category and are, therefore nonplantable.

I would defend my "rights" as an anthropologist, however, to contend for purposes of analysis that these species are indeed intentionally managed and planted.

What I had learned most from the dialogue with my Kayapó friend was that my concept of "planting" was much broader than his and that to understand the emic view of the matter, I had to return to a cognitive analysis of Kayapó words and expressions related to seed scattering, tuber replanting, epiphyte spreading, and a potential field of many additional unknown, unnamed, or unimagined (by me) categories.

Consciousness of Reality

Sometimes the "consciousness" of knowledge is just a matter of putting an abstract label on a well-known but unconscious nonverbalized phenomenon. Edward T. Hall in *The Hidden Dimension* (1966) points out how culturally different perceptions of space change social interactions. Living in a Latin culture, for example, one quickly learns that intimate spaces between people are closer than in northern European cultures. Latins are seen as pushy or forward by gringos, while the gringos are considered cold and formal by Latins. Different perceptions of space are to a large extent to blame for these generalized differences in behavior.

Behavior in elevators teaches the conscious observer much about the organization of unconscious interpersonal space. Observers need only a book like Hall's to become sensitized to the well-known but generally overlooked behavior of people in elevators. As the number of users of the restricted space changes, "commonsense" rearrangements of the individuals occur to reestablish the proper interpersonal space.

So it is with the consciousness of native peoples and their management practices. The native can become conscious of certain commonsense acts of management when alerted to the phenomenon by the researcher, as in the case of observing an Indian throwing seeds on the ground and stepping on them.

"Yes, we do that," he may say, "but that's not planting or management." At least the phenomenon is agreed upon by researcher and the people researched, even if the emic and etic terminologies are different. But the informant too learns of the categories being used by the researcher and may modify his own way of looking at his culture.

The distinction between interpretation and reality becomes even more complicated when higher levels of abstraction such as the notion of spirits and mythological beings or forces come into play. Native peoples generally make a point of saying that the forest, for them, is not just an inventory of natural resources but represents the spiritual and cosmic forces that make life what it is. What, then, is the role of metaphysical concepts in management practices?

The Kayapó, for example, believe that old, abandoned village sites are full of spirits. Fear of spirits puts these old sites off limits for many Indians. Only those who deal with spirits—shamans—and special hunting parties go to these sites. Thus these abandoned camps and villages effectively become protected reserves with a high diversity of secondary growth that also attracts many animals. The spirits effectively serve as ecological protective agents.

Consider what the Kayapó call the *pitu* plant (actually a grouping of plants from several botanical families). *Pitu* is one of the few plants that is thought to have a spirit—a very powerful spirit that once killed thousands and thousands of Indians. The spirit of this plant is so dangerous that if people go near it or touch it they will die. It is said, however, that *pitu* can be planted by shamans in specialized, secret medicinal gardens. Fear of coming into contact with *pitu* is sufficient to keep unwanted guests out of these gardens and to guarantee the secrecy of the gardens and their contents.

An etic interpretation of *pitu* maintains that the fear of its spirit functions to protect medicinal gardens and restrict use to specialists of medicinal plants. It would be impossible, however, to find an Indian who would say, "Well, yes, the *pitu* spirit functions in our society as an ecological management agent to protect our medicinal plantations."

So, we return to the initial problem of interpreting the "reality" of native peoples. It is the problem of emic versus etic analyses that has scared biologists and ecologists away from anthropology, which they consider unscientific. Botanists and zoologists do not, after all, have to confirm their scientific analyses with their biological subjects. Having only an etic level of analysis makes scientific investigation much easier. It is important, however, not to confuse scientific "objectivity" with the obscuring of "reality." There is much to learn from the interpretation of indigenous myths, legends, and folk taxonomies, whether or not the methods meet the rigorous standards of some scientists.

A raging debate is occurring now among anthropologists as to whether cultural interpretation can ever become scientific. To the ethnobiologist, the debate is rather inane. Ethnobiologists attempt to use all of the scientific tools that can be borrowed from botany, zoology, geography, pedology, genetics, ecology, and other disciplines. But that does not mean abandoning the quest for an emic view as well. If searching for native "realities" noses a bit too far into metaphysics or "fuzzy science" for some hardliners, then we must conclude that not everyone can or should be an ethnobiologist. We cannot forget, however, that attitudes regarding the nature of "true science" seriously divide the social and natural sciences.

Methodological Barriers

One of the greatest barriers to interdisciplinary scientific investigation is the different time frames for research used by social and natural scientists. Biologists consider a few months to be a reasonable field period, whereas anthropologists think in terms of years in order to master a language well enough to delve into native perceptions of natural resources, concepts of management, mythological forces, and other levels of conscious or unconscious knowledge. To cope with this fundamental difference alone, it is justifiable to develop a hybrid field of ethnobiology that trains students to weigh as equally important the cognitive analyses of semantic fields and the gathering of basic geological and ecological data.

Another fundamental barrier to research into traditional knowledge is the methodological problem of assessing the degree to which knowledge is shared within a society. Even in the smallest of societies, individuals do not know the same things. Scientists who have worked with native peoples have learned this painfully—or ignored it in their data analyses.

The Kayapó and other Ge peoples, for example, have highly specialized knowledge. In the Kayapó village of Gorotire, 26 percent of the population of approximately seven hundred are medicinal curing specialists. Each specialist knows certain types of animal spirits that provoke diseases and can be treated only with a specific array of medicinal plants, magical songs, and curing rituals.

Roughly 15 percent of Gorotire inhabitants can identify and name at least thirty-five species of stingless bees (Meliponidae). The remaining 85 percent have difficulty recognizing more than eight. But some specialists can tick off sixty-five species, including details of their morphological characteristics, nesting habits, flight patterns, seasonal production of honey, and varied uses of their wax, bitumen, pollen, and honey.

To complicate matters, specialists frequently do not agree on details of their knowledge. Two specialists on "fish disease" (*tep kane*), for example, may vehemently disagree about which method of preparing "fish medicine plants" is most effective—or even which plants can be used for which type of the disease.

The Hypothesis-Testing Bridge

These methodological problems can be handled by trying to construct statistically significant survey and analytical methods to describe "typical" Kayapó knowledge, but such endeavors are nightmares for field researchers and result in questionable benefits. If it is the detailed knowledge of biological and ecological knowledge that is of interest, then careful documentation, checking, and cross-checking to find anomalies and contradictions between informants is adequate to advance ethnobiological research. More important, it is adequate to advance hypotheses (Posey 1986).

Most ethnobiological studies have tended to search in native knowledge only for what is already known from science. So we look for categories of plant use, animal behavior, ecological relationships, soil types, and landscapes that already exist in our own knowledge system.

Utilizing indigenous concepts, however, shortcuts or even breakthroughs in scientific investigation can occur through the quite appropriate scientific method of generating and testing hypotheses. No ethnobiologist has ever insisted that traditional knowledge be taken at face value but rather that such statements be used to guide researchers to look for unknown categories of knowledge or unknown relationships, i.e., to propose hypotheses to test indigenous concepts.

In this manner, new species and subspecies of bees have been "discovered" from native bee specialists, interesting active compounds have been isolated in laboratories as a result of ethnopharmacological research with medicinal curers, diets of animals have been analyzed with the help of skilled hunters, pioneering behavioral studies of little-known species have been carried out with the help of native specialists, and soil-plant-animal complexes have been learned from veteran agriculturalists.

The decisions scientists make in proposing hypotheses based upon indigenous knowledge reveal the arbitrary nature of this basic step in the scientific pursuit, since the researcher must frequently discard from his formulations the "unlikely" or "unbelievable" elements of informants' statements. But what is "unlikely" and "unbelievable" more often reflects the researcher's inability to grasp native "reality" than any real scientific criteria. Nonetheless, the proposal and testing of hypotheses provides the methodological and theoretical bridge necessary to link scientific research with traditional knowledge.

Advances Against Archaic Dichotomies

Despite the many problems that complicate the study and use of traditional knowledge, some important advances have been made. Pioneering emic studies of indigenous knowledge and management practices are appearing (Alcorn 1984 and 1989; Boster 1984; Conklin 1957; Carneiro 1978; Johnson 1989; Ribeiro and Kenhiri 1989; Salick 1989; and others) to reveal the sophistication of folk knowledge. More and more evidence is now available to show that what were once considered "natural" landscapes are really "human artifacts" (Alcorn 1981; Anderson and Posey 1985; Frickel 1959; and others). And very slowly the deeply embedded concept of "abandoned fields" is being replaced with data on field successions and their long-term roles in natural resource management (Balée 1989; Denevan and Padoch 1987; Irvine 1989; Posey 1985; and others).

Yet most of these advances continue to be ignored in general scientific literature. Readers continue to be plagued with archaic dichotomies. Plants must be either domesticated or wild, native peoples must be either hunter-gatherers or agriculturalists, and agriculture is considered mutually exclusive of agroforestry. Some years ago I coined the term "nomadic agriculture" to describe the widespread manipulation of "semidomesticated" plants by the Kayapó Indians (Posey 1983). The term was intended to emphasize that most plants used by the Kayapó are not domesticates, nor are they wild, if wild means uninfluenced by human manipulation. More than 76 percent of the species used by the Kayapó that are not domesticates are nonetheless systematically selected for desirable traits and propagated in a variety of habitats. During times of warfare, the Kayapó could abandon their agricultural plots and survive on the semidomesticated species that had been scattered for millennia in known spots throughout the forest and savanna. Old field sites became hunting preserves and orchards, since as younger fields they had been planted by the Indians to mature for such purposes. In other words, agricultural plots were designed to develop into productive agroforestry plots dominated by semidomesticated species, thereby allowing the Kayapó to shift between the activities usually designated as those of either agriculturalists or hunter-gatherers. Such patterns appear to have been widespread in the lowland tropics and make archaic the traditional dichotomies, which, unfortunately, continue to persist.

Future Research and the Discovery of History

Far too many biologists and ecologists still assume ecological systems to be "natural" instead of investigating the historical and prehistorical activities of humans in the region. This ignores a sizable literature documenting the widespread effects of humans on environmental communities that is reflected, among other ways, in the genetic diversity of a region. Future biologists will have to become aware of anthropological, archaeological, ethnobiological, and historical research in order to produce credible results.

Many of the assumed historical processes that produced today's landscapes, are, in fact, current processes as well. The formation of *terra preto do indio* (Indian black earth) can still be observed in Amazonian Indian villages. Kayapó ethnopedological studies (Hecht and Posey 1989) have shown that even some agricultural soils should be considered human artifacts due to their extensive manipulation by indigenous agriculturalists.

Likewise the processes of genetic selection among thousands of species is currently underway throughout the Amazon (Clement 1989; Kerr and Clement 1980; Kerr 1987; Patiño 1963). Domestication, therefore, is not just a historical question but a dynamic one that can be studied today in many Indian villages. Yet we continue to read mostly theory about domestication, as though it were only some archaic process to be encountered in libraries.

Suggestions for the Future

The following suggestions are offered to guide future research into traditional natural-resource management techniques and their applications:

1. Genetic manipulation of flora and fauna by native peoples remains relatively little known. Research into selection procedures, decisions behind selection choices for different species and varieties, inter- and intratribal variations in selection of variables, and evolutionary consequences of differential selective behavior should be systematically studied.

2. More extensive studies should be made to describe how native peoples modify landscapes and environments.

2.1. Fire is immensely important in almost all traditional management systems, yet few details of fire use are available (when to burn, what can or cannot be burned and when, temperature of burn, frequency of burning, protection from burning, products of burning—use of charcoal, ashes, stumps, charred root systems, etc.).

Likewise, little is known of the effects of fire on biological communities (which species are destroyed, which are stimulated, effects of burning on blooming and fruiting timing and production, modification of morphological structures due to burning, etc.).

2.2. Most research to date has focused on indigenous manipulation of forests and agricultural plots. Yet evidence exists showing that other habitats may also be significantly modified. Scrub forests and savannas, for example, are certainly molded by use of fire. Creation of "forest islands" (*apete*) in *campo-cerrado* (scrub-grassland) by the Kayapó shows that ecological communities can even be human-made with a high diversity of useful plants from very distant areas. Hillsides, trails, and even rocky outcroppings are also modified to maximize resource availability (Posey 1985). To get a more complete picture of indigenous adaptation and management, future research should emphasize investigations into these lesser-studied habitats.

2.3. Rivers, streams, and seashores also can be modified by indigenous activities. Although extensive documentation of the dependence of native peoples on fresh and salt water species exists, only scattered evidence (Chernela 1989) is available on how native peoples manage these important resources.

3. Baseline studies are necessary to establish wildlife population

composition, numbers, and carrying capacities. Many researchers have observed that old fields are favored hunting areas for Indians, and I have proposed that where old-fallow management is practiced, wildlife populations may indeed be higher in number and in species diversity. This hypothesis goes against traditional thinking that maintains the innate destructiveness of Indians to wildlife populations. To prove or disprove the thesis, baseline data from uninhabited areas must be available for comparison with managed inhabited areas.

4. Ethnolinguistic studies can provide valuable data on the historical relationships between peoples and the botanical interchanges that occurred between them (see Balée, page 41), or the cognitive geographical maps that link resources with the physical world. Myths, legends, ceremonies, rituals, and songs are filled with ecological and biological information, but very few studies have systematically analyzed their content.

5. More long-term sequence studies of fallows and other managed habitats are needed to understand the ecological transitions that accompany growth. These studies, coupled with knowledge of the natives' uses of resources in the growth sequences, can give a much clearer picture of options for forestation and reforestation schemes.

6. Multidisciplinary research teams should be encouraged, despite the conceptual and time difficulties that separate disciplines. Focusing upon the study of traditional knowledge can indeed be used to reunite the fragmented disciplines of science.

7. Special ethnobiological academic and training programs should be established to develop interdisciplinary fields, combining the methods and techniques of ecology, biology, anthropology, and linguistics. Ethnobiological research centers should also be established to coordinate multidisciplinary research and design programs for the application of traditional knowledge. These centers should include laboratories for the analysis of medicinal compounds, the nutritional value of edible plants, the chemical characteristics of natural fertilizers, and so forth.

Application of Traditional Knowledge

Major international policy shifts by governments, lending institutions, research institutions, private foundations, and industry should occur to support research into traditional knowledge use and its potential applications in solving modern world problems. Applied research centers need to be established so that experimental plots, analytical laboratories, and field research stations can investigate methods of sustained resource management based on indigenous models.

New categories of plant and animal use (as cosmetics, alternative building materials, and other natural products) should be investigated,along with international markets for these products. Pragmatically, if economic value cannot be given to the myriad natural products—and the people who know how to propagate, prepare, and use them—then there is little chance to save the planet's remaining environments and native peoples.

Once the diversity of native products and their market potential are known, it will be possible to design reforestation and forestation projects that are productive in all stages of their development. A real challenge rests with the implementation of forestation programs and forest reconstruction projects that include native peoples as *intellectual* participants in all stages of project planning and implementation.

Intellectual Property Rights

Headway has been made in convincing the world that native peoples have much to teach us about the ecological and biological diversity of the planet. Until there are international agreements to protect the intellectual property rights of indigenous peoples and to compensate them for their knowledge, however, it will be difficult to proceed ethically much further in the application of traditional knowledge. Development of these policies and procedures should be given highest priority.

The Danger of Romanticism

There are lingering Rousseauistic notions that "true" native peoples are totally adapted to their environment and languish in perfect harmony with nature. If natives begin to wear clothes and wristwatches, drive cars, use video cameras, and drink Coke, then they have been polluted and acculturated and have nothing more to tell us.

This notion is both incorrect and dangerous. Even in the most acculturated and devastated societies valuable knowledge remains. *Ca-*

boclos (Amazonian peasants), for example, have an enormous amount to teach us. In fact, *caboclos* have been applying indigenous models to change and modernize situations for centuries. Romantic idealism must not be allowed to blind researchers to the remnants of traditional knowledge that may in fact be close to home.

Finally, those who study traditional knowledge and attempt to find its modern applications are not proposing that the world revert to a tribal existence. We are merely calling out to any and all that will listen to help stop the senseless destruction of the planet's natural resources and the native peoples who best know them. There are options for the survival of humankind in the biosphere, and many of the options lie encoded within the "realities" of native peoples.

LITERATURE CITED

Alcorn, J. B. 1981. Huastec noncrop resource management: Implications for prehistoric rain forest management. *Human Ecology* 9:395–417.

—— 1984. *Huastec Mayan Ethnobotany.* Austin, Texas: University of Texas Press.

—— 1989. Process as resource: The traditional agricultural ideology of Bora and Huastec resource management and its implications for research. In D. A. Posey and W. Balée, eds., *Resource Management in Amazonia, Advances in Economic Botany* 7:63–77 New York: New.

Anderson, A. B. and D. A. Posey. 1985. Manejo de cerrado pelos Indios Kayapó. *Boletim do Museu Paraense Emilio Goeldi, Botanica* 2(1):77–98.

Balée, W. L. 1989. The culture of Amazonian forests. In D. A. Posey and W. Balée, eds., *Resource Management in Amazonia, Advances in Economic Botany* 7:1–21. New York: New York Botanical Garden.

Balée, W. L. and A. Gély. 1989. Managed forest succession in Amazonia: The Ka'apor Case. In D. A. Posey and W. Balée, eds., *Resource Management in Amazonia, Advances in Economic Botany* 7:129–148 New York: New York Botanical Garden.

Boster, J. S. 1984. Classification, cultivation, and selection of Aguaruna cultivars of *Manihot esculenta* (Euphorbiaceae). In G. T. Prance and J. A. Kallunki, eds., *Ethnobotany in the Neotropics Advances in Economic Botany* 1:34–47. New York: New York Botanical Garden.

Carneiro, R. 1978. The knowledge and use of rain forest trees by the Kuikuru Indians of central Brazil. In R. Ford, ed., *The Nature and Status of Ethnobotany,* Anthropology Papers, no. 67, pp. 210–216. Ann Arbor, Michigan: Museum of Anthropology, University of Michigan.

Chernela, J. 1989. Managing rivers of hunger: The Tukano of Brazil. In D. A. Posey and W. Balée, eds., *Resource Management in Amazonia, Advances in Economic Botany* 7:238–248. New York: New York Botanical Garden.

Clements, C. R. 1989. A center of crop genetic diversity in western Amazonia: A new hypothesis of indigenous fruit-crop distribution. *Bioscience* 39(9):624–630.

Conklin, H. C. 1957. *Hanunoo Agriculture: On an Integral System of Shifting Cultivation in the Philippines*. FAO Forestry Development Paper 12. Rome: FAO.

Denevan, W. M. and C. Padoch, eds. 1987. *Swidden-Fallow Agroforestry in the Peruvian Amazon. Advances in Economic Botany,* Vol. 5: New York: New York Botanical Garden.

Frikel, P. 1959. Agricultua dos Indios Munduruca. Antropología. *Boletim Museu Paraense Emilio Goeldi,* 8:1–41.

Hall, E. T. 1966. *The Hidden Dimension*. Garden City, N.Y.: Doubleday.

Hecht, S. B. and D. A. Posey. 1989. Preliminary results on soil management techniques of the Kayapó Indians. In D. A. Posey and W. Balée, eds., *Resource Management in Amazonia, Advances in Economic Botany* 7:174–188. New York: New York Botanical Garden.

Irvine, D. 1989. Succession management and resource distribution in an Amazonian rain forest. In D. A. Posey and W. Balée, eds., *Resource Management in Amazonia, Advances in Economic Botany* 7:223–237.

Johnson, A. 1989. How the Machiguenga manage resources: Conservation or exploitation of nature? *Advances in Economic Botany* 7:213–222. New York: NYBG.

Kerr, W. E. 1987. Agricultura e seleções geneticas de plantas. In B. Ribeiro, ed., *SUMA Etnologica Brasileira,* Petropolis, Brazil: Vozes/Finep.

Kerr, W. E. and C. R. Clement. 1980. Praticas agricolas com consequencias geneticas que possibilitariam aos Indios da Amazonia uma melhor adaptação as condições ecologicas da região. *Acta Amazonica* 10:251–261.

Patiño, V. M. 1963. *Plantas cultivadas y animales domesticos en America equinoccial*. Cali, Colombia: Impresa Departmental.

Posey, D. A. 1983. Indigenous knowledge and development: An ideological bridge to the future. *Ciencia e Cultura* 35(7):877–894.

—— 1985. Indigenous management of tropical forest ecosystems: The case of the Kayapó Indians of the Brazilian Amazon. *Agroforestry Systems* 3(2):139–158.

—— 1986. Topics and issues in ethnoentomology with some suggestions for the development of hypothesis-generation and testing in ethnobiology. *Journal of Ethnobiology* 6(1):99–120.

Ribeiro, B. G. and T. Kenhiri. 1989. Rainy seasons and constellations: The Desana economic calendar. In D. A. Posey and W. Balée, eds., *Resource Management in Amazonia, Advances in Economic Botany* 7:97–114 New York: NYBG.

Salick, J. 1989. 7 189–212. Ecological basis of Amuesha agriculture, Peruvian upper Amazon. In D. A. Posey and W. Balée, eds., *Resource Management in Amazonia, Advances in Economic Botany.* New York: NYBG.

THREE

People of the Fallow: A Historical Ecology of Foraging in Lowland South America

WILLIAM BALÉE

Ecological studies often presume that the habitats of lowland South American foragers are somehow "natural" or pristine. Modern foragers (hunter-gatherers) are etched in anthropological minds as being the few remaining people of the earth who use no agriculture. The underlying assumption is that hunter-gatherers exploit "wild" resources over whose reproduction and distribution human beings have little influence (Foley 1987:75). Since hunting and gathering was the only means of subsistence throughout most of human evolution, research on modern foragers putatively can help elucidate patterns of resource use by preagricultural hominids (Foley 1987:76–77; Lee and DeVore 1968).

These assumptions about foragers derive from cultural ecology and evolutionary ecology. In both paradigms, a strategic ahistoricity has prevailed. Such guiding assumptions about foragers in lowland South America are at best partly mistaken and at worst, gross distortions of historical processes. In addition, historical ecology is a more powerful tool for explaining patterns of resource use by lowland South American foragers than is either cultural ecology or evolutionary ecology. In anthropology, widely held views on resource use by tropical South American foragers date from the work of Julian Steward, if not earlier. In his major theoretical work, Steward envisioned cultural ecology as being "an heuristic device for understanding the effect of environment upon culture" (1955:30). In his writings on

South American Indians, not only was Steward uninterested in the possible effects of culture upon the environment, but he also showed no concern with the possible interpenetration of the two. On these questions, modern evolutionary ecology offers no quibble with Steward. Evolutionary ecology accepts the primacy of environment over culture and the treatment of the two as alienated subject and object (Balée 1989). This ultimately derives from reductionism in evolutionary biology itself, wherein "organism" and "environment" represent segregated, noninteractive entities of research (Levins and Lewontin 1985:134).

Steward's adherents codified the primacy of environment over culture in terms of "environmental limiting factors." Steward (1949:609; Steward and Faron 1959:293, 453–454) had specified that soil exhaustion "required" village movement and that the "need" to relocate periodically swiddens and/or the village itself precluded "community stability" of tropical forest Indians. This anticipated, of course, Meggers' theory (1954; 1971) that Amazonian soils had restrained cultural development in lowland South America, given that community stability is a prerequisite for the development of complex society. Meggers interpreted extinct prehistoric Amazonian chiefdoms in the archaeological record, for example, as being the relics of introduced civilization, which could not have survived as such, in the tropical forest (Roosevelt 1987:161).

Steward (1947; 1949) proposed an essentially evolutionary typology of South American Indian means of economic production (Willey and Sabloff 1980:175). Steward classified foragers as "marginal tribes" who mainly inhabited zones that were unfavorable for agriculture (Steward 1947; Steward 1949; Steward and Faron 1959). Steward (1947) suggested that foragers (or trekkers) living in lands favorable to agriculture (those of the "tropical forest tribes") had borrowed certain cultural traits from horticulturalists. I would not deny that lowland South America, including Amazonia, supplies evidence for preceramic, prehorticultural society. But Steward's "marginals" were mostly known from colonial and postcolonial times, and many were coinhabitants of the tropical lowlands with horticultural peoples. The presence of foragers in arable habitats probably may be best explained by nonenvironmental considerations (Lyon 1985:3). Such considerations belie an evolutionary typology of modern lowland South American Indians.

The alienation of culture from environment in cultural ecology and evolutionary ecology, in attempting to establish a historical approach to foraging in the tropical lowlands of South America, obscures the fact that foragers use not merely pristine forests, but agricultural zones, especially old fallows and the culturally dependent biological resources therein encountered. Some foragers of the tropical lowlands, moreover, apparently regressed from an earlier phase of horticultural society, because of historical and sociopolitical forces. It seems unlikely, hence, that study of their patterns of resource use would reflect pre-Neolithic patterns, even in South America. In addition, since these foragers depend on the resource concentrations of old fallows, the habitats they exploit must be qualitatively and quantitatively distinct from those of pre-Neolithic peoples of the tropical lowlands.

Agricultural Regression in Lowland South America

Lowland South American Indian societies affiliated with the Tupi-Guarani language family are remarkably unalike (Viveiros de Castro 1986:106). The Tupinambá and Guarani societies of late prehistoric and early colonial times displayed paramount chiefdoms (Balée 1984a; Clastres 1973; Dean 1984). But many others such as the Hetá, Aché, and Avá-Canoeiro exhibited no social units more complex than nomadic bands (Kozák et al. 1979; Clastres 1972; Toral 1986). The Tupinambá occupied villages with an average population of 625 people. The modern Tupi-Guarani speaking Ka'apor, in contrast, have average village sizes of only 33 (Balée 1984a:254; 1988a:167), while Guajá bands number only between 5 and 15 persons (Balée, 1988b:48). Although the Ka'apor prescribe the ritual consumption of tortoise meat for girls at puberty, menstruating women, and men and women in the couvade (Balée, 1985), the linguistically related Guajá actually forbid this meat to such ritually restricted persons. The list of infrastructural, structural, and superstructural differences between Tupi-Guarani societies, therefore, is indeed long. But, in addition to broadly defined activities like "gathering" and "hunting," these otherwise disjunct societies share, or shared in the past, a key characteristic: the intensive cultivation of plants.

Food plants aboriginally cultivated by Tupi-Guarani peoples of the Atlantic littoral, for example, included manioc, bananas, maize, peanuts, capsicum peppers, pumpkins, sweet potatoes, and pineapples (Benitez 1967:27; d'Evreux 1864:74; Léry 1960:113, 115, 158, 160, 162; Lisboa 1967:97, 99, 100, 101, 115, 121; Sousa 1974:89, 95, 105–106, 110; Staden 1930:141, 176; Vasconcellos 1865:131, 133). The inventory of cultivated species was high, with numerous varieties of many of these crops. There were some twenty-eight named varieties of manioc (Métraux 1928:65–67). The Tupinambá also had access to semidomesticated tree species such as *Hymenaea* and *Inga* (Sousa 1974:101; Vasconcellos 1865:133).

Given that no chronicler reported a Tupi-Guarani society of early colonial times to be lacking horticulture, it is curious that several Tupi-Guarani societies of the present lead exclusively foraging existences. These include the Hetá, Aché, Avá-Canoeiro, Guajá, and Setá. Yet historical and "inferential" linguistic evidence (Sapir 1949:430–432) suggests that each of these groups regressed from an earlier phase of horticultural society (Martin 1969; Clastres 1973:271; Lathrap 1968). The Aché of Paraguay, for example, possess a Tupi-Guarani cognate for maize (*waté*) (Clastres 1968:51–53). Evidently, the Aché cultivated maize in the eighteenth century (Métraux and Baldus 1946:436; Clastres 1972:143). The Avá-Canoeiro, who number some thirty-five people, did at one time cultivate maize, the term for which, *avaši,* is cognate in Tupi-Guarani (Rivet 1924:177; Toral 1986). The trekking Sirionó of Bolivia also planted some maize, in spite of their mischaracterization as foragers (Holmberg 1969), the word for which is *ibaši,* also a cognate (Allyn Stearman, letter, 1987). The Hetá, who spoke a language phonologically and lexically derived from Guarani (Rodrigues 1978), evidently "practiced some plant cultivation" before they became total foragers (Kozák et al. 1979:366). The Guajá language has cognates for cashew (*akayu*), annatto (*araku*), pineapple (*nana'i*), yam (*kara*), chili peppers (*kiki*), and maize (*wači*), all traditional Tupi-Guarani cultigens, but none of which the Guajá recently cultivated. The Setá, a now extinct society of southern Brazil, were hunter-gatherers whose language had Tupi-Guarani cognates for manioc and maize (Loukotka 1929:374, 392).

Sociopolitical, not evolutionary forces, appear to be responsible for the regression of semisedentary horticultural societies into foraging nomadism (Martin 1969; Clastres 1968; Balée 1988a). These forces such as epidemic disease, slave raids, and colonial warfare, also devastated the paramount chiefdoms of the Tupinambá of the Atlantic littoral. In the emergence of true foraging societies, these forces were acting on societies already of a much smaller scale. The Guajá (called Uayás) apparently lived in settled villages in the 1760s in the lower Tocantins River (Noronha 1856:8–9). The lower Tocantins saw intensive warfare between colonial militia, their Indian mercenaries, and uncontacted Indian societies after this time (Balée 1988a). By 1872, the clandestine Guajá, now of the upper Gurupi to the east, were described as being "persecuted" by all surrounding indigenous groups and as lacking horticulture (Dodt 1939:177). According to a Ka'apor informant, the Ka'apor would not have permitted the Guajá to establish permanent settlements, Ka'apor warriors also being more numerous than those of the Guajá (Balée 1988b).

The forests of the Héta and Aché, respectively, in Paraná state, Brazil and eastern Paraguay, experienced a long history of missionization, epidemic disease, and slave raids (Métraux 1948; Clastres 1972:143–144). In the region of the Héta (between the Piquiri and Ivai rivers in Serra dos Dourados), there are numerous archaeological sites dating from thousands of years ago (Chmyz and Sauner 1971). Jesuit missions (*reducciones*) were founded between 1576 and 1630 in the region (Maack 1968:41), where Guarani Indians were concentrated. After 1632, however, gold and slave seekers from São Paulo, called *bandeirantes,* destroyed these missions (Chmyz and Sauner 1971:10; Maack 1968:42–43). The Guarani, who fled into Paraguay, abandoned their village sites. Gé-speaking Indians (the Kaingáng) subsequently occupied these settlements (Maack 1968:44). These newcomers, in turn, later became enemies of the Héta and may have prevented the Héta from maintaining a sedentary life-style (Kozák et al. 1979:366). Full-time hunting and gathering does not appear to be an option that a horticultural society freely chooses. Warfare, epidemic disease, and depopulation evidently can make horticulture and semisedentarism less rewarding than foraging for some societies.

The process of agricultural regression itself has been fairly unexplored. It does not seem to be often an abrupt but rather a gradual transition from sedentarism through trekking to complete nomadism. Dependence on given plant species diminishes over time, until only one or two cultigens are left and finally none. One of these final cultigens appears to be maize. Although maize cultivation is closely associated with the sedentary pre-Columbian Tupi-Guarani chiefdoms, it is also ironically seen among modern trekking peoples. The Araweté, who until recently trekked for an average of six months a year (Viveiros de Castro 1986:271), are highly dependent on maize instead of starchy tubers. Viveiros de Castro (1986:26) appropriately noted that, with respect to the Araweté, "maize aggregates people; it is practically the only force which does so. Innumerable other forces work toward dispersion." Maize aggregated the Araweté during planting and harvest seasons. In the other seasons, people tended to forage in the forest. Among several macro-Gé peoples such as the Botocudos, Tapuias, central Kayapó, and the Akwẽ-Xavante, the maize harvest united society. Otherwise the tribesmen dispersed themselves into small groups for gathering, hunting, and fishing (Galvão 1979:245; Maybury-Lewis 1967:47). Maize dependence combined with trekking also occurs with the Panoan-speaking Amahuaca of eastern Peru (Carneiro 1985) and the linguistically isolated Hoti of southern Venezuela (Coppens 1975:68; Coppens and Mitrani 1974:36). Several trekking groups traditionally cultivated other food plants, especially bananas and sweet manioc, one or more of which may have been calorically more significant than maize in the diet, but little or no bitter manioc (Coppens and Mitrani 1974:136; Smole 1976:119).

Manioc is, of course, the caloric staple of most semisedentary (nontrekking) lowland South American Indians. Brochado (1977:57) found that of 553 indigenous peoples of the tropical forest that cultivate manioc, 478 (86.4 percent) employ it as a primary food source and 75 (13.6 percent) use it only as a supplementary resource. For 60 (80 percent) of these latter groups, the dietary staple is maize, bananas, peanuts, or sweet potatoes. In particular, trekking peoples seem to choose maize rather than bitter manioc as a staple because of its relatively low start-up costs, simple processing requirements, low transport costs, and fast maturation time (Balée 1985; Galvão, 1979:245). When trekking peoples do cultivate manioc, it tends to be only the sweet variety, as with the Hoti (Coppens and Mitrani 1974:136) and the Yanoama of Venezuela (Smole 1976:119). Sweet manioc demands no elaborate detoxification nor installations in order to be made fit for human consumption (Balée 1985:489–490). Brochado (1977:64) observed that of the 75 groups that employ manioc as a dietary supplement, the nontoxic varieties outnumber the toxic ones almost two to one (Lathrap 1970:53).

One can reasonably propose that of the two staple cultigens, maize and bitter manioc, maize disappears last from a trekking people's horticultural repertoire, prior to their becoming true foragers (see Figure 3.1). That the Guajá language has retained a cognate for maize but not one for bitter manioc (called *tarəmə*) supports this hypothesis. This is precisely the case, moreover, with the Aché (Clastres 1968:53).

In the transition from sedentarism through trekking to nomadism, it is logical to assume that there is a growing economic dependence on plants that are not cultivated per se by the groups undergoing this transition. The species that tend to become primary botanical resources, however, are not readily characterized as being "wild." Rather, these plants, in their dominance and/or frequency, appear to be vegetational artifacts of an agricultural past, sometimes of other societies.

Artifactual Resources of Foragers

What is an artifact? Two basic definitions, implicitly or explicitly, are common in the literature. One is based on purposeful human activity, whereby an artifact is any object "made or unmade by *deliberate* human action" (Childe 1956:11, emphasis mine). The other is broader, entailing no necessary "consciousness" on the part of human actors. In this second definition, an artifact is "any material expression of human cultural activity" (Spaulding 1960:438). This definition seems more appropriate in considering use of biological re-

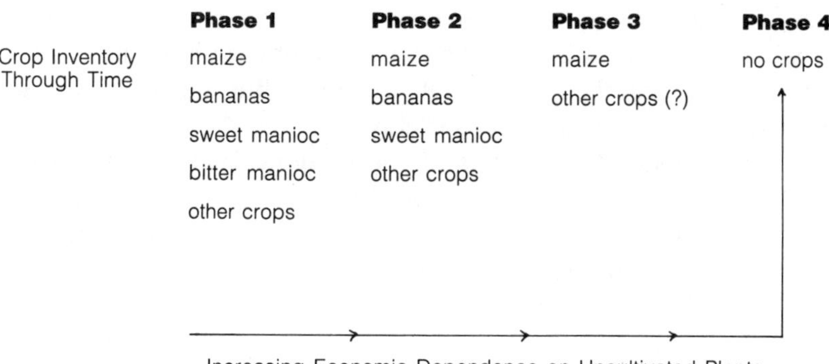

Figure 3.1

sources by lowland South American Indians. It is true, however, that some contemporary Indians use prehistoric artifacts that were produced by "deliberate human action." In North America, for example, the Navajo use Anasazi ruins for gathering medicinal plants and clay, and as sites for curing ceremonies (Holt 1983). In South America, the Yanomamo and Araweté recently felled trees with prehistoric stone axheads, the origins of which they believe to be divine (Chagnon 1977:46; Viveiros de Castro 1986:152). But many groups, including the Araweté, use an anthropogenic resource, *terra preta do Índio,* which is better understood as an artifact in the broad, materialistic sense that Spaulding gave it. The formation of *terra preta do Índio* in hearths and garbage pits over hundreds or thousands of years is clearly no result of "deliberate human action." The Araweté, Kuikuru, Mawé, Mundurucu, Xikrin-Kayapó, and many other Indians prefer to plant nutrient-demanding crops in this most fertile cultural horizon (Balée 1989; Carneiro 1983:67; Frikel 1959, 1968; Smith 1980:562).

Another artifact of this sort concerns disturbance indicator and late successional plants. In their dominance and/or frequency, these form artifactual landscapes in Amazonia (Anderson and Posey 1985; Balée 1989; Balée and Campbell 1989; Posey 1985), Central America (Gomez-Pompa, Flores, and Sosa 1987), and no doubt elsewhere. Although not all of these landscapes, such as anthropogenic savannas of northwestern and central Amazonia (Huber et al. 1984; Smole 1976; Whitten 1979; Prance and Schubart 1977), proffer very important plant resources to indigenous peoples, many others do. These old, artifactual landscapes, which often cover archaeological sites, appear to be especially useful to modern foragers of the tropical South American lowlands. These foragers, for sociopolitical reasons, could not remain sedentary horticulturalists, and thereby could no longer produce any artifactual landscapes of their own. But they could use the artifactual landscapes of other societies.

Descriptions of the habitats of the foraging Héta of Paraná and Aché of Paraguay point to past agricultural management. First, recall that the region was at one time occupied by agricultural Indians, the Guarani, whose population at contact has been estimated at 1,500,000 with a population density of 4 persons per square kilometer (Clastres 1973). This is higher than the population density of modern Amazonia and twenty times higher than the average Indian population density (0.2 persons per square kilometer—see Hames 1983:425) of lowland South America at present. The region was subsequently colonized, became the scene of much colonial warfare, and was later in large part abandoned. It is possible that much of the "primary" forest of the region never recovered or was at least irrevocably modified by past disturbance, in spite of what has been described often as "primary" forest cover (Hill and Hawkes 1983:143; Maack 1968:217; Fischer 1987:16). The forests the Héta exploited, like the Héta themselves, regrettably no longer exist. When they did, no systematic inventories or studies of forest recovery were made (Paulo Sodero Martins, pers. comm., 1988). General studies of the vegetation of the Serra dos Dourados, habitat of the Héta, nevertheless, were carried out. The geographer Maack (1968:215–217) mentioned some nine species that were "predominant" in the "rain forest" (*mata pluvial*) that typified the region of the Héta. These included *Acrocomia sclerocarpa* Mart., *Acacia polyphylla* DC., *Cassia multijuga* H. & P., and *Jacaratia spinosa* A.DC., all of which are well known late successional species. The mucajá palm (*Acrocomia sclerocarpa*) in the central zone of eastern Paraguay, near the Aché and less than 200 km from the Serra dos Dourados, is strongly associated with agricultural fallows. Markley (1956:12–13) noted that this palm germinates best in areas that have been used for horticulture. He added, "It is not surprising, therefore, that the most intensively and thoroughly cultivated farmlands of the Central Zone contain the greatest concentration of palms" (1956:13). Evidently, *Acrocomia* is a long-lived late successional species that, through the agency of human beings, can grow in what otherwise appear to be "virgin" forests. Markley is worth quoting further with respect to a seemingly dense forest with *Acrocomia* present, at a place called Curuguaty:

> A subsequent study of the history of Curuguaty revealed that the area had been a center of agriculture off and on from pre-Columbian times until after the war of the Triple Alliance (1865–70). The settlement of the area, its agricultural development, its subsequent decline and the return of the land to what now resembles virgin forest explain the apparently anomalous occurrence of the mbocayá [*Acrocomia*] palm. (Markley 1956:15).

In the Amazon, *Acrocomia* often occurs in *terra preta do Índio* and other fallows. It has been described as "semi-domesticated" (Huber 1900:21; Balée 1988b:8). *Acrocomia* was an important dietary item for the Héta Indians. They consumed its proteinaceous kernels, a flour made from the palm heart (the only flour they made from any plant), and the weevil larvae that bored into the fruit (Loureiro Fernandes 1964:40–41; Kozák et al. 1979:379).

Continuing with Maack's list, *Acacia polyphylla* is known to be a late successional species elsewhere in lowland South America (Balée and Campbell 1989; Huber 1909:162). *Cassia multijuga,* according to many herbarium labels I have seen at the Museu Goeldi, is by far most commonly found in secondary forests from eastern to western Amazonia (Lisboa, Maciel, and Prance 1987:55). The bamboo (*Bambusa guadua* H. & P.) stands of the Héta habitat referred to by Maack (1968:217) and Kozák et al. (1979:367) are also usually found in areas previously disturbed by slash-and-burn agriculture (Balée 1989; Sombroek 1966:188–189). The Héta used bamboo blades as the plaits for their sieves (Loureiro Fernandes 1964:41).

Maack also noted a "wealth" of lianas, epiphytes, aroids, bromeliads, and orchids (1968:217) in the Héta forest. Liana forests (*mata de cipó*) appear to be late successional in Amazonia (Balée 1989; Balée and Campbell 1990) if not elsewhere in the Neotropics (Al Gentry, letter, 1987). The authors of the habitat descriptions of the Héta Indians observed two other palms, *Euterpe edulis* Mart. and *Syagrus romanzoffiana* (Cham.) Glass. *Syagrus* (called *jerivá* in Brazil) was the "most common" species in the Héta habitat (Kozák et al. 1979:367–368; Loureiro Fernandes 1964:39; Maack 1968:217). Both these palms are now very commonly cultivated as ornamentals in the major cities of southern Brazil (Pinheiro and Balick 1987:28). *Jerivá* is extremely varied in habit and occurs in a wide variety of climates, soils, and elevations (Barbosa Rodrigues 1898:15; 1899:8). Its wide dispersion may in fact have been related to human influence, given its frequent occurrence in swiddens and fallows (Barbosa Rodrigues 1898:16). In the 1980s, *jerivá* is seen almost only in disturbed areas (Dalmo Giacometti, pers. comm., 1988). The Héta used *jerivá* palms for many utilitarian ends. Palm fruits, in general, were described as constituting the "basis of the diet" of the Héta (Loureiro Fernandes 1964:40), of which those of *jerivá,* which is perennial, were most important (Kozák et al 1979:367–368; Loureiro Fernandes 1964:40). The Héta consumed the mesocarp, the kernel, and the palm heart of *jerivá.* They used the spathe as a receptacle. All basketry, rope, infant-carrying straps, sleeping mats, and fishing line derived from the fiber and/or sword leaves of this palm (Loureiro Fernandes 1959:25; 1964:41–43; Kozák et al. 1979:395).

The frequent presence of one final species, feral oranges (*Citrus sinensis* [L.] Osb.), supplies incontrovertible evidence for prior human modification of the landscape in which the Héta were first reported to be living in the 1950s (Maack 1968:217). The orange trees were planted by Jesuit missionaries in the sixteenth and seventeenth centuries. Although some orange groves appear to have expanded since then, these would not have existed in the habitat of the Héta were it not for their having been an agricultural introduction of the past. The orange trees, together with the many late successional species, signify that the habitat of the Héta should hardly be described as "primeval" (Kozák et al. 1979:357)."

Feral orange trees exist as well in the habitat of the Aché, having been also originally planted by Jesuits of the sixteenth and seventeenth century missions (Hill et al. 1984:125; Métraux and Baldus 1946:436; Kaplan and Hill 1985:229; Vellard 1934:240). It seems inaccurate from this evidence alone, even if the orange trees were subsequently dispersed by nonhuman agents, to describe their habitat as "primary forest" (Hill and Hawkes 1983:143; Bailey et al. 1989:65). Some question exists as to whether feral oranges were edible to the Aché. Clastres (1972:156) indicated that they were not, but others (Hill et al. 1984:112; Kaplan and Hill 1985:229; Vellard 1934:240) listed the oranges among the Aché's gathered foods.

In addition to feral orange trees, Vellard (1934:228) listed numerous trees of the Aché habitat that are well known late successional species, such as *Cedrela, Piptadenia, Astronium, Cordia,* and *Ceiba*. He also noted the presence of *Jacaratia*, the fruits of which both the Aché and Héta ate (Vellard 1934:240; Loureiro Fernandes 1964:39, 42). This species (*Jacaratia spinosa* A. DC.) is the ecologically most important species on a one-hectare fallow of forty years in the habitat of the Ka'apor Indians (Balée, unpublished data; Balée and Gély 1989). It does not dominate in "primary" forest.

Although the Aché obtained most of their calories from game animals and not plants (Kaplan and Hill 1985:228; Clastres 1972:160; Vellard 1934), the most important botanical resource, both as food and for making many tools and utensils, was *jerivá* (*Syagrus romanzoffiana*), as it was for the Héta. *Jerivá* supplied the Aché with edible buds, flour, and weevil larvae in addition to bow wood, bowstring, roofing thatch, matting, fans, and cases for keeping feather art (Clastres

1972:161; Métraux and Baldus 1946:436; Vellard 1934:232, 240). The preferred campsites of the Aché, moreover, were groves of *jerivá* palms (Clastres 1972:150). The disturbance indicator, *Acrocomia* (*mucajá* palm), a food resource of the Aché, is also present (Hill et al. 1984).

Other gross indicators of past agricultural perturbations of the Aché habitat include bamboo forests and liana forests (Baldus 1936:750; Hill and Hawkes 1983:143; Vellard 1934:229). Regardless of whether the Aché would have been best described as a society of honey gatherers (Vellard 1939) or a society of hunters (Clastres 1972; Kaplan and Hill 1985), it is not unreasonable to hypothesize that they are also a people of old fallows. These fallows harbor artifactual resources upon which they directly and indirectly depended.

The campsites and principal botanical resources of the foraging Guajá Indians of northern Maranhão, Brazil, evoke striking similarities to those of the Héta and Aché. In particular, as with the Héta and Aché, the Guajá display a very high dependence on palms for food, shelter, clothing, tools, and utensils (see table 3.1; Balée 1988b:50–51). The economically most important palms in Guajá subsistence, moreover, are well known disturbance indicators. These include *babassu* (*Orbignya phalerata* Mart.), *inajá* (*Maximiliana maripa* [Corr. Serr.] Drude), and *tucuma* (*Astrocaryum vulgare* Mart.). These species often occur in enclaves of two to three hectares, surrounded by dense *terra firme* forest. *Babassu*, when present, is usually the most dominant species in such enclaves. The Guajá refer to this forest as "*babassu* grove" (*wa'i'ɨ-tu*). *Babassu* groves are the exclusive sites of their camps (Balée 1988b), as are *jerivá* groves for the campsites of the Aché. The *babassu* palm (*wa'ĩ'ɨ*) supplies edible kernels (*ha'ĩ*) and mesocarp flour (*tarəmə*). They extract the kernel by placing the fruit on a small concave rock (*ita*) and then crack the fruit open with another smooth rock. They eat the kernels raw. To obtain flour, young fruits are placed directly in hot ashes, which soften the exocarp. The exocarp is peeled off and the mesocarp, now doughy, falls off in irregular plates, which are then eaten by hand. Although consuming *babassu* mesocarps is not common in Amazonia, the Araweté, Apinage, and Guajajara do so (Viveiros de Castro 1986:164; May et al. 1985:125), especially in the failure of domesticated staples. May et al. (1985:125) pointed out that for rural families, as well as for some indigenous peoples of the *babassu* zone south of the Amazon River, the *babassu* mesocarp flour

Table 3.1
Economic Palms in Guajá Society

Scientific Name	Guajá Name	Uses	Part Used
Astrocaryum munbaca Mart.	*yu* (3483)	food	seed
A. vulgare Mart.	*takamã* (3481)	bowstring, hammock, women's skirt, foreskin string, infant carrying strap, rope, string for beads	sword-leaf fibers
Bactris humilis (Wallace) Burret	*wə̃'ə̃* (3355)	arrow shaft	petiole
Bactris maraja Mart.	*mariawa* (3385)	food	mesocarp
Bactris major Jacq.	*kiripiri-hu* (3490)	food	mesocarp
Bactris sp. 2	*kiripirim' ɨ* (3525)	food	mesocarp
Euterpe oleracea Mart.	*pinuwã-pihun* (3281)	food	mesocarp
Geonoma baculifera (Poiteau) Kunth.	*yowo'ĩ* (3347)	arm band	seed
Maximiliana maripa (Corr. Serr.) Drude	*inaya' ɨ* (3377)	food, bowl	mesocarp, seed, spathe
Oenocarpus distichus Mart.	*pinuwa' ɨ* (3144)	food	exocarp
Orbignya phalerata Mart.	*wa'ĩ' ɨ* (3376)	food, thatch	kernel, mesocarp, young leaves

Note: Numbers in parentheses below Guajá names for palms refer to collections numbers in the series Balée with vouchers at the New York Botanical Garden and duplicates at the Museu Paraense Emílio Goeldi.

serves as a "substitute" for manioc flour during manioc shortages (Balée 1988b:52). The Guajá harvest *babassu* fruits on the ground. Like the fruits of *jerivá,* so important to the Aché and Héta, *babassu* fruits are available year round, with young fruits appearing at the beginning of the rainy season (December–January). In addition to using *babassu* for food, the Guajá use the leaves as roofing thatch.

Babassu palm groves in the habitat of the Guajá are always indicative of prior agricultural occupations. In the upper Turiaçu River region, where I have done research with the Guajá, all *babassu* forest enclaves were former Ka'apor Indian agricultural settlements and fields (Balée 1988b), attested to by Ka'apor oral history and the remains of ceramic manioc griddles.

Fruits of the *inajá* palm (*Maximiliana maripa*) provide oil, protein, and calories in the Guajá diet. The Guajá heat the exocarp, which they then peel off and discard. They boil the mesocarp to make a porridge (*mika*) that they serve on an inajá spathe (*hawera-kutai*). They also eat the *inajá* kernels. *Inajá,* like *babassu,* frequently occurs in fallows in Amazonia (Balée 1988b:48; Pesce 1985:66; Schulz 1960:222; Wessels Boer 1965:152).

The Guajá make all garments and rope from the young sword-leaf fibers (*tikwira*) of *tucuma* palm (*Astrocaryum vulgare*). Infant-carrying straps, hammocks, women's skirts, bowstrings, string for tying the foreskin, rope, and string for arm bands all derive from *tucuma* fibers (Balée 1988b:50). In the habitat of the Guajá, the *tucuma* palm occurs only on former sites of Ka'apor agricultural occupation, with or without the presence of *babassu* (Balée and Gély 1989). In a forty-year-old fallow of the Ka'apor, also mentioned above, only 20 kilometers from the Guajá of the upper Turiaçu River, *tucuma* was the second ecologically most important species. *Tucuma* is one of two species of *Astrocaryum* taken to be a "good" indicator of prior human settlement in Surinamese noninundated habitats; further, it is "never" found in undisturbed forest there (Wessels Boer 1965:132).

Since *babassu, inajá,* and *tucuma* occur on and indicate prior settlement sites of the horticultural Ka'apor, the Guajá, who camp within *babassu* groves and who intensively exploit these palm species therein, have adapted not merely to a "natural" environment, but to the vegetational artifacts of another society.

Discussion

The foraging Guajá, Héta, and Aché appear to have undergone a kind of parallel historical regression from prior agricultural societies to exclusively foraging ones. Although the specific botanical resources they utilize vary (less so between the Héta and Aché), it is remarkable that all three groups depend heavily on palms that appear to dominate in old agricultural fallows. In each case, the principal habitats they exploit exhibit plant indicators of past human disturbance. These "hunter-gatherers," in other words, heavily rely on agriculture (Bailey et al. 1989; Headland 1987). This does not mean that they use a priori domesticated crops, even though the Aché (Vellard 1934:241) and the Guajá (Araújo Brusque, 1862:15) were historically known to raid occasionally the swiddens of other Indians and settlers for maize, sweet potatoes, and the like. Such raids entailed high risks, since at least in the Guajá case, when the other Indians caught them in their swiddens, they would "beat or kill them" (Araújo Brusque 1862:15). For the Guajá, Aché, and Héta, raiding swiddens was clearly a far less usual occupation than harvesting the fruits of old fallows. Sometimes these fruits were domesticates, as with the feral oranges of the Héta and Aché and yams and lime trees that the Ka'apor planted many years ago and which the Guajá now harvest in old fallows (Balée and Gély 1989). But for all three foraging societies the economically most important plant resources were palms.

The foraging Warao Indians of the Orinoco delta demonstrated a major dependence also on a palm, the *moriche* palm (*Mauritia flexuosa* L.f.) (Heinen and Ruddle 1974). Even though some monospecific groves of *moriche* palm in swamp forests of the Warao habitat seemed to be anthropogenic (Heinen and Ruddle 1974:124), many were no doubt natural groves unrelated to human activity. The palm groves of the Aché, Héta, and Guajá Indians, however, appear to be anthropogenic, considering, in addition, many other late successional species found therein. The Aché, Héta, and Guajá also underwent agricultural regression, about which no evidence exists with respect to the Warao (Wilbert 1969:19).

Not a general theory of foraging, therefore, but a general theory of agricultural regression and the utilization by foragers of old fallows in lowland South America follows from this information. Because of epidemic disease and/or defeat in warfare, some sedentary agricultural societies became oriented toward a trekking life-style. Some eventually became full-time foragers, with population densities considerably lower than that of surrounding agricultural peoples. This is certainly so with the Aché, Héta, and Guajá, who are (or were) circumscribed by far more dense agricultural populations, such as those of the Guarani, Mbyá, Kaingáng, Ka'apor, and Tenetehara. In other words, agricultural regression appears to be associated with a concomitant drop in population density. It is unlikely that very small bands of five to fifteen people could have maintained or opted for agriculture ex post facto. First, they would be more liable to defeat in warfare and the subsequent loss of their swiddens, orchards, and other agricultural investments. Second, with only one or at most two small swiddens, appropriate for such a tiny village size, the group would run the risk of generalized crop failure. Agricultural societies such as that of the Ka'apor tend to spread this risk, insofar as many small families of a village have their own swiddens. If one fails (due to insect pests, overgrazing by peccaries, and so on), this does not ipso facto become a village-wide crisis (Balée 1984b). In addition, no one family keeps the seeds and/or cuttings of all the crops grown by the Ka'apor. The largest Ka'apor settlements, incidentally, tend to be the longest-lived, perhaps because of ecological risk management and a well established system of economic reciprocity between many households (Balée 1984b).

The smaller a society gets, in contrast, the more nomadic it becomes. In late September 1987, an isolated family of Araweté Indians came to the attention of the Brazilian National Indian Foundation (FUNAI). They were living some 200 km distant from the Araweté village of 155 people, having become "lost" in 1976, at the time the FUNAI authorities were first making contact with the Araweté and when introduced epidemic disease killed some sixty Araweté (Viveiros de Castro 1986:181). Early in September 1987, Kayapó Indians attacked this isolated family of seven people, killing one man and one boy. The Kayapó raiders took captive two women and two children. One Araweté man escaped into the forest. The hostages were taken to the Kayapó village of Cateté (in the basin of the Itacaiunas River, between the Tocantins and Xingu rivers). Once apprised of the situation, the FUNAI authorities, with two Araweté Indian interpreters from the Araweté village, departed in search of the missing Araweté man. They also entered into negotiations with the Kayapó for the release of the four hostages, which eventually took place.

The search party made contact after a few days with the missing Araweté man, convincing him to return with them. All the surviving Araweté of this isolated group were later returned to live among their far more numerous kinsmen again (the missing man had a sister in the Araweté village). I interviewed these survivors, their interpreters, and members of the FUNAI search party when they arrived in the city of Marabá on the Tocantins River in transit to the Araweté village. At the site of the raid stood a house in a small swidden. The only crop in the swidden was maize. The Araweté traditionally grow a reasonable number of species, including maize, manioc, papaya, cotton, annatto (*Neoglaziovia variegata* L.) and yams (Viveiros de Castro 1986:150–153). The isolated family was subsisting, according to the remains found at the site, from maize, *babassu* mesocarps, and Brazil nuts. These last two botanical resources are often present because of past agricultural legacies (Balée 1989; Ducke 1946:8). Although the Araweté traditionally make their hammocks from cotton, the hammocks of the isolated family were woven from the bark of an uncultivated tree (*Couratari quianensis* Aubl.) Whereas the Araweté usually fashion their bowstrings from the cultivated bromeliad *Neoglaziovia variegata*, the bowstring of the missing man was from *Astrocaryum vulgare* (*tucuma* palm) sword-leaf fibers. Evidently, this group of seven people had become so small that they could not maintain an inventory of more than one crop, maize. As pointed out above, maize appears to be the last cultigen a trekking society loses before becoming true foragers. These Araweté seemed to have been living on the edge between trekking and foraging.

Even though agriculture becomes increasingly implausible for a group approaching band size (between about fifteen and fifty persons or less), a very small number of people can survive exclusively by foraging—even a number smaller than that of a band. I met the "lost" Araweté in Marabá while I myself was in transit to a hitherto unknown group of Indians consisting of only two men between the ages of thirty and forty who were living some 190 kilometers west-northwest of Marabá on a tributary of the Tapirapé River. I was participating in a FUNAI mission with the objective of identifying the cultural and linguistic affinities of these Indians and determining if there were more of their kind in the vicinity (there were none). Their language was determined to be a new Tupi-Guarani language (Moore and Maciel 1987). Although a substantial list of Tupi-Guarani cognates for cultivated plants was elicited from the two men (Balée 1987; Moore and Maciel 1987), their campsites evinced no horticulture. Based on material encountered at their only two known campsites, the botanical resources they were subsisting on were *babassu* fruits and Brazil nuts. Evidently, they had lived in a horticultural village in the past, but after their kinsmen died (apparently in a massacre by other Indians), they alone, who were probably teenagers at the time, could not keep horticulture viable (Balée 1987). The two orphans turned instead to a major reliance on plant species that dominate in very old fallows, depending on artifactual resources. Later, the two men were transferred to the Asurini do Xingu village, where they were living as of October 1988.

It is probably not merely the small size to which a trekking people can be reduced that leads to the inevitability of a foraging life-style. An intervening variable is, no doubt, the frequent uneveness of adult sex ratios in such tiny populations, given statistical odds. In a group with few or no women left, as with the orphans of the Tapirapé River, a previously learned sexual division of labor in horticulture would cease to be meaningful, as would horticulture itself.

The resource concentrations of old fallows are, to foragers, substitutes for the caloric vitality, to horticulturalists, of young swiddens. Whereas these old fallows could not alone permanently support a horticultural village without being supplemented by new clearings and plantings, they can maintain for indefinite periods small, nomadic foraging bands. The orphans of the Tapirapé, like the Héta, Aché, and Guajá, relied on agriculture without raising any crops. A cultural past, in other words, was embodied in the food they gathered.

The principal problems concerning the use of biological resources by lowland South American Indians therefore are not evolutionary, but historical. The artificiality of culture/environment dichotomies and of evolutionary typologies with regard to lowland South America should by now be apparent. Cultural ecology and evolutionary ecology offer no theories for explicating three seemingly paradoxical facts: 1) the "natural" habitats of various foragers are "cultural"; 2) many foraging societies began in a historical process of agricultural regression, buoyed by sociopolitical forces; and 3) many modern foragers (or "hunter-gatherers") use agriculture, as is readily documented in their profound dependence on old fallows. These facts are not paradoxical, however, within the framework of a historical ecology.

LITERATURE CITED

Anderson, A. B. and D. A. Posey. 1985. Manejo de cerrado pelos indios Kayapó. *Boletin Museu Paraense Emílio Goeldi* 2(1):77–98.

Araújo Brusque, F. C. de. 1862. *Relatório apresentado à assemblea legislativa da Província do Pará na primeira sessão da XIII Legislatura*. Belém, Brazil: Typographia Frederico Carlos Rhossard.

Bailey, R. C., G. Head, M. Jenike, G. Owen, R. Rechtman, and E. Zechenter.

1989. Hunting and gathering in tropical rain forest: Is it possible? *American Anthropologist* 91:59–82.

Baldus, H. 1936. Ligeiras notas sôbre duas tribos Tupi da margem paraguaya do Alto Paraná (Guayaki e Chiripá). *Revista da Museu Paulista* 20:749–756.

Balée, W. 1984a. The ecology of ancient Tupi warfare. In R. B. Ferguson, ed., *Warfare, Culture, and Environment*, 241–265. Orlando, Fla: Academic Press.

——. 1984b. The persistence of Ka'apor culture. Ph.D. dissertation, Columbia University.

——. 1985. Ka'apor ritual hunting. *Human Ecology* 13(4):485–510.

——. 1987. Relatório etnológico sobre os últimos dias da Frente de Atração do Rio Tapirapé (Municipio de Marabá, Estado do Pará) em 22–29 de outubro de 1987. Brasília, Brazil: *Fundacão Nacional do Indio.*

——. 1988a. The Ka'apor Indian wars of lower Amazonia, ca. 1925–1928. In R. R. Randolph, D. M. Schneider, and M. N. Diaz, eds., 155–169. *Dialectics and gender: Anthropological approaches.* Boulder, Colo.: Westview Press.

——. 1988b. Indigenous adaptation to Amazonian palm forests. *Principes* 32(2):47–54.

——. 1989. The culture of Amazonian forests. In D. A. Posey and W. Balée, eds. *Resource Management in Amazonia, Advances in Economic Botany* 7:1–21.

Balée, W. and A. Gély. 1989. Managed forest succession in Amazonia: The Ka'apor case. In D. A. Posey and W. Balee, eds. *Resource Management in Amazonia, Advances in Economic Botany* 7:129–158. New York: New York Botanical Garden.

Balée, W. and D. G. Campbell. 1990. Evidence for the successional status of liana forest (Xingú River basin, Amazonian Brazil). *Biotropica* 22(1) 36–47.

Barbosa Rodrigues, J. 1898. *Palmae Mattogrossenses, novae vel minus cognitae.* Rio de Janeiro, Brazil: Typographia Leuzinger.

——. 1899. *Palmae Novae Paraguayensis.* Rio de Janeiro, Brazil: Typographia Leuzinger.

Benitez, J. P. 1967. *Formación social del pueblo Paraguayo.* Asunción, Paraguay: Ediciones Nizza.

Brochado, J. P. 1977. *Alimentação na floresta tropical.* Caderno No. 2. Pôrto Alegre, Brazil: Universidade Federal do Rio Grande do Sul.

Carneiro, R. L. 1983. Cultivation of manioc among the Kuikuru. In R. B. Hames and W. T. Vickers, eds., *Adaptive Responses of Native Amazonians*, 65–111. New York: Academic Press.

——. 1985. Hunting and hunting magic among the Amahuaca of the Peruvian Montaña. In P. Lyon, ed., *Native South Americans: Ethnology of the Least Known Continent*, 122–132. Prospect Heights, Ill.: Waveland Press.

Chagnon, N. A. 1977. Yanomamo: *The Fierce People.* 2nd ed. New York: Holt, Rinehart and Winston.

Chmyz, I. and Z. C. Sauner. 1971. Nota prévia sobre as pesquisas arqueológicas no Vale do Rio Piquiri. *Dédalo* 7(13):7–35.

Childe, V. G. 1956. *A Short Introduction to Archaeology.* London: Frederick Muller.

Clastres, P. 1968. Ethnographie des indiens Guayaki (Paraguay-Brésil). *Journal de la Société Americanistes* 57:9–61.

——. 1972. The Guayaki. In M. G. Bicchieri, ed., *Hunters and Gatherers Today,* 138–174. New York: Holt, Rinehart, and Winston.

——. 1973. Eléments de démographie amérindienne. *L'Homme* 13(1–2):23–36.

Coppens, W. 1975. Contribución al estudio de las actividades de subsistencia de los Hotis del Río Kaima. *Boletin Indigenista Venezolano* 16(12):65–78.

Coppens, W. and P. Mitrani. 1974. Les indiens Hoti. *L'Homme* 14(3–4): 131–142.

Dean, W. 1984. Indigenous populations of the São Paulo–Rio de Janeiro coast: Trade, aldeamento, slavery and extinction. *Revista de História,* n.s., 117: 3–26.

d'Evreux, Y. 1864. Voyage dans le nord du Brésil fait durant les années 1613 et 1614. Leipzig: Librairie A. Franck.

Dodt, G. 1939. Descripção dos rios Paranaíba e Gurupy. *Coleção Brasiliana,* Vol. 138. São Paulo, Brazil: Cia Editora Nacional.

Ducke, A. 1946. Plantas de cultura precolombiana na Amazônia brasileira: Notas sôbre as espécies ou formas espontáneas que supostamente lhes teriam dado origem. *Boletim do Instituto Agronomico do Norte* 8:1–24.

Fischer, G. R. 1987. *Manejo sustentado de florestas nativas.* Joinville, Brazil: G. R. Fischer.

Foley, R. 1987. *Another Unique species: Patterns in Human Evolutionary Ecology.* Hong Kong: Longman Group UK Ltd.

Frikel, P. 1959. Agricultura dos índios Mundurukú. n.s., *Antropologia (Boletimdo Museu Paraense Emilio Goeldi),* n.s., Antropologia 4:1–35.

——. 1968. Os Xikrin. *Publicações Avulsas do Museu Goeldi* 7:3–119.

Galvão, E. 1979. *Encontro de sociedades: Indios e brancos no Brasil.* Rio de Janeiro, Brazil: Editora Paz e Terra.

Gomez-Pompa, A., J. Salvador Flores, and V. Sosa. 1987. The "pet-kot": A man-made tropical forest of the Maya. *Interciencia* 12(1):10–15.

Hames, R. B. 1983. Settlement pattern of a Yanomamo population bloc. In R. B. Hames and W. T. Vickers, eds., *Adaptive Responses of Native Amazonians,* 393–427. New York: Academic Press.

Headland, T. N. 1987. The wild yam question: How well could independent hunter-gatherers live in a tropical rain forest ecosystem? *Human Ecology* 15(4):463–491.

Heinen, H. D. and K. Ruddle. 1974. Ecology, ritual, and economic organization in the distribution of palm starch among the Warao of the Orinoco delta. *Journal of Anthropological Research* 30(2):116–138.

Hill, K. and K. Hawkes. 1983. Neotropical hunting among the Aché of eastern Paraguay. In R. B. Hames and W. Vickers, eds., *Adaptive Responses of Native Amazonians,* 139–188. New York: Academic Press.

Hill, K., K. Hawkes, M. Hurtado, and H. Kaplan. 1984. Seasonal variance in the diet of Aché hunter-gatherers of eastern Paraguay. *Human Ecology* 12(2):145–180.

Holmberg, A. 1969. *Nomads of the Long Bow.* Garden City, New York: American Museum of Natural History Press.

Holt, H. B. 1983. A cultural resource management dilemma: Anasazi ruins and the Navajos. *American Antiquity* 48(3):594–599.

Huber, J. 1900. *Arboretum amazonicum*. Belém, Brazil: Museu Paraense de História Natural e Etnographia.

——. 1909. Matas e madeiras amazónicas. *Boletim do Museu Goeldi (Museu Paraense) de História Natural e Ethnographia* 6:91–225.

Huber, O., J. A. Steyermark, G. T. Prance, and C. Alés. 1984. The vegetation of the Sierra Parima, Venezuela–Brazil: Some results of recent exploration. *Brittonia* 36(2):104–139.

Kaplan, H. and K. Hill. 1985. Food sharing among Aché foragers: Tests of explanatory hypotheses. *Current Anthropology* 26(2):223–246.

Kozák, V., D. Baxter, L. Williamson, and R. L. Carneiro. 1979. *The Héta Indians: Fish in a Dry Pond*. Anthropological Papers of the American Museum of Natural History, vol. 55, part 6. New York: American Museum of Natural History.

Lathrap, D. 1968. The "hunting" economies of the tropical forest zone of South America: An attempt at historical perspective. In R. B. Lee and I. DeVore, eds., *Man the Hunter*, 23–29. Chicago: Aldine.

——. 1970. *The Upper Amazon*. London: Thames & Hudson.

Lee, R. B. and I. DeVore. 1968. Problems in the study of hunters and gatherers. In R. B. Lee and I. DeVore, eds., *Man the Hunter*, 3–12. Chicago: Aldine.

Léry, J. de. 1960. *Viagem à terra do Brasil*. Translated from the French (1586) by S. Milliet. São Paulo, Brazil: Livraria Martins Editora.

Levins, R. and R. Lewontin. 1985. *The dialectical biologist*. Cambridge, Mass.: Harvard University Press.

Lisboa, C. de. 1967. *História dos animais e árvores do Maranhão*. Lisbon: Publicações do Arquivo Histórico Ultramarino e Centro de Estudos Históricos Ultramarinos.

Lisboa, P. L. B., U. N. Maciel, and G. T. Prance. 1987. Perdendo Rondônia. *Ciência Hoje* 6(36):48–56.

Loukotka, C. 1929. Le Setá: Un nouveau dialecte Tupi. *Journal de la Societe des Americanistes* 21(2):373–398.

Loureiro Fernandes, L. 1959. The Xetá: A dying people of Brazil. *Bulletin of the International Committee on Urgent Anthropological and Ethnological Research* 2:22–26.

——. 1964. Les Xetá et les Palmiers de la foret de Dourados: Contribution á l'ethnobotanique du Paraná. In *VI Congrés International des Sciences Anthropologiques et Ethnologiques*, Tome II, 39–43. Paris: Musée de L'Homme.

Lyon, P. J. 1985. Editor's note. In P. J. Lyon, ed., *Native South Americans*, 3. Prospect Heights, Ill.: Waveland Press.

Maack, R. 1968. *Geografía física do Estado do Paraná*. Curitiba, Brazil: Banco de Desenvolvimento do Paraná, Universidade Federal do Paraná e Instituto de Biologia e Pesquisas Tecnológicas.

Markley, K. S. 1956. Mbocayá or Paraguay cocopalm: An important source of oil. *Economic Botany* 10(1):3–32.

Martin, M. K. 1969. South American foragers: A case study in cultural devolution. *American Anthropologist* 71(2):243–260.

May, P. H., A. B. Anderson, M. M. Balick, and J. M. F. Frazão. 1985. Subsistence benefits from the babassu palm (*Orbignya martiana*). *Economic Botany* 39:113–129.

Maybury-Lewis, D. 1967. *Akwẽ-Shavante Society.* Oxford, England: Clarendon Press.

Meggers, B. J. 1954. Environmental limitation on the development of culture. *American Anthropologist* 56:801–824.

——. 1971. Amazonia: Man and culture in a counterfeit paradise. Chicago: AHM Publishing Corp.

Métraux, A. 1928. La civilisation matérielle des tribus Tupi-Guarani. Paris: *Librairie Orientaliste Paul Geuthner.*

——. 1948. The Guarani. In J. H. Steward, ed., *Handbook of South American Indians.* Vol. 3, *The Tropical Forest Tribes,* 69–94. Washington, D.C.: Smithsonian Institution.

——. and H. Baldus. 1946. The Guayaki. In J. H. Steward, ed., *Handbook of South American Indians.* Vol. 2, *The Marginal Tribes,* 435–444. Washington, D.C.: Smithsonian Institution.

Moore, D. A. and I. L. Maciel. 1987. *Relatório da consulta lingüística à frente de atração Rio Tapirapé.* Xerox, Fundação Nacional do Indio Brasília, Brazil.

Noronha, J. M. de. 1856. Roteiro da viagem da cidade do Pará até as últimas colonias dos dominios portugueses em os rios Amazonas e Negro. In *Noticias para a história e geographia das nações ultramarinas,* Vol. 6. Lisbon.

Pesce, C. 1985. *Oil Palms and Other Oilseeds of the Amazon.* Edited and translated by D. V. Johnson. Algonac, Michigan: Reference Publications.

Pinheiro, C. V. B. and M. J. Balick. 1987. Brazilian palms. In *Contributions from the New York Botanical Garden,* Vol. 17. Bronx, New York: The New York Botanical Garden.

Posey, D. A. 1985. Indigenous management of tropical forest ecosystems: The case of the Kayapó Indians of the Brazilian Amazon. *Agroforestry Systems* 3:139–158.

Prance, G. T. and H. O. R. Schubart. 1977. Notes on the vegetation of Amazonia I: A preliminary note on the origin of the open white sand campinas of the lower Rio Negro. *Brittonia* 30(1):60–63.

Rivet, P. 1924. Les Indiens Canoeiros. *Journal de la Société des Americanistes* n.s., 16:169–181.

Rodrigues, A. D. 1978. A lingua dos indios Xetá como dialeto Guarani. *Cadernos de Estudos Lingüisticos* 1:7–11.

Roosevelt, A. 1987. Chiefdoms in the Amazon and Orinoco. In R. D. Drennan and C. A. Urike, eds., *Chiefdoms in the Americas,* 153–185. Lanham, Md.: University Presses of America.

Sapir, E. 1949. Time perspective in aboriginal American culture: A study in method. In D. Mandelbaum, ed., *Selected Writings of Edward Sapir,* 389–462. Berkeley, Calif.: University of California Press.

Schulz, J. P. 1960. *Ecological Studies on Rain Forest in Northern Surinam.* Amsterdam: N.V. Noord-Hollandsche Uitgevers Maatschappij.

Smith, N. J. H. 1980. Anthrosols and human carrying capacity in Amazonia. *Annals of the Association of American Geographers* 70(4):553–566.

Smole, W. 1976. *The Yanoama Indians: A Cultural Geography.* Austin, Tex.: University of Texas Press.
Sombroek, W. G. 1966. *Amazon Soils: A Reconnaissance of the Soils of the Brazilian Amazon Region.* Wageningen, Netherlands: Center for Agricultural Publications and Documentation.
Sousa, G. S. de. 1974. *Notícia do Brasil.* São Paulo, Brazil: Departamento de Assuntos Culturais do MEC.
Spaulding, A. C. 1960. The dimensions of archaeology. In G. E. Dole and R. L. Carneiro, eds., *Essays in the Science of Culture in Honor of Leslie A. White,* 437–456. N.Y.: Crowell.
Staden, H. 1930. *Viagem ao Brasil.* Translated from the German (1557) by A. Lofzren. Rio de Janeiro: Officina Industrial Graphica.
Steward, Julian H. 1947. American culture history in the light of South America. *Southwestern Journal of Anthropology* 3:85–107.
——. 1949. South American cultures: An interpretative summary. 669–772. In *Handbook of South American Indians.* Vol. 5, *The Comparative Ethnology of South American Indians.* Washington, D.C.: Smithsonian Institution.
——. 1955. *Theory of Culture Change: The Methodology of Multilinear Evolution.* Urbana, Ill.: University of Illinois Press.
Steward, Julian H. and L. C. Faron. 1959. *Native Peoples of South America.* New York: McGraw-Hill.
Toral, A. 1986. *Situação e perspectivas de sobrevivência dos Avá-Canoeiro.* Xerox, São Paulo, Brazil: Centro Ecumênico de Documentação e Informação.
Vasconcellos, S. de. [luu3] 1865. *Chronica da Companhia de Jesus no Estado do Brasil.* Lisbon: A. J. Fernando Lopes.
Vellard, J. 1934. Les indiens Guayaki. *Journal de la Société des Americanistes* 26:223–292.
——. 1939. *Une Civilisation du miel.* France: Gallimard.
Viveiros de Castro, E. 1986. *Araweté: Os deuses canibais.* Rio de Janeiro, Brazil: Jorge Zahar.
Wessels Boer, J. G. 1965. Palmae. In J. Lanjouw, ed., *Flora of Suriname,* Vol. 5, Part 1. Leiden, Holland: E. J. Brill.
Wilbert, J. 1969. Textos folkloricos de los indios Waraos. *Latin American Studies, Vol. 12.* Los Angeles: University of California.
Willey, G. R. and J. A. Sabloff. 1980. *A History of American Archaeology.* 2nd ed. San Francisco: W. H. Freeman.
Whitten, R. G. 1979. Comments on the theory of Holocene refugia in the culture history of Amazonia. *American Antiquity* 44(2):238–251.

FOUR

Traditional Productive Systems of the Awa (Cuaiquer) Indians of Southwestern Colombia and Neighboring Ecuador

JORGE E. OREJUELA

The department of Nariño in southwestern Colombia has an extraordinarily rich biological legacy and is home to a sizable population of indigenous people. Accelerated rates of primary forest loss and ecological decline threaten the resource base and consequently the continued viability of the indigenous Awa (Cuaiquer) population. A small nature reserve, La Planada, was established in 1982 to promote nature conservation and appropriate community development in the region (Cobo 1982). In 1988 a large biosphere reserve was proposed for inclusion in UNESCO's Man and the Biosphere (MAB) program, comprising both the entire Awa territory and non-Indian areas inhabited by *campesino/colono* people. The involvement of the local populations of this as well as other tropical areas in conservation and in the promotion of sound ecological and economic natural-resource-use activities is essential to their long-term well-being. However, unless major changes in land use occur, neither the resources nor the traditional rural communities have a chance of survival.

The intermittent expansions and contractions of vegetational lifezones during the Quaternary period resulted in significant interactions between the lowland Chocoan (Pacific) and Northern Andean biotas (Haffer 1967; 1974). The Western slope of the Andes between Ecuador and Colombia therefore acquired an enormously rich biological heritage with many endemic organisms that occur nowhere else on earth. This important biogeographic region, part of the Southern

Choco Pleistocene Refuge (Gentry 1982) includes well-developed mangroves, major river systems, and mountainous terrain with mostly humid and pluvial forest formations in the tropical and subtropical regions. Some of the upper-elevation forests (1,200 to 2,800 meters) are typical cloud forests. The characteristic feature of the region is the extremely high annual precipitation, which varies between 4,000 and 8,000 millimeters per year. Fifteen thousand Awa people live on the Pacific slope (twelve thousand in Colombia and three thousand in Ecuador). Their settlements are dispersed over a mostly forested area of more than 1 million hectares, straddling the international border between the Mira, Guiza, and Telembi drainage systems (Binational Plan Awa Project 1987). The Awa are descended from people who drew upon primary and secondary forests for food, medicine, fuel, and fibers and created appropriate agricultural production systems in the world's wettest rain forest. Understanding the ecological constraints imposed by the environment (high rainfall, low nutrient content of soils, steep hillsides, complex timing of phenological events in a heterogeneous environment) helps to explain why the Awa are dispersed thinly across this complex spatiotemporal system (Martinez et al. 1984; Cerón 1986). In clearings within primary and secondary forests the Awa developed in isolation, settling remote riverine systems in a fragile but uncontested environment. The isolation of the Awa resulted in the fortunate outcome that neither their land nor its resources were coveted or expropriated by either the highland or lowland tribes, by Spaniards during the conquest or colonial times, or by recent wealthy landowners or large-scale developers (MacDonald 1986). The region's difficult access and relative desolation has not attracted, thus far, either permanent guerrilla groups or well-established coca growers or traffickers. In fact, the Awa have been little known even to local Indian organizations. The present-day territories of the Awa, however, are far from isolated, since they are surrounded on all sides by human settlements. Territorial expansion is not possible, and loss of occupied territory is taking place rapidly as new roads are opened up or old ones are improved and a corresponding influx of colonists and land speculators continues.

Campaigns exist on the national and international level to prevent loss of Awa territories and the wildlife within them. These include ongoing programs both in Ecuador (the Awa Plan's technical unit, formerly the Tobar Donoso project, which manages the Awa Ethnic Reserve [Villareal 1988]), and in Colombia (the La Planada Nature Reserve's efforts to combine nature conservation and study with rural development activities [Orejuela et al. 1983; Samper and Orejuela 1988]). For slightly longer than one year the two countries have been working together, using the guidelines of UNESCO's MAB program in the development of the proposed Awa International Biosphere Reserve, with terms of reference for the control of the Awa lands agreed upon by the two countries' interagency committees and Awa representatives (Binational Plan Awa Project 1987).

The traditional knowledge of the Awa in regard to the management of natural resources is indeed an extremely valuable asset that must be tapped and at the same time preserved. With the goal of achieving sustainable utilization of the rich but fragile resource of the western slopes of the Andes, it is appropriate to look at Awa production systems.

Awa Productive Systems

Most semi-isolated Awa settlements, like the one at Pialapí (where most of this study's observations were made), consist of about 150 persons. The settlement is a social unit with a discrete community organization evident especially in *mingas,* or communal work groups that are effectively used to improve trails, construct bridges, and so forth. Within the settlements, the agricultural-production and landholding units are family groups of five to seven people (Osborn 1969–72). Within the family unit the elders control the organization of work groups, which initiate large endeavors like clearing additional forest; the elders also own seeds. However, the male members of the family (i.e., the sons) have the right to harvest part of the crop. The group of brothers is the core of the work group (Osborn 1969–72). Inheritors of hunting/gathering and shifting agricultural traditions, the present-day Awa have developed three integrated production systems:

1. A type of subsistence (shifting) agriculture known as slash-mulch.
2. A rather intensive agroforesty production system that includes the maintenance of several crops and animals in sizable house gardens (Osborn and Melo, 1986; Acosta, 1988).
3. Hunting/gathering activities in primary and secondary forests.

Several factors determine the relative contribution of each production mode. The fertility of the soil influences the duration of farming and fallow periods, the preponderance of agriculture versus hunting/gathering, and the practice of slash-mulch agriculture versus cattle-raising. Naturally, soil fertility is often related to population density and to the intensity and manner of land use (Goodland 1985). The ever-present concern of the Awa is therefore the maintenance of soil fertility, and their production systems attempt to keep a balance between nutrient losses and gains.

The availability of game and fish is also important; as game and fish stocks are reduced, agriculture gains preponderance. This situation prevails in and around Pialapí and other communities in the subtropical zone, while in the tropical region agriculture is less intensive, but hunting, fishing, and gathering activities are more significant.

Subsistence Agriculture

Subsistence farming is by far the most important occupation of the Chocoan/Andean populations. Good farmland is scarce, for most of the soils are highly leached and infertile. As in most humid tropical areas a type of shifting agriculture prevails. Today the basic crops represent a mixture of Old- and New-World plants. Indian maize and Old World plantains are the main staples of the diet and the main source of starch foods. The native sweet manioc and other American tubers, the New World *chontaduro* (Bactris gasipaes), and the Old World sugarcane are other important elements of the diet (West 1957; Osborn 1986). Awa shifting cultivation essentially involves the cutting of natural vegetation and the mulching of this material to prepare a temporary agricultural field. Once yields decline after several cycles of harvest, the field is left fallow and permitted to regrow as an enriched secondary forest. The system is employed effectively by the Awa in the growing of both maize and, to a lesser degree, beans. A system similar in general form is used to clear lots to raise cattle. The woodlot is slashed, some mulching is done, and grass is planted; when the grass is fully grown, cattle are introduced into the fields.

Maize Maize is cultivated by means of the slash-mulch method. First seeds are broadcast into tall second-growth vegetation, which is then cut and permitted to rot over the seeds. The sprouting maize takes root in the shallow humus layer on the surface and appears through the rotting mulch in a few days' time. The ears are ready for harvest from five to eight months after planting, depending on the variety (Martinez et al. 1984). Usually two crops of corn are raised in different plots per year. One is planted in January and the other in August, with both growing periods coinciding with the months of least rainfall. The Awa family usually maintains three or four maize fields on hillsides surrounded by forest. In Pialapí one site was found to have been fallow for only three years, without reduction in yields. This fallow time is significantly lower than that reported in the literature for other Awa regions, where a five to seven years' fallow was the norm. Successive cropping in the same field, however, appreciably lowers the yield (J. M. Guanga, 1986–88, pers. comm.). Evidently the fallow time in this production unit allows for the contribution of adequate amounts of nutrients to the field. The production of maize was estimated at about 100 to 150 kilograms per hectare. If all the corn were sold, it would provide the family with almost the equivalent of a minimal annual wage (Acosta 1988; Acosta and Jaramillo 1988).

Beans and Other Crops Like maize, short-cycle varieties of red and white beans are planted using the slash-mulch system. The Awa also have a "permanent" or perennial type of bean (a shrubby vine type with year-round production) that is part of the house garden. Intercropping of beans and maize is common, with dry stalks used as poles for the beans. A system of *roza* (slash of primary or secondary forest) to plant grass (*gramalote,* or *Axonopus* sp.) for raising cattle is also practiced by the Awa, with households typically having two to three fields of about 2 hectares each. Cattle commonly are raised *a medias* (in partnership) with an outsider *campesino,* frequently a *compadre* partner.

The fallow period of shifting agriculture can be subjected to improvements depending on the time available and the size of the family of the production unit. Such improvement can be accomplished with the planting of selected species that will yield important byproducts and enrich the soil. The Awa occasionally improve fallows by leaving certain trees in the maize field, including fast-growing arboreal and shrubby legumes like *Inga.* Other plant species left in the field serve to attract game animals; these include fruit trees like *Brosimum, Ficus,* and various palms. These second-growth sites are visited periodically by the Indians to trap spiny-rats or to cut firewood. If the fields are reasonably close, the Awa take pigs to the recently harvested area to feed on crop residues, to turn over the soil, and to enrich it with manure. In reality, the field is not abandoned in the true sense of the word, but rather enriched and tapped for products on a regular but seasonal basis.

Agroforestry Practices in Awa House Gardens

Agroforestry is a technique that combines the production of annual crops and tree species simultaneously or sequentially on the same parcel of land. The typical polyculture of crops, palms, and fruit trees grown by the Awa around their farmhouses in a heterogeneous mixture is actually a very efficient agroforestry pattern in excellent accordance with the local environment (Holt and Ferguson 1986). Several of the elements of the garden are perennial species; some, like plantains and bananas, sweet manioc, and sugarcane, are dependent on long-cycle planting. The maintenance of this system demands constant attention; it is a rather intensive polyculture.

In planting gardens of .5 hectare up to 2 hectares, the Awa choose from a wide assemblage of crop species. Plantains, manioc, and sugarcane are planted in virtually all fields. Fruit trees including *caimito* (*Chrysophyllum*), *madroño* (*Rheedia*), *guayaba* (*Psidium*), *guabo* (*Inga*), *chontaduro* (*Bactris*), avocado (*Persea*), *Citrus* spp., and cacao (*Theobroma*) are commonly planted. Tubers, transplanted ornamentals (*Hibiscus*, anthuriums, and orchids), and timber species are optional plants. Tomatoes, herbs, medicinal plants, a few coffee bushes, tree tomatoes, *lulo* (*naranjilla*), *aji* (chili) peppers, and fish-stunning plants (*barbascos*) are regularly found in Awa gardens.

The combination of plants, their maturation schedules, and spatial arrangements determine the management alternatives for the Indian farmer. A brief discussion of some of the most important plants of the garden follows, and a composite list of plant species appears in Table 4.1.

Plantains Today plantains are considered to be the main element of the diet of the Awa. Five varieties were observed under cultivation in Pialapí: the large-fruited *hartón*, *dominico*, *enano*, *chiro*, and *maqueño*. Since plantains bear no viable seed, they are dependent on man for reproduction, which is carried out by planting the rootstock, or *colino*. After planting the *colinos*, the surrounding bush is cut down, forming

Table 4.1
Principal Plants and Animals Used by the Awa Indians of Southwestern Columbia

Family	Scientific Name	Common Name	Uses
Plants			
Actinidiaceae	*Saurauía*	*moquillo*	edible fruit
Acanthaceae	*Trichanthera gigantea*	*nacedero*	live fence
	Fittonia	*contra*	medicinal
Amaranthaceae	*Alternanthera*	*escansel*	medicinal
Anacardiaceae	*Mangifera indica*	mango	edible fruit
	Camnosperma panamensis	*sajo*	construction
Annonaceae	*Annona muricata*	*guanábana*	edible fruit
	Rollinia mucosa	cherimoya	edible fruit
	Xylopia	*rayado*	construction
Araceae	*Caladium/Anthurium*	*anturio*	ornamental
	Colocasia	*papa cum*	edible tuber
	Xanthosoma jacquinii	*cum*	edible tuber
Aristolochiaceae	*Aristolochia*	*guaco. contra*	medicinal
Bignoniaceae	*Tabebuia guayacan*	*guayacán*	construction
	Crescentia cujete	*totumo*	utilitarian
Bixaceae	*Bixa orellana*	*achiote*	food coloring
Bombacaceae	*Ceiba pentandra*	ceiba	construction
	Matisia alata	*sapote*	edible fruit
	Ochroma pyramidale	*balso*	construction/ utilitarian
	Quararibea sp.	*sapotillo/castaño*	construction
Boraginaceae	*Cordia alliodora*	*laurel*	firewood/ construction
Bromeliaceae	*Aechmea*	*pita*	utilitarian
	Ananas comosus	pineapple	edible fruit
Burseraceae	*Dacryodes*	copal	construction/ medicinal
	Protium	*anime*	construction
	Trattinnickia	*pulgande*	construction
Cannaceae	*Canna* sp.	*achira*	utilitarian
Caricaceae	*Carica papaya*	papaya	edible fruit
	Jacaratia spinosa	*papayuelo*	edible fruit
Chenopodiaceae	*Chenopodium* sp.	*paico*	medicinal
Compositae	*Matricaria* sp.	*manzanilla/ chamomillo*	medicinal
	Pollaslesta klugii	*kinde*	construction
Convolvulaceae	*Ipomoea batatas*	*camote*	edible tuber
Cucurbitaceae	*Sechium edule*	*cidra*	edible fruit
Elaeocarpaceae	*Sloanea*	*altaquer*	construction
Ericaceae	*Macleania* spp. *Psammisia*	*chaquilulo*	edible fruit
Erythroxylaceae	*Erythroxylum*	coca	medicinal
Euphorbiaceae	*Alchornea*	*algodoncillo*	construction

Table 4.1
Continued

Family	Scientific Name	Common Name	Uses
Euphorbiaceae	*Croton*	*sangregae*	medicinal
	Hieronyma chocoensis	*motilón*	construction/ firewood
	Manihot esculenta	sweet manioc (*yuca*)	edible tuber
	Sampium utile	*barbasco*	pesticide
Gesneriaceae	*Cremosperma*	*contra*	medicinal
Graminae	*Guadua angustifolia*	*guadua* (bamboo)	construction/ utilitarian
	Gynerium	*caña brava*	construction
	Saccharum officinarum	sugarcane	edible
	Zea mays	corn	edible
Guttiferae	*Clusia*	*incienso*	firewood
	Rheedia madruno	*madroño*	edible fruit
	Vismia ferruginea	*mancharropa*	construction
Hippocastanaceae	*Billia colombiana*	*cariseco achiotillo*	construction/ firewood
Humiriaceae	*Humiriastrum rocera*	*chanul*	construction
Labiatae	*Salvia*	salvia/*menta*	medicinal
Lauraceae	*Aniba perulitis*	*jigua./malde*	construction
	Beilschmiedia	*jigua*	construction
	Nectandra	*palealte./ gualpante*	construction/ game attractants
	Ocotea	*chimbusté/ chachajo*	construction
	Persea americana	avocado	edible fruit
Lecythidaceae	*Eschweilera*	*tete*	construction/ firewood
	Lecythis	*aray*	edible nuts
Leguminosae	*Erythrina edulis*	*chachafruto*	edible fruit/ fodder
	Erythrina poeppigiana	*mambla./porró*	fodder
	Platymiscium	*caobo*	construction
	Inga edulis	*guaba*	edible fruit/ multiple use
	Inga spectabilis	*guaba lomomachete*	edible fruit/ firewood
	Inga spp.	*guabo*	firewood
	Tamarind indicus	*tamarindo*	edible fruit/ firewood
	Phaseolus vulgaris	bean	edible fruit
	Pithecellobium	*chipero*	edible fruit/ firewood

Table 4.1
Continued

Family	Scientific Name	Common Name	Uses
Malpighiaceae	*Banisteriopsis*	*barbasco*	roots stun fish
Malvaceae	*Hibiscus*	*resucitao*	ornamental
Maranthaceae	*Calathea* spp.	*bijao/chinula*	utilitarian/ thatch
	Unknown	*tetera*	utilitarian/hats
Melastomataceae	*Miconia*	*amarillo*	construction/ firewood
Meliaceae	*Carapa guianensis*	*tangare*	construction/ canoes
	Cedrela odorata	*cedro*	construction/ utilitarian
	Guarea	*pialde*	construction
	Guarea sp.	*piaste*	construction/ firewood
Moraceae	*Artocarpus*	*fruta de pan*	edible fruit
	Brosimum utile	*sande*	edible fruit/ fodder
	Castilla ulei	*caucho*	construction/ utilitarian
	Cecropia	*yarumo*	game attractant/ firewood
	Ficus	*higuerón*	game attractant
	Perebea	*cabecita*	poisonous bark
	Poulsenia armata	*damajagua*	utilitarian
	Pourouma	*uva*	edible fruit
	Maclura tinctoria	*moral*	utilitarian
Musaceae	*Musa paradisiaca*	*plátano*	edible fruit
	Musa	banana	edible fruit
Myristicaceae	*Otoba*	*cuángare*	construction/ firewood
	Virola	*chalvinde*	construction/ medicinal/ game attractant
Myrtaceae	*Eugenia*	*pomorroso*	edible fruit/ firewood
	Myrcia	*arrayán*	edible fruit/ firewood
	Psidium guajava	*guayabo*	edible fruit/ firewood

Table 4.1
Continued

Family	Scientific Name	Common Name	Uses
	Psidium sp.	*guayabillo*	edible fruit/ game attract
Olacaceae	*Minquartia*	*guayacán*	construction
Palmae	*Attalea*	*táparo*	edible fruit
	Bactris gasipaes	*chontaduro* (peach palm)	edible fruits/ multiple purpose
	Catablastus	*gualte*	construction
	Euterpe	*naidí*	palmito/ heart-of-palm
	Geonoma	*gualte/chaler*	construction
	Orbignya	*corozo*	construction
	Phytelephas	*tagua*	utilitarian
	Prestoea	*palmito*	heart-of-palm
	Socratea	*pambil*	construction
	Wettinia	*gualte*	construction
Passifloraceae	*Passiflora ligularis*	*granadilla*	edible fruit
Piperaceae	*Piper*	*cordoncillo*	medicinal
	Peperomia	*contra*	medicinal
Rubiaceae	*Cephaelis*	*cascarilla*	utilitarian
	Cinchona	*cascarilla/quina*	medicinal
	Genipa americana	*jagua*	utilitarian
	Coffea arabica	coffee	food/ medicinal
Rutaceae	*Citrus sinensis*	*naranja*	edible fruit
	Ruta graveolens	*ruda*	medicinal
Sapotaceae	*Chrysophyllum cainito*	*caimito*	edible fruit
	Pouteria sp.	*sapote*	edible fruit
	Mammea	*mamey*	edible fruit
Solanaceae	*Solanum quitoensis*	*lulo/naranjilla*	edible fruit
	Cyphomandra betacea	tree tomato	edible fruit
	Datura sp.	*borrachero/ floripond*	medicinal
	Capsicum frutescens	*aji/chile*	edible fruit
	Lycopersicum	tomato	edible fruit
	Cestrum sp.	*gallinazo*	construction/ firewood
	Nicotina tabacum	*tabaco*	utilitarian/ medicinal
Sterculiaceae	*Guazuma ulmifolia*	*guácimo*	firewood
	Theobroma cacao	cacao	edible fruit
Theaceae	*Freziera*	*chimbuste*	construction
Tiliaceae	*Apeiba*	*peine de mono*	construction

Table 4.1
Continued

Family	Scientific Name	Common Name	Uses
Umbelliferae	*Arracacia xanthoriza*	*arracacha*	edible roots
Ulmaceae	*Celtis*	*gallinazo*	construction
	Trema	*isinde*	construction
Verbenaceae	*Verbena*	*verbena*	medicinal
	Aloysia	*cedrón*	medicinal
Animals			
Amphibia			
Dendrobatidae	*Dendrobates histrionicus*	poison-arrow frog	utilitarian (poison)
Reptilia			
Iguanidae	*Iguana iguana*	common iguana	food
Aves			
Tinamidae	*Tinamus/Crypturellus*	tinamou	food
Anatidae	*Cairina moschata*	Muscovy duck	food (poultry/ eggs)
Cracidae	*Penelope/Chamaepetes Crax.*	guan	food
Phasianidae	*Odontophorus, Rhynchortyx*	wood quail	food
Columbidae	*Columba* spp.	pigeon	food
	Geotrygon spp.	dove	
Psittacidae	*Ara*	parrot	food
	Amazona		
	Pionus		
	Pionopsitta		
Ramphastidae	*ramphastos*	toucan	food
Mammalia			
Didelphidae	*Didelphis marsupialis*	oppossum	food
Cebidae	*Alouatta*	howler monkey	food
	Ateles	spider monkey	
	Cebus	white-faced monkey	
Myrmeco-phagidae	*Tamandua aetradactyla*	anteater	food
Bradypodidae	*Bradypus*	three-toed sloth	food
	Choloepus hoffmani	two-toed sloth	food
Dasypodidae	*Dasypus novemcinctus*	nine-banded armadillo	food
Leporidae	*Sylvilagus brasiliensis*	cotton-tailed rabbit	food

Table 4.1
Continued

Family	Scientific Name	Common Name	Uses
Sciuridae	*Sciurus granatensis*	red-tailed squirrel	food
	Microsciurus	dwarf squirrel	food
Erethizontidae	*Coendu prensilis*	prehensile-tailed porcupine	food
Dasyproctidae	*Agouti paca*	paca	food
	Dasyprocta punctata	agouti	food
Echimydae	*Hoplomys gymnurus*	spiny rat	food
	Proechimys semispinosus	spiny rat	food
Ursidae	*Tremarctos ornatus*	spectacled bear	food
Tayassuidae	*Tayassu pecari*	collared peccary	food
Cervidae	*Odocoileus virginianus*	white-tailed deer	food
	Mazama americana	brocket deer	food

a rotting mulch through which the shoots emerge. Harvest takes place after nine to twelve months. An Awa family consumes between four and six stems of fruit every week, indicating clearly their dependence on this crop. *Chiro* plantains are generally fed to farm animals. With good cultivation and mulching, plus fertilization of the soil with manure, a good plantain field will yield for years without replanting. Frequently, however, the cycle lasts between seven and ten years.

Bananas are usually grown in the same manner as plantains but are of secondary importance. Several varieties are cultivated.

Root Crops The most important root crop is sweet manioc, or *yuca* (*Manihot esculenta*). Fairly sizable hillside plots, often intercropped with plantains, are seen in the Awa garden, where they are cultivated by the slash-mulch process. At maturity, between six and eight months after planting, the long, slender roots are dug and prepared by boiling or frying. The Awa never grate the tubers to make *fariña,* as is done in the Amazon.

The native aroid tubers, similar to taro, locally known as *papa china* or *papa cum* (*Colocassia* and *Xanthosoma*) are rich in starch and are eaten boiled or baked. *Arracacha* (*Arracacia xanthoriza*) is also intercropped with *yuca*, hot peppers, and *cum.* The roots have a high starch content.

Sugarcane Since its introduction by the Spaniards in colonial days, sugarcane has become an important subsistence crop grown by Indian farmers in small plots (.5 to 1 hectare) in the house garden. The main products made of the cane juice, which is extracted in homemade sugar mills (*trapiches*) are *guarapo*, or fermented juice; *chapil*, or homemade *aguardiente* (brandy); and *panela*, the block, raw brown sugar, prepared by cooking the juice in copper caldrons. This sweet concentrated sugar is the most important carbohydrate source.

Sugarcane is grown by the Awa and along the Pasto-Tumaco road. The system begins with slashing, clearing debris, and sowing the shoots on hillsides. The field is then weeded after two to three months, the leaves are removed and worked into the ground together with the suckers. The first cutting of mature, ripe stalks comes one year after planting.

Three varieties of sugarcane are common in the Awa area. The purple variety is the best juice producer. A container called a *paila* is used to make *panela*. The *cachasse*, or supernatant debris, produced by the boiling process is removed, along with some impurities, by employing the bark of *balso* (*Ochroma*), which releases a sticky sap to which the *cachasse* adheres. Waxy laurel is used to thicken the product just prior to pouring over the hardwood block molds. The *panela* bricks are wrapped in *bijao* (*Calathea*) leaves for transport. In the Awa territory the most common mills are still made of wood and they are drawn by horses. A few Indians are now using metal presses, which are more efficient in the extraction of the juice.

Chontaduro (Peach Palm) Probably the cultivated plant most characteristic of indigenous cultivation in the region is *chontaduro* (*Bactris gasipaes*). This spiny-trunked palm is grown in clumps in gardens and in fallows. It yields large quantities of yellow fruits of mealy pulp rich in starch, protein, carotene, and other important dietary elements (West 1957). *Chontaduro* is harvested in Pialapí in the drier months, the largest crop coming in August and a smaller one in January. *Chontaduro* is a multi-purpose plant that provides, in addition to the fruit, construction material, thatch, and heart-of-palm. The seeds are rich in oil and are consumed by several animals.

Guadua (Bamboo) Bamboo is another multiple-use plant; it provides excellent construction material when used in round form or as *esterillas* (cut slats), and is used for musical instruments (including the resonant chambers of the marimba and, when filled with *achiras* (canna lily seeds), the *guasa*). Stakes for crops and fences for animal pens are some of the many other uses of this *Gramineae.*

Fruit Trees The Awa use a wide selection of native American fruit trees in their gardens. These include: papaya (*Carica papaya*), *madroño* (*Rheedia chocoensis*), *caimito* (*Chrysophyllum caimito*), *pomarroso* (*Eugenia* sp.), *guayaba* (*Psidium guajava*), *guamo* (*Inga edulus, I. spectabilis*), avocado (*Persea americana*), *achiote* (*Bixa orellana*), *aji* or chili (*Capsicum* spp.), tomato (*Lycopersicum*), tree tomato (*Cyphomandra*), and *lulo* or *naranjilla* (*Solanum quitoensis*). Less frequently grown in Awa gardens are the annonaceous *guanábana* (*Anona muricata*) and cherimoya (*Rollinia*): *zapote* (*Matisia cordata*), *mamey* (*Mammea*), cacao (*Theobrama*), *aray* (*Lecythis*), melons and pineapples, *uvilla* (*Pourouma*), *sande* (*Brosimum*), *moquillo* (*Saurauia*), and *uvo de monte* or *chaquilulo* (*Ericaceae*). Among the Old World fruits, the breadfruit (*Artocarpus*) is common in the lowlands. Coffee and citrus are also grown in small quantities.

Medicinal Plants Like many forest groups, the Awa rely on a variety of forest products for their traditional healing practices, and they maintain medicinal plants in their gardens. Among these are the *contra,* used against snakebites (of the Piperaceae, gesneriads, and some melastomes (Holm-Nielssen and Barford 1984). A Pialapí *curandero* (shaman) pointed out *guaco,* a vine used against *ambucama* (*Bothrops* spp.) bites; this plant is probably *Aristolochia,* reputed for its antibothropic powers. *Dacryodes* produces a resin used to heal wounds, and *panela,* the raw sugar, is commonly used as an excellent wound-healing substance.

Several euphorbs including *Sapium,* sangregao (*Croton*), and certain ferns are said to relieve stomach pains. In many Awa gardens it is common to see *barbasco* (*Lonchocarpus* sp.), the fish-stunning plant, under cultivation. When pressed, *barbasco* produces a milky substance, similar in action to rotenone, which is probably the active ingredient in most *barbascos. Chutún* is a widespread generic name for several ailments of the Awa, which include symptoms not unlike those of influenza. These conditions are treated with magical plants (perhaps *Banisteriopsis*) by shamans. (Detailed studies in the field of ethnobotany will undoubtedly yield important results.)

Other Plants of Awa Gardens Besides food and medicinal plants, there are usually a few herbs and spices like *cidrón* (*Aloysia*), and coriander (*Coriandrum sativum*) in Awa gardens. Other useful cultivated trees or shrubs include *totumo* or gourd tree (*Crescentia*) and *bijao* (*Calathea*), grown rather extensively between permanent crops such as sugarcane and plantains. *Bijao* is maintained for the repair of roofs of sugar mills and kitchens and for wrapping all sorts of items like *panela*, chickens, *chontaduro*, chicken coops, and *avio* (food taken on journeys or excursions to distant cornfields). *Tetera* is a maranthaceous plant whose processed stem is used in the manufacture of hats similar to Panama hats. *Cosedera* (*Casearia*) is a shrub whose stems are cut, peeled, pounded, washed, and dried to make fibers or wound into multiple-use string that is frequently worked into *jigras*, or fine bags, made by women. Men are responsible for making *canastos*, or burden baskets, used for many purposes. Other types of fine bags are made from a bromeliad (*Aechmea*). The Indian tumpline for carrying burden baskets is made from the bark of *balso* (*Ochroma*). In the lowlands, the moraceous tree *damajagua* (*Poulsenia armata*) still is used occasionally to make skirts and sleeping mats by pounding the bark. *Catangas* are baskets used to collect freshwater shrimp (*Macrobrachium*).

Large aroids like elephant-ear *rascadera* (*Anthurium* sp.) are the umbrellas of the Awa, and thin vines (*bejucos*) are used to tie all kinds of items.

Among the preferred ornamental plants, the Awa use anthuriums, *Hibiscus*, and rain-of-gold orchids (*Oncidium*), canna lilies, crotons, and *francesinas*.

Animal Husbandry The plant production of house gardens is complemented by animal husbandry on a modest scale, with pigs, chickens, guinea pigs, and ducks most frequently raised. The chickens roam in the garden looking for food and are given cooked plantains, *cum*, and sweet corn. In some families, fairly sizable populations of up to thirty-five chickens are maintained, providing eggs almost every day. Pigs normally are rustic varieties resistant to diseases and inclement weather. They are allowed to roam the gardens near the house, and in the interharvest period they are taken to the subsistence plots, where they fatten on crop residues. Near the house, they are fed cooked plantains, *cum*, and sugarcane shoots. Pigs are usually sold in town; Awa rarely eat them. Guinea pigs (*cuyes*), common in the subtropical Awa communities, represent an influence from the highlands and provide an important source of protein. Muscovy ducks (*Cairina moschata*), hardy domesticated lowland ducks, are frequently kept in substantial numbers by the Awa. Some wild fowl like *pavas* (*Penelope* spp.) and wood quail (*Odontophorus*) are maintained in pens until an occasion to consume them appears. Horses are kept as burden animals and to operate the sugar mills. Most of the cows are maintained in the fields far away from the house, where they are fattened for sale.

Use of Forest and River Ecosystems

A variety of forest and river products are tapped by the Awa through hunting, fishing, and gathering activities as well as small-scale timber exploitation for various purposes; most of these products are exploited for subsistence.

Hunting The Awa derive a significant part of their diet from hunting. This activity is particularly important for the more isolated settlements in the lower-elevation tropical forest. Today the animals most hunted for meat are the large rodents agouti or *guatin* (*Dasyprocta punctata*) and, to a lesser extent, the paca or *guagua* (*Agouti paca*). Because these species consume hard nuts from trees like *Dypterix* and some palm trees, the Awa usually hunt them in locations where these are found. A good variety of Echimyid rodents also are commonly hunted with ingenious traps baited with *yuca* and maize. Among these, *Proechymys semispinosus* and *ratón puyoso* (*Hoplomys gymnurus*) are of great importance. This type of hunting is done on a cyclical basis, which allows sufficient time for the recovery of wild populations. Ungulates have been somewhat depleted, possibly through overhunting mostly by groups of outsiders. The collared peccary (*Tayassu tajasu*) can still be found and is sought for its fine meat. The small brocket deer (*Mazama americana*) is still plentiful, and the much rarer white-tailed deer and pudu are also hunted. The lowland tapir (*Tapirus bairdii*) is now extremely rare and little known, even to the Indians.

Many populations of animals regularly hunted by Indians, blacks, and mestizos have been drastically reduced to the point that several of them have become rare, and their status is now highly vulnerable. Victims of this type of overexploitation, even by native Awa populations, are monkeys (the black howler, *Allouatta palliata*, and the black spider monkey, *Ateles seniculus*), the spectacled bear (*Tremarctos orna-*

tus), and tapir (*Tapirus bairdii*). Many of the small mammals such as the oppossum (*Didelphis*), the two-toed sloth (*Choeleopus*), the three-toed sloth (*Bradypus*), anteater (*Tamandua*), nine-banded armadillo (*Dasypus*), and the prehensile-tailed porcupine (*Coendu*) have also been significantly reduced through hunting. Perhaps the only still-abundant species are the squirrels (*Sciurus* and *Microsciurus*) and the cotton-tailed rabbit (*Sylvilagus*). Large snakes such as boas and venomous types such as the fer-de-lance (*Bothrops*) and bushmaster (*Lachesis*) are seldom hunted for meat, although the venomous snakes are killed whenever they are found. The variety and density of these snakes is a significant cause of mortality in the Awa territory, and a significant portion of the traditional curative medicine is devoted to counteracting snakebites. Other reptiles commonly killed are the common iguana (*Iguana iguana*) which is hunted for its meat, as are the large birds like the cracids (guans [*Penelope* spp.] and curassows [*Crax*]) and other large birds like toucans (*Ramphastos*), *Amazona* parrots, pigeons (*Columba*) and wood quail (*Odontophorus*).

Several animals are killed to fend off crop destruction; these include the small flocking parrots (*Pionus* and *Aratinga*), coatis (*Nasua*), and oppossums as well as animals that prey on chickens, such as weasels (*chucuros* or *comadrejas* [*Mustela*]). Since felids and bears can become accustomed to preying on farm animals, the puma (*Felis concolor*), ocelot (*Felis pardalis*), jaguar (*Felis onca*), and bear (*Tremarctos ornatus*) are also killed. Bears are killed with alarming frequency when they come down to feed on the tender maize crop. This is particularly the case at higher elevations, where Awa agriculture is more intensive and has resulted in a reduction of habitat available to bears.

The Awas' comprehensive knowledge of the natural ecosystems is evident in their hunting practices. They are, for example, aware of feeding associations of birds and mammals at fruiting trees. Their knowledge of the phenology of trees, their acute powers of observation, and excellent site memory of the animals that are attracted to exploit either the crop of fruit in the canopy or the fruits dropped on the ground, and of the respective guilds of predator associations are particularly notable. For example, Awa hunters know well that dropped *Otoba* and *Virola* seeds attract large rodents and that these in turn are the regular prey of ocelots; *guayabillo* (*Psidium* sp.) and *ficus* spp., on the other hand, have abundant fruit crops that are consumed by a large variety of both diurnal and nocturnal birds and mammals, including ocelots, which will climb trees to feed on fruit. Whenever the Awa wish to hunt large game animals they build platforms in *fruteros* (fruiting trees). Other trees used as natural bait include lauraceous species (*Nectandra, Ocotea,* and *Persea*), sapotaceous species (*Quararibea* and *Matisia*), lecythids (*Lecythis* and *Eschweilera*), moraceous trees (*Ficus, Brosimum,* and *Cecropia*), and a variety of palm trees (*Geonoma* and *Wettinia*), which may occur in solid stands.

A traditional hunting technique that tends to disappear with the introduction of shotguns is the blowgun, or *bodoquera,* used to kill large and small game species with poisoned darts. Two types of venom are used: an animal-derived poison from the *kokoe,* or poison-arrow frogs (*Dendrobates histrionicus*), and a plant-derived poison from the Moraceous *Perebea* sp. The latex of the tree is boiled and concentrated before use as a cardiac poison (the trans-Andean equivalent of curare).

Fishing The lowland Awa rely as much on fishing as on farming for subsistence, and a significant portion of their dietary protein is obtained from fish. Although fishing is also practiced by the upper-elevation Awa, it is not as important to them. The most abundant freshwater fish are the characins, which comprise more than 50 percent of South American species. The *sabalo* and *sabaleta* (*Brycon* spp.), the *mojarra,* and the *saña* (catfish) are the most common fish consumed by the Awa. Fishing tools and techniques consist mostly of traps designed by the Awa themselves (now widely used throughout the Chocoan lowlands). The corral is a small, rectangular enclosure constructed of bamboo stems or palm strips driven into the mud along the banks of the large rivers, including the Mira, Nulpe, and San Juan. A sliding guillotinelike gate weighted with a stone is tripped by the observant fisherman during high-water seasons when *sabalo* regularly migrate upstream (West 1957).

For catching freshwater shrimp (*Macrobrachium*) and catfish, a basket trap (*catanga*) made of palm slats and baited with *chontaduro* is secured with lianas along the muddy banks. The Awa also use harpoons, with nondetachable tips to catch various fish. At higher elevations, where fish is less abundant, the Awa have developed cultigens to kill fish (*barbascos*) in man-made channels or dammed portions of the rivers, since the use of *barbasco* requires quiet waters. The stunned fish float to the surface where they are picked up into baskets. Recently both the Awa and recent immigrants have been using dynamite, causing drastic reductions in fish populations.

Gathering of Forest Products Aside from the subsistence activities of farming, hunting, and fishing, the Awa gather a vast number of forest products for use in construction and as firewood, medicine, food, and fiber. In a development parallel to the great plant diversity of their forests, the Awa have found many important uses for them.

House Construction The basic Indian house is a roofed platform constructed on beams over tree trunks and elevated above the ground as protection from animals and dampness. More remote Indian dwellings generally lack walls, but many Indians are now building their rectangular houses with walls and room partitions. Some Awa in the lowlands build their houses on low, floodable sites close to rivers, but most build them on high, well-drained hillsides or terraces. To reach the raised floor, between 1.5 and 2.5 meters above the ground, the Indians use a notched hardwood log or chicken ladder. Indian houses are variable in size, but medium-size houses predominate (6 × 5 meters) to accommodate the average family of five to seven members.

Floors are made of split-trunk slats (or hand-sawn boards) taken from a variety of palms, depending upon elevation. In the higher elevations *gualte* (*Geonoma*) is preferred, while in the lower elevations *barrigonas* (*Catablastus, Wettinia, Socratea,* and *Bactris*) are common. The roof is thatched with fronds of different palms, among them *Orbignya, Euterpe, Jessenia, Prestoea* and *chalar* (*Geonoma*). For other kinds of construction, including kitchens, which often occupy a *culata* or extension of the main dwelling, the roof is generally made of *bijao* (*Calathea*). The basement pilings and the floor beams are always made of hardwoods like *guayacan* (*Minquartia guianensis*), *chanul* (*Humiriastrum procera*), or *jagua* (*Nectandra*). When a house has walls, these are usually constructed of regular boards using softwoods and bamboo or palms split into slats. For long, straight, round poles, the Awa have a wide selection including *sajo* (*Camnosperma*), *tangare* (*Carapa*), *kinde* (*Pollalestra*), *motilón* (*Hyeronyma*), *mancharropa* (*Vismia*), *malde* (*Aniba*), and *achiotillo* or *cariseco* (*Billia*). For furniture and cabinets they prefer cedar (*Cedrela odorata*).

Firewood Of paramount importance as firewood species is the leguminous *guabo* (*Inga* spp.), a fast-growing, multiple-purpose tree.

Other fine species for firewood are the Myrtaceous *arrayán* (*Myrcia*), *guayabo* (*Psidium*), *pomarosso* (*Eugenia*), various Melastomes such as *amarillos* (*Miconia*), and Lauraceous species including *Ocotea*, *Beilschmeidia*, and *Nectandra*. Other species commonly used are *pialde* (*Guarea*), *cariseco* (*Billia*), *motilón* (*Hyeronyma*), *gallinazo* (*Celtis*), *guácimo* (*Guazuma*), *tete* (*Eschweilera*), and *sapotillo* (*Quararibea*). Many of these species are also used for other purposes including construction of sheds for sugarcane mills, eating utensils, and construction of canoes.

Miscellaneous Products Other forest products are gathered by the Awa to supplement their income. During Holy Week, they usually gather heart-of-palm (*Euterpe* and *Prestoea*), which they sell in town. The extraction of *anturios* (*Anthurium andreeanum*) is done throughout the year by several families to the point that stocks now are rare in the wild. Another important product from the forest that the Awa women process is the vine locally known as *tetera* (a Maranthaceous plant), which is made into fine hats. The material is sold in town and local women do the final manufacturing of the hats.

Awa Productive Systems and the Biosphere Reserve

A salient feature of traditional agroecosystems is their maintenance of high levels of species and genetic diversity. This characteristic represents a multifaceted strategy aimed at promoting diversity in the diet and in sources of income, thereby stabilizing production and reducing risk, reducing the incidence of plant plagues and diseases, using labor efficiently, and intensifying production with limited resources while maximizing returns at low technological levels (Altieri 1983; Altieri, Letourreaud, and Davis 1983).

Awa traditional agroforestry helps preserve life-support systems through sustainable forms of land use, and it enhances genetic and ecological diversity in a benign mosaic landscape of forest, fallow, and gardens. With low population densities and labor allocation dependent on family production units, the general impact on native ecosystems appears to be kept within the ecological and present technological carrying capacity. This situation resembles that of a closed system, within the defined territory of the social settlement, in which wasteful competition between individuals is reduced and the environment is well preserved.

The opening of the system through invasion by *colonos* of different backgrounds is leading in many areas to the overexploitation of natural resources and to the loss of cultural integrity (McNeely and Pitt 1985). Formerly self-contained and self-sufficient, the Awa population is becoming increasingly dependent on outside markets, both for trade and assistance. Cultural disruption and environmental degradation have been occurring as a result of the opening of Awa society to outsiders. Moreover, in the last ten to fifteen years, both cultural and natural degradation have accelerated, resulting in territorial restriction and assimilation of the indigenous people into a market economy. The adoption by the Awa of outsiders' production systems such cattle-raising, based on rather extensive conversion of forests to grazing fields, and the progressive dependence on single (or few) crops for meeting cash demands, may indeed be pushing the Awa close to exceeding the carrying capacity of the environment and may lead to the loss of the ecological and cultural equilibrium that probably existed until just a few years ago, or that may still be present in the more remote Awa lands. The substitution of the Awa's harmonious agricultural systems for the simplified and unsustainable methods of immigrants is a serious process which needs careful consideration. The knowledge of the Awa is needed to help reverse these adverse tendencies and to again contribute to the maintenance of life-support systems, genetic diversity, and critical habitats, and to the technical restructuring of production systems (Leff 1985). It would seem that the reestablishment of semiclosed systems could promote the preservation of traditional patterns together with the conservation of natural resources. The recognition of the rights of indigenous communities over their living resources—the land they occupy—and their full participation in management will go a long way toward the preservation of both culture and environment (de Klemm 1985).

The MAB biosphere reserve system was initiated to demonstrate the relationship between conservation and development patterns sensitive to cultural idiosyncracies. To achieve these complementary purposes, biosphere reserves are subdivided into rather distinctive zones: a *core* of relatively intact critical ecosystems surrounded by *buffer* zones with examples of harmonious landscapes resulting from traditional land and natural-resource-use patterns, and *areas of experimentation* in methods of sustainable development and restoration. This region intergrades with others of more intensive human activities such as small-scale industry (Oldfield 1987).

Each biosphere reserve includes, in essence, a gradation of intensity of land-use forms which vary from *slight* (i.e. subsistence hunting/fishing in the core or critical ecosystem) through *intermediate* intensity activities (such as use of secondary forests and fallow cultivation in the buffer zones), and *intensive* uses (like monocultures and cattle-raising in the zones of influence outside the reserves).

The Awa production systems embody, on a microscale, the concepts of the biosphere reserves. That is, there is within each family production unit, a core or protected forest unit (critical ecosystem) that provides important services to other productive areas in the form of microwatershed protection, prevention of erosion and nutrient loss, provision of nutrients through leaf fall and through the action of the root systems that assist in the weathering of soils. In addition, in this core area the Awa harvest a variety of forest products on a sustainable basis. There are buffer zones of intermediate-intensity agriculture using slash-mulch methods under natural and man-made conditions, maize fields, fallows of different ages, and secondary forests frequently enriched by the planting of selected species or the leaving of some important plant species during the clearing of primary forest. Finally, there are year-round intensive agroforestry practices which combine horticultural and animal husbandry activities. In Awa gardens the intensive use of resources is well adapted to the regenerative capacity of the soil. The spatial distribution and production schedules of the permanent, semipermanent, occasional, and annual plants included in the polycultural Awa agroecosystem combined with animal production is an excellent example of sustainable agriculture.

The ultimate goal of the Binational Awa Biosphere Reserve is to improve relationships of the various human groups of the area among each other and with their natural resource base, so as to ensure long-term prosperity for all (Orejuela 1986; Binational Plan Awa Project 1987). Immediate objectives include wildland conservation, recognition of communal ownership of traditional Awa lands (where feasible) and the creation of anthropological reserves (excluding new settlers from Awa territories), cultural conservation and recuperation, and the broad implementation of appropriate development programs with emphasis on education, health, community organization, and sustainable agroforestry activities. Scientific research will focus on the potential economic use of natural resources and on devising technological improvements compatible with existing systems. Environmental education and training for sustainable development are central components of the project.

Ultimately, it is expected that in the coming years the interdependency of the Awa Indians and the *campesino/colono* segment of the population, can lead to mutual benefits and respect. The long-term occupation by the Awa of their traditional territories should be recognized and assured through appropriate legislation. Awa traditional agroecological knowledge should be incorporated into new, much needed models of tropical agriculture. Teams of scientists from outside the community and Awa agriculturalists should conduct studies to share knowledge about the complex natural environment, with the purpose of improving the livelihood and security of all the human groups of the region. This type of intercultural, interdisciplinary study will provide new ideas to bridge the gap between the immediate wants and needs of local people and the long-term objectives of wildland conservation.

ACKNOWLEDGMENTS

This account is based on observations made in the community of Pialapí, Colombia, the closest Awa settlement to the La Planada Nature Reserve. Personal visits to the area and frequent conversations with various members of the community allowed me to provide this fragmentary description of their production systems. The experience of my colleagues at La Planada, mainly those of biologist Guillermo Cantillo and agronomist Carlos Acosta provided important insights. This treatment does not in any way do justice to the Awas' impressive holistic knowledge of their environment.

I want to express my deep gratitude to the entire community of Pialapí, the finest neighbors La Planada could wish to have. My debt of appreciation goes especially to the Guanga family (José María, Angel, and Pedro) and to Don Demetrio Guanga for sharing so generously their experience and friendship with us, the folk of La Planada.

LITERATURE CITED

Acosta, C. 1988. Informe de actividad: Sistemas de producción agropecuaria Awa (Cuaiquer) en Pialapí, Ricuarte, Nariño. Paper presented to the Reserva Natural La Planada/Fundación para la Educación Superior. Unpublished.

Acosta, C. and M. F. Jaramillo. 1988. Informe de actividades del proyecto: Etnohistoria y etnografía de los sistemas de aprovechamiento de los recursos naturales en tres comunidades Awa del sur-occidente de Nariño. Fundación para la Educación Superior—FES, Cali, Colombia.

——. 1988. Propuesta de: Investigación sobre la etnobiología Awa (Cuaiquer) en comunidades de diferentes densidades. Paper presented to the Fundación para la Educación Superior, Cali, Colombia.

Altieri, M. A. 1983. *Agroecology: The Scientific Basis of Alternative Agriculture.* Berkeley, Calif.: Division of Biological Control, University of California.

Altieri, M. A., D. K. Letourneaud, and J. B. Davis. 1983. Developing sustainable agroecosystems. *Bio-Science* 33:45–49.

Binational Plan Awa Project. 1987. Plan de ordenamiento para el Proyecto Binacional Awa (Colombia-Ecuador). Binational Plan Awa Project, La Planada Nature Reserve, Nariño, Colombia. Working Document.

Cobo, A. 1982. Plan for the environment and natural resources area. Fundación para la Educación Superior. Nariño, Columbia. Appendix to a Memorandum from R. M. Wright to the Program Committee of the World Wildlife Fund on Finca "La Planada," Nariño, Colombia.

Cerón, B. 1986. *Los Awa-Kwaiker.* Quito, Ecuador: Abya.

De Klemm, C. Culture and conservation: Some thoughts for the future. In J. A. McNeely, and D. Pitt, eds., *Culture and Conservation: The Human Dimension in Environmental Planning,* London: Croom Helm.

Gentry, A. H. 1982. Phytogeographic patterns as evidence for a Choco refuge. In G. T. Prance, ed., *Biological Diversification in the Tropics,* New York: Columbia University Press.

Goodland, R. 1985. Tribal Peoples and Economic Development: The human and ecological dimension. In J. A. McNeely and D. Pitt, eds., *Culture and Conservation: The Human Dimension in Environmental Planning,* 13–31. London: Groom Helm.

Haffer, J. 1967. Speciation in Colombian birds west of the Andes. *American Museum Natural History Novitates,* no. 2294, pp. 1–57.

——. 1974. Avian speciation in tropical South America with a systematic survey of the toucans (Ramphastidae) and jacamars (Galbulidae). Publication Nuttall Ornithological Club No. 14. Cambridge, Mass.

Holt, C. and D. Ferguson. 1986. *Sustainable Development Designs for the Awa Territory of Ecuador (Interim Report).* Canberra, Australia: Rainforest Information Center.

Holm-Nielssen, L. and A. Barford. 1984. Las investigaciones etnobotánicas entre los Cayapas y los Coaiqueres. *Misc. Antropol. Ecuatoriana* 4.

Leff, E. 1985. Ethnobotany and anthropology as tools for a cultural conservation strategy. In J. A. McNeely and D. Pitt, eds., *Culture and Conservation: The Human Dimension in Environmental Planning,* 259–268. London: Croom Helm.

MacDonald, T. 1986. Anticipating "colonos" and cattle in Ecuador and Colombia: Awa-Cuaiquer land and resource management project. *Cultural Survival Quarterly* 10(2):33–36.

Martinez, Y., H. Martínez and J. Parra. 1984. *Diagnóstico Preliminar de la Comunidad Cuaiquer*. Pasto, Colombia: ICA.

McNeely, J. A. and D. Pitt. 1985. *Culture and Conservation: The Human Dimension in Environmental Planning*. London: Croom Helm.

Oldfield, S. 1987. *Buffer-zone Management Techniques*. Gland, Switzerland: International Union for the Conservation of Nature.

Orejuela, J. E. 1986. Documento de recomendación a la Corporación Autónoma Regional de Nariño para el establecimiento de la Reserva de Biosfera Cumbal y Chiles (Región Awa). La Planada, Ricaurte, Nariño.

——. 1987. La reserva natural La Planada y la biogeografía Andina. *Humboldtia* 1(1):117–148.

Orejuela, J. E., J. Barborak, D. Glick, and A. Echeverri. 1983. Plan Operativo de Manejo y Desarrollo de la Reserva Natural La Planada. Cali, Colombia: Fundación para la Educación Superior. Internal document.

Osborn, A. 1969–72. Alliance at ground level: The Kwaiker of southern Colombia. *São Paulo Revista de Antropología* 12(2):211–315.

Osborn, A. and F. Melo. 1986. *Estudio para el conocimiento y rescate de formas autóctonas de atención al niño: Los Kwaikeres de Nariño*. Bogotá, Colombia: Instituto Colombiano de Bienestar Familiar.

Samper, A. and J. Orejuela. 1988. La Planada: A private nature reserve for nature conservation and community development. Cali, Colombia: Fundación para la Educación Superior.

Villareal, C. 1988. *Términos de referencia para el ordenamiento del Proyecto Awa*. Quito, Ecuador: Unidad Técnica Ecuatoriana del Proyecto Awa.

West, R. 1957. *The Pacific Lowlands of Colombia: A Negroid Area of the American Tropics*. Social Science Series, no. 8. Baton Rouge, Louisiana: Louisiana State University Press.

FIVE

Resource Use, Traditional Technology, and Change Among Native Peoples of Lowland South America

HILLARD KAPLAN AND KATE KOPISCHKE

Traditional Amazonian peoples employ a broad array of strategies and technologies in the acquisition of food. Prior to contact with Western society and the associated introduction of metal-based technologies, Amazonian peoples were limited, and some still are, to technologies produced from raw materials found in the nearby environment. With the exception of stone axes, the principal traditional technologies are derived from plant products, particularly palms, and include the bow and arrow, blowgun, spears, digging sticks, clubs, traps, snares, dams, and plant poisons.

As groups begin the process of contact with Western society, traditional subsistence practices are supplemented by techniques involving new technologies. Among the most important technological introductions are shotguns, machetes, axes, fishhooks and fishline, nets, flashlights, dogs, and dynamite. In most cases, new technologies do not completely replace traditional technologies but rather add to the repertoire of resource-acquisition techniques.

Though such contact can bring certain material advantages to indigenous peoples, it also generally involves massive mortality due to disease. Also, contact frequently leads to the deterioration and occupation of native homelands by colonists and large corporations. The economic, social, and personal consequences of contact are often dramatic and complex and usually occur before peoples possess the political influence to battle the forces they confront.

The food-acquisition strategies of three different communities in Peru and Paraguay—1) an Aché community, Chupa Pou, located in the subtropical forests of eastern Paraguay; 2) the Machiguenga community of Tayakome, located along the central portion of the Manu river in southeastern Peru; and 3) the Machiguenga community of Yomiwato (also called Quebrada Fierro), located near the headwaters of a small tributary of the Manu River—demonstrate the variety and flexibility of traditional Amazonian subsistence practices and show how traditional peoples adjust to varying environmental challenges and to the availability of new technologies. The base of knowledge about these communities facilitates much-needed discussion of the research and policy implications for conservation and development among newly contacted, highly traditional people. Variability in the extent to which native peoples engage in resource conservation, the environmental impact of access to new technology, and how increased interaction with modern state societies affects the economic, social, and physical welfare of the people are important foci of this discussion.

Methods

The three Aché and Machiguenga communities have been studied for varying time periods. The most extensive research was conducted among the Aché of Paraguay. The first field team of May–July 1980 focused on subsistence practices of members of the Chupa Pou community and included Kim Hill and Kristen Hawkes. From October 1981 to April 1982, another field session was conducted by Hill, Hawkes, A. Magdalena Hurtado and Hillard Kaplan. Similar data were collected from September 1984 to April 1985 by Hill, Kaplan, Kevin Jones and Heather Dove.

Research in the Machiguenga communities was in progress during the writing of this paper. The field session began in July 1988 and continued until June 1989. The authors collected data in Tayakome and Yomiwato, communities previously studied by Hurtado and Hill from June to October 1986.

Methods employed during each of these field periods, developed during the Aché project, are very similar. Data discussed in this paper are derived from "focal follows" of individuals and by weighing food brought back to settlements and sleeping sites. During follows, the investigator maintains an ongoing record of all the individual's activities and weighs any food he or she acquires. Special emphasis is placed on timing all aspects of food acquisition (search, pursuit, travel, tool manufacture, etc.), so that the productivity of alternative techniques can be measured. In order to obtain a larger sample of food acquired, homes (in the case of the Machiguenga) and sleeping sites (in the case of the Aché) were also sampled throughout the day by an investigator who weighed all food brought in. The caloric equivalents and the macronutrient constituents of foods are determined using both published tables and laboratory analyses (see Hill et al. 1984 for details). Unless otherwise noted, all descriptions of resource acquisition are based on direct observation.

The Study Populations

The Aché

Until about the mid-1970s, inhabitants of Chupa Pou were full-time hunter-gatherers who traveled nomadically throughout a large part of eastern Paraguay (see Hill 1983 for a history of contact). Between 1970 and 1978, bands of Aché were contacted primarily by a frontiersman who convinced already-contacted Aché to live and work on his ranch. Following contact, the Aché were introduced to swidden horticulture and missionized by Catholic and Evangelical missionaries. In 1978, Chupa Pou was established on a plot of land bought for the Aché by the Verbo Divino Catholic order. During the study period, about two hundred Aché lived at Chupa Pou, practicing a mixed economy of swidden horticulture, with hunting and fishing near the settlement and extended hunting-and-gathering treks. They complemented their basic subsistence strategies with small amounts of wage labor and some selling of crops. Fewer than 5 percent of the calories consumed were derived from food they did not produce (Hawkes et al. 1987).

Aché resource use can be analyzed both in terms of the traditional hunting-and-gathering life-style and the present mix of foraging, horticulture, and wage labor. Knowledge of traditional Aché foraging lifeways has been derived from interviews and through direct observations of foraging trips. Foraging trips that generally last from one to three weeks in many ways resemble precontact subsistence patterns. Informant reports, historical accounts (Bertoni 1941), other ethnographic observations (Clastres 1972), and Hill's observation (Kaplan and Hill 1985) of a band's foraging practices within weeks of first contact in 1978, indicate that techniques for food acquisition are the same and band composition is similar in size and kin relatedness. The composition of the diet also is similar; virtually all food eaten during trips is derived from foraging, and almost none is brought back to the permanent settlement. Perhaps the most important differences between these foraging trips and precontact patterns are that the range covered is much smaller and there are no longer "time-outs" for important ritual events and club fights. Massive mortality at contact probably has had an effect on group membership and on the age-structure of foraging bands. In addition, some children are left behind at the settlement during foraging trips. This means that the average number of child-dependents per family on current trips is about one fewer than in precontact bands. Informants also report that women pounded more palm fiber for starch in the past, probably because they had more children to feed.

The Machiguenga

Tayakome, a village currently numbering fifty-seven inhabitants, was established by Summer Institute of Linguistics missionaries in the mid-1960s. Two contacted Machiguenga from the Urubamba River system traveled to uncontacted Machiguenga settlements at the headwaters of the right-bank tributaries of the Manu River. They convinced many people to settle in Tayakome, midway along the main watercourse of the Manu. A school was established and goods were distributed, either free of charge or in exchange for animal skins.

In 1973, Manu National Park was established. Park rules prohibited the Machiguenga from using shotguns and selling skins. Virtually all outsiders, with the exception of park officials, were prohibited from entering the inhabited regions of the park. The Summer Institute of Linguistics established an alternative settlement along the Camisea River, outside the park, and about half the inhabitants of Tayakome followed to this new site.

During the early years of the park, Tayakome decreased in size, with residents dispersing along the Manu and its tributaries. The dispersal was due to a variety of factors, including attacks from neighboring Yaminahua groups and conflicts within the community. The Machiguenga also were returning to their traditional settlement pattern, living in dispersed extended-family groups of fifteen to twenty people separated by one to several hours by foot or canoe. In 1981, a schoolteacher was sent to Tayakome by Dominican Catholics. He convinced many of the dispersed families who lived along the Manu to resettle in Tayakome so that their children could attend school. Although the extended families now live closer together in the village, they are still economically independent of one another and spatially separate. Virtually all Western goods are obtained as gifts from the Catholic Church and from anthropologists or biologists working in the park. Except for occasional and brief trips out of the park, the residents of Tayakome have no contact with the Peruvian monetary economy. All food eaten in Tayakome is produced or acquired there. No food is sold.

Yomiwato, known to outsiders as Quebrada Fierro, is inhabited by one hundred people in two major settlements: the community proper with seventy-five and an upstream settlement with twenty-five. About half of the adults living in Yomiwato in 1988–1989 had never left the headwaters and had not lived with missionaries; the remainder lived in Tayakome at contact but later returned to their traditional habitat. They were settled in dispersed, extended-family groups until a schoolteacher was sent in 1984 to centralize the community. Centralization was underway before the teacher arrived, however, as a defense against Yaminahua attacks. Nevertheless, families even now maintain separate residences away from the community in which they spend a substantial portion of their time.

Because of Yomiwato's remote location (five days by motorized canoe from the nearest road), access to Western goods is even more limited than in Tayakome. Each family may own a plate, a spoon, a metal pot, a blanket, one to two knives, a machete, an ax, and fishhooks and fishing line. Almost all these goods are donations from the Catholic Church and from visiting investigators. There is virtually no involvement in the monetary economy.

Food Acquisition: Results

The Aché

Hunting and Fishing Aché land use is extensive. According to demographic data and informant reports, nearly eight hundred people covered a range of about 12,000 square kilometers prior to peaceful contact (Hill 1983). On current foraging trips the Aché are extremely mobile, moving camp almost every day (as is reported in historical accounts; see Bertoni 1941). At the start of each day, men leave camp in a line and are followed by women and children. After walking for about half an hour, men fan out in search of game while women (usually accompanied by a man who breaks the trail) walk together in a direction determined earlier by the group. The men attempt to space themselves far enough apart so they do not overlap areas covered, but close enough to call to one another when a cooperatively hunted resource is spotted. This pattern of mobility is associated with diets that emphasize hunting and male provisioning. Meat provides 59.6 percent of Aché calories, and men provide 87 percent of the total diet. Table 5.1 (taken from Kaplan and Hill 1985 and Hill et al. 1984) shows the composition of the diet obtained on foraging trips observed in 1981–1982.

About half of the major game animals are hunted cooperatively, providing some 66 percent of the total meat calories obtained. These include white-lipped peccaries, capuchin monkeys, pacas, and coatis. The remainder—deer, collared peccaries, tapir, agoutis, and armadillos—are pursued by solitary hunters. Meat is pooled for consumption. In fact, meat is so widely shared that hunters and their families receive no special share of what they acquire; 90 percent is consumed by individuals outside their nuclear families. It is not clear, however, whether this pattern of sharing is due to cooperative acquisition or to the reduction of risk associated with a diet that emphasizes meat (see Kaplan and Hill 1985; Kaplan et al. 1990 for a detailed treatment of Aché food sharing).

Aché hunters use two major types of technology: the bow and arrow and the machete. Sometimes shotguns are used (although all the data cited above and in Table 5.1 were obtained from cases in which no shotguns were present). Bows and arrow points are made of palm wood, and arrow shafts are made of cane. Monkeys, peccaries, tapir, deer, and sometimes coatis are taken with bow and arrow. Techniques used in solitary hunts of tapir and deer involve stealth and positioning before the animal notices the hunter and flees. For cooperatively acquired animals, such as monkeys and peccaries, the hunter who encounters the game waits for others to arrive before pursuing the animal. Cooperative hunting is designed to prevent animals from escaping (see below, and Hill and Hawkes 1983, for further description of cooperative hunting techniques). Bows and arrows provide about 58 percent of the meat calories.

Table 5.1
Aché Diet: October–May 1981–82

Resource	Technique	Kg/ consumer/ day	Cals/kg Raw Wt.	Cals/ consumer/ day	% of diet
Hunted					
White-lipped peccary (*Tayassu pecari*)	cooperative, bow and arrow	.31	1,620	495	12.9
Monkey (*Cebus apelle*)	cooperative, bow and arrow	.37	1,290	483	12.6
Deer (*Mazama americana*)	solitary, bow and arrow	.16	1,095	174	4.6
Collared peccary (*Tayassu tajacu*)	solitary, bow and arrow/ hand	.07	1,620	121	3.2
Paca (*Agouti paca*)	cooperative, hand	.24	1,620	381	9.9
Coati (*Nasua nasua*)	cooperative, hand	.09	1,620	143	3.7
Armadillo (*Dasypus novemcinctus*)	solitary, hand	.29	1,290	375	9.8
Other (birds, snakes, fish	various	.09	1,275	112	2.9
Total hunted	—	1.62	—	2,284	59.6
Collected					
Larvae (various spp.)	chop	.02	3,100	774	1.9
Palm fiber (*Arecastrum romanzol-fianum*)	chop, pound	1.09	336	368	9.6
Palm heart (*Arecastrum romanzol-finanum*)	chop	.10	595	59	1.5
Oranges (*Citrus sinensis*)	collect	.39	297	116	3.0
Other fruit	collect	.35	550	193	5.1
Bee honey (from *Apis melifera*)	chop	.23	2,673	622	16.3
Other honey (various spp.)	chop	.04	2,673	111	2.9
Total collected	—	2.22	—	1,543	40.3
Total hunted and collected		3.84	—	3,827	99.9

The remaining 42 percent of the meat is obtained with hand-based hunting techniques, often assisted by machetes first obtained through theft some forty years before peaceful contact. Armadillos are often captured by hand when found above ground. They typically are chased into hiding. If the animal takes shelter in a small hole or log (because it cannot reach its burrow), the hunter will grab its tail, hold it down and break its neck with his hand. Occasionally, the hunter will chop its back open with a machete. If the animal manages to flee into its burrow, it is dug out by loosening the dirt with a machete, bow stave, or pointed stick. The process appears to be fastest with a machete, but sufficient data to calculate comparative efficiency are lacking.

In hunting coatis, men who spot an animal call the others to the site. The area around the tree where the coati troop is hiding is then cleared of underbrush. Clearing is usually done today with machetes, but in the past, when machetes were not available, the work was done by hand. An experiment conducted with hand clearance of forest underbrush in Paraguay (Kaplan 1985) indicates that hands are about half as efficient as machetes, and since the bulk of time spent coati hunting is devoted to clearing, hand-clearing could reduce efficiency by about 35 to 40 percent. After the area is cleared, the hunters make noise by shaking branches. As the coatis jump from the trees, hunters grab the fleeing animals by their tails and smash their heads against the ground or against a tree. Women sometimes capture juveniles and subadults (see Bertoni 1941 for a similar account of precontact groups).

Pacas are almost always taken with minimal technological assistance. When a paca is found in its burrow, the hunter calls several other men to the site. Each positions himself at an escape hole. If there are more escape holes than hunters, the extras are blocked with logs. One man then drives a log noisily into the main entrance while the others drop onto the escape holes with their hands covering them. When the animal flees, it is grabbed by its neck and suffocated with the knee. According to Hill and Hawkes (1983), men acquire .27 kilogram of meat per hour of hunting using hand-based kill techniques.

Aché rarely fish, and fishing accounts for less than 2 percent of the total meat acquired. Although infrequent, fishing expeditions are done cooperatively using a hand-based technique. When a lagoon is found that appears to be full of fish, Aché men cut saplings with machetes and build a large weir. The weir spans the width of the lagoon and extends from the bottom to about one foot above the water. The far end of the lagoon is blocked off and the weir is rolled in that direction, progressively trapping the fish in a smaller and smaller area. Virtually all men, women, and children over the age of five participate in rolling the weir. The fish attempt to jump over the weir as they become trapped, and the Aché grab them by hand and break their necks. If conditions are favorable, it is possible to gather 200 to 300 kilograms of fish using this technique. However, the usual take is much less. As with clearing in coati hunts, this technique could be practiced without machetes but would require about twice the setup time, or about two to three hours.

During our study, shotguns were used only occasionally. Most men (88 percent) did not own a shotgun, and of those who did, few could afford shells. Although our sample was weighted toward foraging trips where no shotguns were used, Hill and Hawkes (1983) showed that shotgun hunting is more efficient and targets larger game. Aché bow hunters acquired .53 kilogram of meat per hour of hunting, while shotgun hunters acquired 1.60 kilograms per hour. In addition, shotgun hunters ignored monkeys and small birds. Species weighing over 20 kilograms account for 87 percent of the take of shotgun hunters and only 24 percent of bow hunters (Hill and Hawkes 1983).

Collecting Gathering accounts for the remainder of the calories acquired on Aché foraging trips (40.3 percent). Modern technology has had an interesting effect on patterns of gathering. Axes—acquired by theft at least forty years before peaceful contact—are used to get honey, a resource that accounts for 19.1 percent of the calories consumed and almost half of the gathered calories. The bulk of honey taken is produced by stinging bees first introduced by Europeans. When honey is found, a man or several men fell the tree and chop a hole in it to extract the honey, destroying the hive in the process. The Aché report that before metal axes were available, a man would climb the tree, hold on with his legs and one hand, and chip at the hive with a stone ax. They also report that many hives were ignored because they were too difficult to reach.

Palm products are next in caloric importance, providing 11.1 percent of the calories. Pounding the fiber of palms for starch is the main economic activity of women (see Hurtado et al. 1985 for a detailed treatment of women's work). To extract the starch, they fell the palm with an ax, chop out a section of its outer ring about 2 meters long, and pound the pulp with the back of the ax head. It generally takes a woman about six minutes to fell the palm, another ten to fifteen minutes to expose the central portion and about an hour to pound the fiber. If only stone axes were available, felling time would probably be about twenty to thirty minutes, exposing the central portion about a half hour, and pounding the fiber another hour. These estimates are based on data obtained from felling trees of similar diameter with a stone ax (Kaplan 1985). The experiment suggests that what could be obtained in about eighty minutes with a metal ax would require about two hours with a stone ax, a 67 percent gain of efficiency.

Fruits that drop or are shaken to the forest floor are also collected. Depending upon the tree, fruits may be taken by shaking branches or by cutting them off with a machete. When limbs are cut from a tree, no attempt is made to protect the life of the tree.

Gathering is more seasonally variable than hunting (Hill et al. 1984). With the exception of oranges (a post-Hispanic introduction), which are available from April through September, most fruits are only in season for one to four weeks. Although honey can be found throughout the year, it is most abundant and provides the most calories in October and November (Hill et al. 1984). In contrast, hunted foods and their associated capture techniques vary little throughout the year (Hill et al. 1984).

The Machiguenga

Horticulture All known Machiguenga groups reportedly practice slash-and-burn horticulture. During the sample period in Tayakome and Yomiwato, garden products provided 77 percent and 72 percent of the total calories consumed respectively (see Tables 5.2 and 5.3). "Main" gardens are cleared close to home, and many families also maintain smaller gardens from an hour to a day's travel away. At these secondary gardens, families construct shelters ranging from makeshift huts to well-built structures that may last up to ten years. These gardens often are strategically placed along travel routes and in areas that provide opportunities for acquiring faunal and floral resources. They also are placed near abandoned living sites where old trail sys-

Table 5.2
Machiguenga Diet in Yomiwato: October–December 1988

Resource	Technique	Kg/ consumer/ day	Cals/kg Raw Wt.	Cals/ consumer/ day	% of diet
Hunted					
Collared peccary (*Tayassu tajacu*)	dog, bow and arrow	.08	1,620	127	4.3
Collared peccary (*Tayassu tajacu*)	without dog, bow and arrow	.03	1,620	52	1.8
Tapir (*Tapirus terrestrius*)	dog, bow and arrow	.06	1,620	99	3.4
Monkeys (esp. *Ateles, Lagothrix*	bow and arrow	.02	1,290	20	.7
Paca (*Agouti paca*)	dog, bow and arrow	.01	1,620	18	.6
Agouti (*Dasyprocta/Myoprocta*)	dog, bow and arrow	.02	1,095	25	.8
Birds	bow and arrow	.03	1,275	39	1.3
Total hunted	—	.25	—	380	12.9
Fish					
Various spp.	*barbasco* poison	.28	960	267	9.0
Boquichico (*Prochilodus* sp.)	bow and arrow	.02	960	17	.6
Various spp.	hook and line	.03	960	24	8
Total fish	—	.33	—	308	10.4
Collected					
Larvae (esp. *Calardra palmarum*)	chop, collect	.02	3,100	61	2.1
Turtles	collect	.04	450	18	.7
Palm (heart, various species fruits, nuts)	chop, collect	—	—	28	.9
Honey (sp. unknown)	collect	.003	2,673	10	.3
Fruit (spp. unknown)	collect	.01	509	7	.2
Bird eggs	collect	.003	1,700	6	.2
Total collected	—	—	—	130	4.4
Domestic					
Manioc (*Manihot esculenta*)	harvest	1.42	490	1,410	47.7
Plantains (*Musa paradisiaca*)	harvest	.61	860	527	17.8

Table 5.2
Continued

Resource	Technique	Kg/ consumer/ day	Cals/kg Raw Wt.	Cals/ consumer/ day	% of diet
Bananas (*Musa* spp.)	harvest	.07	676	49	1.7
Tubers (*Dioscorea* sp.)	harvest	.11	800	92	3.1
Fruits (papaya, peanuts, avocado, sugarcane)	harvest	—	—	54	1.8
Chicken	slaughter	.004	1,275	5	.2
Total domestic	—	—	—	2,137	72.3
Total wild and domestic	—	—	—	2,955	100.0

tems can be reopened easily, the terrain is familiar, and mature fruit trees and planted palms can be harvested.

Secondary gardens and the dispersed settlement pattern in the headwaters allow for extensive mobility without having to carry many possessions. Families, and especially parts of extended families, may retreat to secondary gardens to forage in areas with lower hunting pressure, exploit resources that are seasonally abundant, and leave behind social conflicts. They often move from one secondary garden to another, visiting friends and relatives in other locations. Along the way, they eat garden products and acquire faunal resources to share with hosts.

In spite of this mobility, gardens are associated with longer-term occupation of sites than was found traditionally among the Aché before settlement at the mission. This lowered mobility has important effects on hunting and gathering. Garden homes are a central place to which foragers return at night.

According to informant reports, Machiguenga now living in the Manu drainage area acquired metal axes some seventy years ago—well before the first face-to-face contact with whites—through trade with contacted Machiguenga living in the Urubamba River drainage. Some elders report that their fathers did not have metal axes when they were young boys, and others maintain that in their area a single

Table 5.3
Machiguenga Diet in Tayakome: August–September 1988

Resource	Technique	Kg/ consumer/ day	Cals/kg Raw Wt.	Cals/ consumer/ day	% of diet
Hunted					
Collared peccary (*Tayassu tajacu*)	without dog, bow and arrow	.11	1,620	178	6.7
Collared peccary (*Tayassu tajacu*)	with dog, bow and arrow	.00	1,620	0	0.0
Capybara (*Hydrochaeris hydrochaerris*)	with dog, bow and arrow	.03	1,620	41	1.6
Birds (various spp.)	bow and arrow	.02	1,275	22	1.8
Total hunted	—	.16	—	241	9.2
Fish					
Various spp.	*barbasco* poison	.004	960	4	.2
Boquichico (*Prochilodus* sp. and other spp.)	bow and arrow	.1	960	91	3.5
Various spp.	hook and line	.2	960	184	7.0
Total fish	—	.3	—	279	10.6
Collected					
Larvae (various spp.)	collect	.01	1,820	14	.5
Turtle eggs	collect (beaches)	.02	2,290	50	1.9
Fruits (spp. unknown)	collect	.01	320	3	.1
Total collected	—	.05	—	67	2.5
Domestic					
Manioc (*Manihot esculenta*)	harvest	1.8	990	1,819	69.1
Plantains (*Musa paradisiaca*)	harvest	.2	860	186	7.1
Bananas (*Musa* spp.)	harvest	.04	676	24	.9
Papayas (*Carica papaya*)	harvest	.05	320	15	.6
Total domestic	—	2.09	—	2,044	77.7
Total wild and domestic	—	2.60	—	2,631	100.0

ax was shared by many families over large distances. No one has seen a garden cleared with a stone axe, and all agree that the Machiguenga did not clear primary forest before they had metal axes. According to informants, gardens were made in small patches of balsa wood trees (*Ochroma* sp.) along the rivers' edges. Much, if not all, of the clearance was done by hand without stone axes, simply by breaking the trees or knocking them over.

The metal ax has allowed Machiguenga gardeners to take factors other than ease of clearance into consideration when choosing plots. They select sites less susceptible to flooding and conveniently located near their houses. Informants report that contemporary gardens are much larger than those of the pre–metal ax period and also are larger than those currently cleared by their uncontacted relatives, who still share very old, dull axes. It is likely that in the past the Machiguenga were more mobile and more dependent on foraged foods.

Manioc and plantains are by far the most important garden crops. In Tayakome, manioc provides 89 percent of the calories derived from gardens, with plantains contributing 9 percent during the recent study period. In Yomiwato, manioc and plantains provide 66 percent and 25 percent of the garden calories respectively. These figures conform closely to those obtained by Hurtado and Hill (Hurtado et al. 1986), although they found that plantains provided some 40 percent of garden calories in Yomiwato, suggesting some year-to-year variability.

Gardens have a somewhat predictable life cycle. They are cleared and burned during the dry season (June–September) and planted just after burning. The first crops planted include maize, peanuts, papaya, and manioc, together with a small number of pineapples, squash, tubers, sweet potatoes, and *barbasco* plants. The maize, peanuts, and manioc are intercropped. Peanuts are the first to be harvested about three months after planting and are followed by the corn. The manioc at this point is still small enough that it does not block sunlight from reaching the corn and peanuts. As the manioc grows, many varieties of plantains and bananas are added to the garden, especially in Yomiwato. After nine to twelve months, the manioc is ready to harvest. A garden is systematically harvested, replanted with manioc, and weeded at least once a year. As the second crop of manioc is harvested, little effort is made to weed and to replant the manioc. The plot then becomes increasingly dominated by the plantains, bananas, and other fruit trees, especially during the third and fourth years. It is then nearly abandoned and only periodically visited to harvest slower-growing fruit trees such as avocado and peach palm.

There is substantial variation in the effort that individuals put into gardens. Men do almost all the clearing, burning, weeding, and initial planting. Both men and women harvest, although women do more of the harvesting and replanting of manioc. However, some men have no gardens and others have several large ones in different stages of growth. This variation is associated with a complicated pattern of sharing rights to gardens, reciprocal exchanges with faunal resources, freeloading, and long-term work-exchange relationships that require more extensive investigation.

Fishing Fishing is of great economic importance to the Machiguenga of Tayakome and Yomiwato, especially in the dry season, when fish provide 58 percent and 74 percent of the total meat consumed in Tayakome and Yomiwato respectively. However, it decreases in importance during the wet season, at least in Yomiwato, providing only 11.5 percent from mid-November to January.

The three principal techniques of fishing employed by the Machiguenga are bow and arrow, *barbasco* poison, and hook and line. Bow-and-arrow fishing is done either from a canoe or from the shore. Fisherman look for shallows, especially along beaches, where fish feed and can be seen easily. The primary target is *boquichico* (*Prochilodus* sp.), which weighs between .3 and 2 kilograms. The bow-and-arrow technique is quite successful when rivers are clear. Fishermen almost never return empty-handed if they fish for more than about two hours, and often they are successful fishing from the main port in the community. Bow-and-arrow fishing, however, is restricted to the dry season (June to October), when rivers are clear, or to brief dry spells during the wet season. For example, bow-and-arrow fishing provided 33 percent of the fish taken in Tayakome from August 15 to September 15, and 18 percent in Yomiwato during October. In contrast, from mid-November to January, it provided no fish in Yomiwato.

Fish also are taken with *barbasco* poison. The roots of the *barbasco* plant (*Lonchocarpus* sp.) are dug from the ground and carried to lagoons and streams both big and small. They are then pounded with stones or clubs so the fluid poison can be drained into the water. The poison affects gill function in fish, and generally within ten minutes to half an hour they begin to swim and float on the surface of the water. They can then be collected easily by hand, hit on the head with a club or machete, or shot with bow and arrow. *Barbasco* fishing contrasts with bow-and-arrow fishing in that its scale, returns, and participants are highly variable. Unlike bow-and-arrow fishing, which is done by a single adult or adolescent male, *barbasco* fishing can involve everyone in the community, including children above the age of about six. In Yomiwato, nearly the entire community participated in *barbasco* expeditions along the main, 25-meter-wide waterway. These events generally lasted three to six hours, yielding returns of about 6 kilograms per person, including children. On other fishing trips, small amounts of *barbasco* were used in tiny streams by one or two individuals, yielding .5 to 1 kilogram of fish. *Barbasco* fishing also is restricted to waterways not flooded by rains, but unlike bow-and-arrow fishing, it can be performed in small streams, even during the wet season.

Tayakome and Yomiwato differ radically in their relative reliance on *barbasco*. In Tayakome, *barbasco* fishing provided slightly more than 1 percent of the total fish taken. In Yomiwato, it provided 90 percent of the fish in October and 58 percent from mid-November to January.

Hook-and-line fishing is a recent introduction that coincides with direct contact. Unlike axes, dogs, and chickens, which are passed from contacted to uncontacted Machiguenga, fishhooks (probably because they are so quickly lost) did not become integrated into the subsistence pattern until a regular supply was obtained through direct contact. However, once introduced, hook-and-line fishing had a dramatic impact on the Machiguenga economy. Hook-and-line fishing can target all sizes of fish available in waterways and can be done both at night and during the day. Lines can be left unattended, tied to the shore, a boat, or a pole buried in the river bottom. In addition, although it is generally less productive in the rainy season, it is still possible to catch fish with hook and line during the wet months. In Tayakome, hook-and-line fishing provided 66 percent of the total fish taken, including during the dry season when other techniques are also effective. In Yomiwato, it provided about 5 percent during the dry and 40 percent in the wet.

Hunting The communities of Tayakome and Yomiwato are surrounded by an intricate trail system that, within about a 5- to 10-kilometer radius of the community, allows them to forage mostly on cut trails rather than through unbroken forest. Most hunting is done on day trips away from the settlement or a secondary garden house.

For this reason, the Machiguenga have a thorough knowledge of their environment and the many diverse resources it contains. Unlike the Aché, who engage almost exclusively in generalized searches and opportunistic foraging (i.e., exploiting whatever resources may be encountered), Machiguenga frequently engage in focused trips to sites such as salt licks, tree trunks known to house collared peccaries, paca holes, and fruit trees currently being visited by animals. Information from one day's hunt is used on subsequent hunts. This creates a mixed pattern of known resource targets and opportunistic foraging.

Dogs play an extremely important role in Machiguenga hunting. Although dogs probably were a post-Hispanic introduction, they were integrated into Machiguenga subsistence well before direct contact with whites through the same informal trade routes that provided axes. However, unlike axes, dogs can multiply. It is likely that they reached the most remote settlements very early.

Dogs and humans are a formidable predatory team, especially for certain game species. They are particularly effective in hunting paca and collared peccaries. One or two men generally leave camp with between one and three dogs. The dogs run circles around the hunters as they walk the trails, frequently flushing peccaries, pacas, armadillos, and agoutis from their foraging sites. Alternatively, the men will find tracks that appear fresh, and send the dogs off sniffing. Dogs pursue the game as the hunters lag behind. They maintain contact with shouts and are answered by barks. The dogs either grab and kill the game or chase it to a hole or hollow log, from which the hunters can extract it. At times the dogs lose the trail, and the hunters must find it again. In Yomiwato, dogs played a role in capturing 66 percent of hunted foods during the late dry and early wet seasons.

Dogs are a hindrance in some resource pursuits. Because they bark when they spot game, some animals—particularly birds, monkeys, and tapir—escape before hunters have a chance to shoot. Dogs actually decrease in importance during the mid and late wet seasons, when monkeys become fat and are favored as prey items.* Dogs are also left at home when men engage in targeted search for tapirs. However, if a tapir is shot but not killed, men may return the next day with dogs to track and capture the wounded animal, which can no longer outrun them.

Technologies and resource exploitation strategies are frequently

*Tables 5.2 and 5.3 do not include dietary data obtained during the wet season after this paper was written.

combined in a single day's activity. Bow-and-arrow fishing is often combined with hook-and-line fishing. A common technique in Tayakome during the dry season is to travel upstream fishing the beaches with bow and arrow, then float downstream fishing the deep spots with hook and line and the shallows with bow and arrow. Emphasis is shifted on the basis of which technique seems most successful on a given day. In Yomiwato, *barbasco* fishing is frequently combined with hook-and-line fishing and with hunting. Men generally carry their fishing rigs, and sometimes *barbasco* root, when they hunt. In this way they can exploit distant streams for fish and can search for game while they travel to the stream. They can also buffer the high-risk strategy of hunting (which is only successful on about 25 percent of all days) with the low-risk strategy of fishing (which almost always produces some fish, even though sometimes takes are quite small).

Collecting In Manu, there are some twenty-five to thirty species of palm. Approximately ten of these provide foods regularly eaten by the Machiguenga. Fruits, nuts, growing shoots, and larvae all are taken from the palms. Unlike the Aché, Machiguenga do not extract starch from palms, perhaps because gardens supply an abundant source of carbohydrates. Calorically, larvae are the most important resource taken from palms. In Yomiwato, they provided some 8 percent of the faunal calories consumed, mostly in the form of fat. Larvae are taken during focused trips, especially by women, who travel to known sites of rotting palms that contain larvae. Larvae are sometimes taken by men during hunting trips.

In addition to palm products, seasonally available fruits also are exploited. The data cited in tables 5.2 and 5.3 suggest that, overall, fruits do not contribute significant calories to the diet (although their micronutrient contribution to the diet is unknown).

Variability in Resource Use and Its Determinants

The most dramatic difference in resource use between the Aché and the Machiguenga is the practice of swidden horticulture. Although it is possible in the distant past that the Aché practiced some horticulture or trade with horticulturalists, it is clear they were at one time isolated, nomadic hunter-gatherers. In contrast, the Machiguenga have practiced horticulture for hundreds, perhaps thousands of years. Historical accounts indicate that horticulture was practiced prior to the Spanish conquest and the introduction of metal technology (Camino 1977).

The reasons for this difference between the Aché and the Machiguenga have yet to be determined. It is possible that the contrast is due to differences in the climate and geomorphology of their two habitats. During the rainy season in eastern Peru the rivers and their tributaries flood and disturb the vegetation, creating stands of soft-wooded secondary growth. The stands of secondary growth can be cleared rapidly by hand and are suitable for gardening, even though they are susceptible to flood damage. In eastern Paraguay, however, the rainy season is not dramatic and flooding is not as pronounced. Detailed studies on vegetation patterns are necessary to determine the distribution of soft-wooded areas that can support efficient horticulture unassisted by metal axes. An experiment conducted on primary forest clearance with stone axes indicated that even in areas with no large trees, clearing with stone axes takes at least ten times longer than metal ax- and machete-assisted clearance. Under these conditions, hunting and gathering may be more efficient than horticulture.

A second reason for the differences between Aché and Machiguenga subsistence strategies may be that multiple niches in South America's tropical forests are being exploited differently. Research has shown that Guarani-speaking horticulturalists occupied eastern Paraguay during the Spanish conquest (Susnik 1979), but their settlements were patchily distributed and may have been limited to specific ecozones. More archaeological information is needed on the distribution of horticulturalists in Paraguay to determine where they lived in relation to the Aché prior to the introduction of metal axes. In Peru, preliminary information concerning the uncontacted Mashco-Piro Indians living in Manu National Park suggests that they may be full-time hunter-gatherers living sympatrically with the Machiguenga (Kaplan and Hill 1984). Perhaps mobility combined with high levels of meat consumption is one viable option in a given niche, and horticulture combined with more stable settlement and lowered meat consumption is another. It is also possible that once a commitment is made to one or the other subsistence strategy, secondary adaptations develop that make switching from one to the other more difficult.

Another possibility is that the history of intergroup relations has played a role in determining the distribution of nomadic hunter-gatherers and less mobile forager-horticulturalists. Fully nomadic people may be less susceptible to domination and extermination by more powerful groups because they are harder to find and less dependent upon territorial control than horticulturalists. In late pre-Hispanic times, the large Guarani villages along the major rivers may have been able to dominate small horticultural settlements, leaving the deep forest for the hunting and gathering Aché. Following the conquest and during the Jesuit period, the Guarani villages were centralized into large settlements (reductions), and many people died from disease and were sold into slavery in Brazil. The depopulation of eastern Paraguay that resulted may have allowed the Aché, who were not settled or captured by the Spanish, to extend their foraging range.

Regardless of the factors underlying these differences, it is clear that the practice of horticulture has important effects on foraging patterns. For example, Aché mobility allows for more frequent encounters with migratory white-lipped peccary bands. Whereas white-lipped peccaries provide about 22 percent of the meat the Aché eat, they play an insignificant role in the Machiguenga diet. Central-place foraging allows the Machiguenga to engage in short, focused trips to known sites such as licks and fruiting trees and to maintain an elaborate trail system that greatly facilitates movement and pursuit of game. Another advantage of central-place foraging is that it allows for accumulation of more possessions. Most Aché possess four to six arrows at any time, while the Machiguenga often have fifteen or more arrows ready and another 250 shafts (which they plant) for later arrow manufacture.

Meat storage is another advantage of lowered mobility. Machiguenga frequently smoke the fish and animals they capture and consume them over a period of several days. The Aché seldom eat meat more than forty-eight hours after it is killed. Smoking meat enables the Machiguenga to have a steadier supply of protein without having to hunt every day.

A ready supply of carbohydrate calories from gardens buffers periods of poor foraging success. This may be the reason Machiguenga men hunt less frequently than and less cooperatively than Aché men. On a given day, only about 10 to 25 percent of Machiguenga men leave for the forest, and therefore fewer men are available for cooperative pursuits. However, it also could be possible that the lack of emphasis on cooperative food acquisition is due to other factors such as the use of dogs and differences in local ecology.

The importance of fish differs among the three communities. The Aché rarely exploit it, while the Machiguenga frequently acquire large or small amounts, depending on the season. This difference may be due to the lower abundance of fish in the watercourses in Aché territory, or to the lack of a steady carbohydrate source, which forces the Aché to acquire large quantities of meat. Fishing also varies between the two Machiguenga communities. Hook-and-line fishing is more successful in Tayakome because it is located on the Manu River. There, fish weighing 5 to 10 kilograms comprise more than 75 percent of the fish taken by hook and line. In Yomiwato, the largest fish taken weigh no more than 2.5 kilograms. On the other hand, the Manu is too large for *barbasco* fishing, and the residents of Tayakome fish with the poison only in small streams. In Yomiwato, the main watercourse is big enough to contain large quantities of fish but small enough to poison efficiently.

In addition to this spatial and cultural variability in resource use, there is year-to-year, seasonal, day-to-day, and even moment-to-moment variation in resource use. Both the Aché and Machiguenga practice a wide array of resource acquisition strategies that are sensitive to changing opportunities for successful pursuits. Specialization in one strategy is apparently not as successful as mixing strategies. Also, multiple resource targets such as hunting on the way to fishing sites enable foragers to share search and travel time among different pursuits. The importance of a mixed strategy may also explain why new technologies generally add to, rather than supplant, traditional technologies.

A key consideration is that efforts to preserve native cultures must rest on the knowledge that cultural practices are not static. Efforts either to halt cultural change or to direct it should be tempered by a respect for native peoples' ability to make their own choices and respond to new opportunities as they see fit. The challenge in forming policies governing reserves and development programs is to involve native peoples in ways that encourage participation by whole communities, including both men and women, rather than relying on the few individuals who do not fear outsiders.

Native Management and Conservation

A commonly held view is that native peoples live in delicate balance with nature and actively manage their environment in order to ensure continued resource productivity. Results of research on the Aché and Machiguenga suggest that while subsistence strategies are closely tied to challenges and opportunities provided by local environments, the tendency to actively manage and conserve resources is not universal.

It is perhaps better to view conservation and resource management as possible, rather than necessary, responses to ecological conditions. The Aché have little interest in conserving resources. They regularly destroy honey-producing beehives, cut down fruiting trees, and wipe out whole monkey troops. This may be because they once were extremely mobile, and until recently bands had overlapping access to foraging territories. The value of conserving resources may be greater when there is a high probability that someone else won't exploit the same patch.

While the Machiguenga also are not avid conservationists, they do rebuild honey hives so they can return to them in the future. Individual extended families tend to exploit the same zones repeatedly and avoid exploiting zones used by other extended families. Central-place foraging with relatively exclusive access may favor attitudes toward conservation of resources.

In general, when the short-term rewards for environmental degradation outweigh the long-term benefits of conservation, especially when those benefits may never accrue to those who conserve, it is unreasonable to expect people to engage in conservation management. If such management is the goal, then conditions must be created that offer short-term rewards for conservation and that ensure long-term rewards for the people and their direct descendents.

Technological Change, Cultural Change, and Environmental Impacts

The histories of both the Machiguenga and the Aché illustrate that the process of contact with Western technologies and resources occurs in stages and is not simultaneous with face-to-face contact. In the case of the Aché, all contacts with non-Aché were hostile until the early 1970s. There also were no contacted Aché who served as channels through which Western technologies could flow. Nevertheless, they were able to obtain metal tools by theft prior to peaceful contact. In addition, honey and oranges—first introduced by the Spanish—spread to the Aché habitat and contributed significantly to their diet. Precontact technological introductions probably had a greater effect on the acquisition of collected resources than on hunted resources. It was only with the postpeaceful-contact introduction of shotguns that hunting was affected. The Aché are more efficient at killing large game with shotguns than with bow and arrow and thus focus on large animals when they hunt.

The Machiguenga, with a larger population than the Aché, participated in extensive trade networks prior to face-to-face contact. It is reported that those in the Manu region acquired metal axes, chickens, and dogs well before contact. The postcontact introduction of hook-and-line fishing, especially in Tayakome, has had a major impact on the diet. It is probably the case that in the past diets contained more land-mammal meat in the wet season, when bow-and-arrow fishing is unproductive.

Environmental impacts of changing resource acquisition practices in response to new technologies are not well understood. Although it is frequently assumed that the introduction of more efficient technology inevitably leads to more intensive resource exploitation, there are few data to support this contention. For example, shotguns enable hunters to target larger game, and since larger animals tend to reproduce more slowly, their populations may be affected by shotgun use. However, with bow-and-arrow hunting large animals often are shot but escape to such great distances before they die that they are not eaten. This means more animals must be shot to satisfy food needs. In addition, the simultaneous introduction of hook-and-line fishing may relieve some pressure on land mammals.

In order to predict the effects of new technologies, not only their effects on efficiency but the factors that determine people's demand for resources must be understood. If the sole purpose of food acquisition is to feed the local population, the increased efficiency of shotguns may have only a slight impact on faunal densities because there is a limit to how much meat people want. For example, the Machiguenga's neighbors, the Piro, who live just outside Manu National Park, acquire more than 90 percent of their mammalian meat with shotguns. Preliminary analyses indicate that mammalian meat-capture per consumer day is only slightly greater among the Piro than among the bow-hunting Machiguenga. It appears that the Piro simply hunt less and spend their extra time in wage labor.

It is possible that the abundance of faunal resources is more affected by human population density than by technology per se. During initial face-to-face contact, native populations are almost invariably decimated by disease. Demographic data on the northern Aché (Hill 1992) indicate that their total population just prior to contact was 700 to 800 people and dropped to about 350 by 1975, five years after the first bands began to come into contact. Most of the contact-related mortality among the Aché was due to illnesses that began as colds (presumably of viral origin) and developed into fatal respiratory diseases. However, if native peoples are given adequate medical treatment during and following face-to-face contact, there is generally a population rebound. In fact, the Aché population is growing at about 5 percent per year. Population growth is lower among the Machiguenga, who have had less access to medicine.

The contact process is also associated with the movement of people onto the lands of newly contacted groups. For example, as farmers and ranchers moved into the Aché range, hostile contacts became more frequent, and many Aché believed leaving the jungle was the only way to avoid being killed. Now, most of the traditional range of the Aché is covered by fields and pastures owned by both individual colonists and large agribusinesses. The land to which they currently hold title is too small to support foraging, and the Aché are forced to exploit increasingly scarce forest lands owned by other people. It is almost certainly the case that population growth due to improved health care, migration to the tropical forests from other areas, agribusiness, and cattle-ranching have a much greater impact on habitats than does the use of introduced food-acquisition technologies.

The occupation and subsequent degradation of traditional ranges often occurs when the newly contacted peoples do not have the political influence to protect their rights. Efforts to give title to traditional peoples or to create reserves must be made prior to development or large-scale migration into the area. In some countries such as Peru, legal provisions allow reserves to be created before face-to-face contact. It is important for anthropologists and conservationists to determine where native peoples are threatened by migration and industrial expansion before environments are degraded and they lose their land.

LITERATURE CITED

Bertoni, M. 1941. Los Guayakies. *Revista de la Sociedad Científica del Paraguay.* (Asunción, Paraguay), vol. 2.

Camino, A. 1977. Trueque, correrias e intercambios entre los Quechuas Andinos y los Piro y Machiguenga de la montaña peruana. *Amazonía Peruana* 1(2):123–140.

Clastres, P. 1972. The Guayaki. In M. Bicchieri, ed., *Hunters and Gatherers Today.* N.Y.: Holt, Rinehart and Winston.

Hawkes, K., H. Kaplan, K. Hill, and A. Hurtado. 1987. Aché at the settlement: Contrast between farming and foraging. *Human Ecology* 15(2):133–161.

Hill, K. 1983. Adult male subsistence strategies among Aché hunter-gatherers of eastern Paraguay. Ph.D. Dissertation, Dept. of Anthropology, University of Utah.

—— 1992. Demography/Life history of Ache Foragers. Hawthorne, N.Y.: Aldine, in prep.

Hill, K. and K. Hawkes. 1983. Neotropical hunting among the Aché of eastern Paraguay. In R. Hames and W. Vickers, eds., *Adaptive Responses of Native Amazonians,* pp. 139–188. New York: Academic Press.

Hill, K., K. Hawkes, A. Hurtado, and H. Kaplan. 1984. Seasonal variance in the diet of Aché hunter-gatherers in eastern Paraguay. *Human Ecology* 12:145–180.

Hill, K. and H. Kaplan. 1989. Population description and dry season subsistence among the newly contacted Yora (Yaminahua) of Manu National Park, Peru. *National Geographic Research* 5(3):317–334.

Hurtado, A., K. Hill, K. Hawkes, and H. Kaplan. 1985. Female subsistence strategies among Aché hunter-gatherers of eastern Paraguay. *Human Ecology* 13:1–28.

—— 1986. Estudio comparativo sobre la ecología humana entre nativos del Parque Nacional del Manu. Unpublished manuscript. Dept. of Anthropology, University of New Mexico.

Kaplan, H. 1985. Tropical forest clearance with a stone ax: Experimental data. Paper presented at the American Anthropological Association Conference, Washington, D.C.

Kaplan, H. and K. Hill. 1984. The Mashco-Piro of Manu National Park, Peru. *AnthroQuest* (Fall).

—— 1985. Food sharing among Aché foragers: Tests of explanatory hypotheses. *Current Anthropology* 26(2):223–245.

Kaplan, H., K. Hill, and A. Hurtado. 1990. Risk, foraging, and food sharing. In E. Cashdan, ed., *Risk and Uncertainty in the Food Supply,* pp. 107–144. Boulder, Colo.: Westview Press.

Susnik, B. 1979. *Los aborigenes del Paraguay.* Tomo 4: *Etnohistoria de los Guaranies.* Museo Etnográfico "Andres Babero," Asunción, Paraguay.

SIX

Neotropical Indigenous Hunters and Their Neighbors: Sirionó, Chimane, and Yuquí Hunting on the Bolivian Frontier

ALLYN MACLEAN STEARMAN

The effect of settler intrusion on sustainable hunting by native Amazonians has been the focus of several recent studies. Research has concentrated on issues such as changes in hunting yields (Saffirio and Hames 1983; Saffirio and Scaglion 1982; Paolisso and Sackett 1985), differential hunting patterns of Indians and colonists (Redford and Robinson 1987), increased market participation as a result of difficulties in maintaining traditional subsistence patterns (Gross, et al. 1979), and problems of conservation of natural environments used by native hunters (Clad 1984; Posey et al 1984; Vickers 1991). Few of these current studies, however, have addressed the indigenous/outsider interface.

In 1970, Brazilian anthropologist Darcy Ribeiro published *Os Indios e a Civilização*, one of the first efforts to explain the nature of relationships between indigenous peoples and the national societies around them (Henley 1978). Ribeiro analyzed the process of social change and acculturation, and in particular, those factors that contribute to ethnic persistence of indigenous populations. He based much of his analysis on a typology of "national fronts," or those alien socioeconomic systems of relationships imposed on indigenous groups. Although Ribeiro presented a theoretically coherent argument, he did not address the complexity of local situations. As Paul Henley notes in his review of Ribeiro's work, "Where both diachronic and synchronic variation in the type of front have to be taken into account, the study of the effects of contact on a given indigenous group can be extremely complex" (1978:100). Henley also points out in several contexts that the successful survival or extinction of many ethnic groups is often determined by these complexities of variation at the local or regional level (1978:100–104).

Indigenous/outsider relationships and their effect on hunting-derived subsistence in a regional setting are diverse and complex. The variability of contact situations and intervening factors may affect both the social and natural environments involved, and contrary to the commonly held view, episodes of settler incursion are not necessarily essentially analogous and inherently threatening to indigenous foraging activities. Data from three case studies of indigenous peoples who inhabit contiguous regions in lowland Bolivia—the Sirionó, the Chimane, and the Yuquí—each of which depends on hunting and fishing for a major portion of its diet in addition to sharing its foraging territory with local mestizo populations—provide an opportunity to explore contact and the wide range of situations it engenders.

Data Collection

I studied the Sirionó in 1982 during a first visit from June to September, when various settlements were surveyed in an attempt to ascertain the most appropriate site for more in-depth research. At this time, only one settlement, Ibiato, was still ethnically Sirionó. During this same period, I also located and made contact with the Yuquí, a small group of foragers culturally related to the Sirionó but much more recently acculturated.

From 1983 to 1984 and 1987 to 1988, extended research among both the Sirionó and Yuquí collected data related to foraging activities. This fieldwork centered on both quantitative and qualitative investigation of subsistence patterns of the two groups. Data collection methods included records of hunting, fishing, and gathering kept on a daily basis for a period of fifty-six days in 1984 (October to December) for the Yuquí (men only), and for a period of eighty days in 1987 to 1988 for both the Sirionó (September to November) and the Yuquí (February to May) that included all providers. Surveys among non-Indian neighbors were conducted using prepared interview guides and probed such areas as use of natural resources (hunting, fishing, gathering, logging, etc.), settlement history, and primary sources of income.

The Chimane segment of the research was conducted on behalf of Conservation International and the National Academy of Sciences of Bolivia to furnish preliminary information to assist in the formulation of the management plan for the Beni Biosphere Reserve, part of UNESCO's Man and the Biosphere Program. A small population of Chimane is located within the boundaries of the reserve and will be affected by future conservation goals. The study was intended to provide an initial understanding of resource use, particularly faunal, by the Chimane and the local mestizo populations that inhabit the reserve. Research was carried out jointly with Kent Redford, a biologist from the University of Florida, and was completed during the month of May 1987. We used a prepared interview guide for this research and interviewed a total of thirty-six households representing approximately 50 percent of the population within the reserve (Redford and Stearman 1989).

The preceding outline of procedures used in data collection makes evident the problem that research among the three groups was carried out in different periods, using different methodologies, and with different research goals. Nevertheless, the overall orientation of this research, that of exploring the various social and economic contexts of people who depend on hunting for a major portion of their subsistence, permits general observations of the processes affecting each of these target populations.

The Sirionó

Location and History

The Sirionó settlement of Ibiato is located 50 kilometers east of Trinidad (14° 45′S, 64° 48′W), the capital of Beni Department. Ibiato lies in the heart of the Beni floodplain, a large area of seasonally inundated savanna broken occasionally by areas of forest bordering waterways and, more typically, forming "islands" (*islas*) of higher ground on the savanna. The area is flat, and the altitude is 236 meters. Annual rainfall averages 1,800 to 2,000 millimeters per year, with the heaviest rains recorded during the months of October to April (Ministerio de Transportes 1975). As is common throughout the lowlands, the region experiences a marked dry season often accompanied by cold fronts, or *surazos,* during which time the temperature has been known to drop to as low as 0 degrees Celsius. Average temperatures, however, normally range between 10 and 30 degrees Celsius (Ministerio de Transportes 1975). The village of Ibiato is located on an artificial mound constructed, and later abandoned, by Moxos Indians during the pre-Columbian era. The mound rises approximately 10 meters from the center of a large savanna that is flooded during three to four months of the year (December to March), making the village an island. Local people term these flooded savannas *yomomos,* meaning large grass lakes.

The population of Ibiato fluctuates throughout the year due to movement of people to and from cattle ranches during periods of seasonal employment and to outlying areas where small homesteads have been built. Generally, however, Ibiato's population remains at about 250 people living in thirty-three dwellings. The total Sirionó population, including the village and its surrounding area, numbers less than 400 individuals. During the 1930s, Protestant missionaries working with the Sirionó solicited from the Bolivian government a parcel of approximately 10,000 hectares, which was designated as an indigenous community for the Sirionó.

The Sirionó achieved prominence in the anthropological literature following the publication in 1950 of Allan Holmberg's *Nomads of the Long Bow.* They were remarkable for their lack of cultural elaboration and a foraging life-style that placed them among those groups defined by Steward and Faron (1959) as "marginal foot nomads." They are most likely remnants of Tupí-Guaraní-speaking people whose presence in Bolivia can be traced to pre-Columbian incursions by Guaraní from present-day Paraguay into Bolivia, probably to capture women and slaves (Stearman 1984). At the time of Holmberg's research, carried out from 1940 to 1941, the Sirionó were the targets of concerted efforts to contact and "pacify" them. This process was being carried out by several groups, including the Bolivian military, the Catholic Church, and a Protestant sect from the United States that settled two bands of Sirionó at the site known to them as Ibiato (Sirionó *Ibi* = land; *ato* = mound, hump). Ibiato is the only Sirionó settlement that has successfully survived assimilation by other lowland indigenous groups or dispersal among the local populations of mestizos (Stearman 1987).

Settlement Patterns of Local Mestizo Peoples

The Beni Department was settled by Europeans and their mestizo descendants migrating from the Santa Cruz region to the south (Sanabria Fernández 1973). The area's natural grasslands, recognized by the early Jesuits and Franciscans as ideally suited to ranching, were rapidly populated by cattle imported from Europe, which over time became well adapted to the harsh conditions of the savanna. As the missions lost social and political hegemony in the region, the area gradually was taken over by settlers seeking to exploit the land, the indigenous residents, and the cattle herds that grazed there (Jones 1980). Because of the annual threat of inundation, only the small islands of forest were used for cultivation, an activity that gradually became the domain of the numerous Indian peoples and poorer mestizos tied to large landholdings through debt obligations. The purpose of these small farms was to provide staples for the ranchers and their workers. Consequently, the land-tenure pattern for much of the Beni, particularly the area where the Sirionó live, is one of large properties held for the purpose of ranching, not farming. Virtually no improvement is made to pastures other than clearing away brush as it encroaches on open grazing land. Fences also are employed infrequently, making branding necessary to differentiate animal ownership. As is typically the case in this type of situation, disputes between ranchers often break out over unbranded calves, rustling, or brand tampering.

Cattle ranching in the Beni requires only a small resident staff of ranch hands to watch over the herds. As the result of a long-standing tradition among ranchers in the Beni, in addition to receiving a small salary, workers are provided with a monthly ration consisting of staples such as lard, flour, coffee, and sugar, along with beef from the ranch, which is slaughtered and salted as needed. This mestizo settlement pattern has enabled the Sirionó to continue what appears to be a successful pattern of subsistence based on hunting and gathering while remaining semisedentary for more than fifty years.*

*The Panare of Venezuela, studied and written about by Paul Henley (1982) and also depicted in an ethnographic film (BBC 1982) appear to be in a situation similar to the Sirionó. They, too, inhabit an area of mixed forest and natural savanna used by mestizos for cattle ranching. Consequently, they have been able to continue to exploit faunal resources while surrounded by ranching interests because of the same social and economic characteristics found in the Beni region of Bolivia.

Siriono Subsistence Patterns

Although Sirionó hunting territory has experienced change and increased settlement since the time when the Sirionó first were settled at the mission of Ibiato, the change has not radically altered the original environment. Ranchers continue to use the savanna without much alteration to its original state, although with human exploitation of the area there has been increased frequency of dry-season fires. Ranch personnel rarely exceed ten individuals, and provisions come from outside sources and from ranch animals. Thus ranch personnel do not generally depend on local natural resources such as game animals for part of their diet. Cattle are distributed sparsely, and overgrazing is rare. This situation has allowed for the preservation of habitats of fauna native to the area.

The Sirionó are virtually the only people in their immediate vicinity who consistently exploit game and fish resources. They are also able to extend their one-day hunting range beyond their own land by having large, flat savannas with good trails or ranch roads to travel over in the dry season and canoes or saddle animals to use during the wet months. With the exception of a few of the older men, all Sirionó hunt with firearms, the 16-gauge shotgun being the weapon of choice. Although some ranchers object to Sirionó hunting on their property, there is little an owner can do to prevent it. Then too, the Sirionó are a convenient and often willing source of labor, and on occasion they provide the service of tracking and killing predators that threaten calves. Although stories circulate among ranchers that the Sirionó also "hunt" cattle, there is little evidence of this occurring with any regularity, and generally the relationship between the two groups is, if not amicable, at least one of mutual tolerance.

Of primary importance to Sirionó hunting success has been the continued presence of large mammals, particularly ungulates such as white-lipped peccaries (*Tayassu pecari*), collared peccaries (*Tayassu tajacu*), tapir (*Tapirus terrestris*), and at least three species of deer (*Blastoceros dichotomus, Mazama americana, Mazama goazoubira.*), which evidently are capable of coexisting quite well with cattle. These game animals provide periodic but consistent sources of large quantities of meat that may be salted and dried for consumption. Between hunts of larger animals, smaller mammals such as armadillos (*Dasypus novemcintus, Priodontes maximus*), which are numerous in the savanna environment, agoutis (*Dasyprocta* spp.), and pacas (*Agouti paca*) are readily caught using dogs. With the help of hunting dogs, the Sirionó often do not require a weapon of any greater sophistication than a large stick or a machete for a kill.

The Beni is also populated by rheas (*Rhea americana*), and while there is disagreement among the Sirionó about regularly exploiting rheas for meat or especially for commercial purposes (the wing feathers can be marketed), they are hunted from time to time by most Sirionó when other game is not available. The controversy arising over whether rheas should be hunted stems from the fact that rhea eggs are an important seasonal food. Each egg weighs about 1 kilogram and provides a meal for a family of five or six. Since rhea eggs are so highly prized and often are found in nests containing thirty or more, many Sirionó claim that the adult birds should not be hunted and recognize that by doing so, they are jeopardizing an important source of nutrition. Eggs also may be sold or traded to local mestizos, bringing the equivalent of US $2.50 each.

Fishing is another important seasonal activity for the Sirionó, providing more than half of their animal protein intake during the dry months. As the water in the *yomomos* begins to evaporate, fish are concentrated in low places and can be captured easily in baskets or by hand. Virtually every pond within a 5 - 7-kilometer radius of the village is completely exhausted of fish by the time the dry season ends. Due to widespread flooding in the Beni during the rainy season, however, these fished-out areas are quickly replenished from surrounding unexploited regions. Hook and line are used for fishing year round in the large and numerous lagoons in the area.

The Sirionó also sell animal skins to local middlemen, usually itinerant merchants. Nonetheless, with the exception of spotted cats and caimans, the primary purpose of hunting is to provide meat. But by killing a game animal that also has a marketable skin, the dual goal of feeding the hunter and his family as well as providing external income is achieved. Throughout the year, the Sirionó market the hides of collared peccaries, deer, capybaras (*Hydrochaeris hydrochaeris*), and an occasional puma (*Felis concolor*), ocelot (*Felis pardalis*), or margay (*Felis wiedii*). Jaguars (*Panthera onca*) now are encountered only rarely in this area, but they are occasionally hunted when their presence has been noted by ranchers. Although selling the hide is illegal, the Sirionó seldom have problems marketing a good jaguar skin through the network of dealers in lowland Bolivia. Yet even in the case of predators, hunted principally for their skins or to protect cattle, many of the older Sirionó also consume the meat.

With the advent of the dry season, the Sirionó intensively hunt caimans, or *lagarto* (*Caiman yacare*). As with dry-season fishing, the Sirionó take advantage of rapidly shrinking sources of water and the tendency of these animals to concentrate in remaining ponds. During this time, caiman hunting is carried out on a daily basis and usually involves small groups of men who work one water hole at a time. Bow staves or long, sharp poles are used to probe for the animals, which often take refuge in the muddy bottom of the pond. The *lagarto* may be stabbed on the dorsal side with the sharpened stave or simply stirred up from the bottom and then killed with a machete blow across the back of the head. Unlike mestizo hunters who abandon the carcass after removing the side strips of hide, the only commercially valuable part of the animal, the Sirionó frequently sever a portion of the tail for consumption. If the *lagarto* is a female, the abdomen will be explored for eggs, considered a great delicacy by the Sirionó.

Although about seven hundred to eight hundred skins may be taken in a season, the Sirionó continue to hunt these animals successfully on an annual basis. Like fish, *lagarto* move easily through open waterways, and once the savannas flood, those in outlying areas quickly take over the available niches near the village. Although the Sirionó virtually exterminate all animals measuring .75 meters or greater* found within a 10-kilometer radius, smaller animals are not usually taken because dealers will not purchase them.

The black caiman (*Melanosuchus niger*), now illegal to hunt in Bolivia, is all but extinct in this region. Only one was captured in 1987 by a Sirionó hunter and it was taken in the middle of a large lake by using a canoe and a firearm. Generally, the Sirionó do not like to hunt these animals, (even when they were plentiful) because of their great size and perceived aggressiveness. The smaller *lagarto* is much less dangerous and can be taken by hand. Black caimans always brought a good price and consequently were pursued by professional mestizo hunters until they were exterminated. *Lagarto*, on the other hand, rarely bring in more than a few dollars per skin, making them, in the eyes of many mestizos, not worth the effort of having to go into rough

*Measurements of caimans are *chaleco* (vest) lengths, the side strips of skin removed from the snout to just behind the vent.

terrain to hunt them. For the Sirionó, who are experienced hunters and who have need of small but steady amounts of cash to purchase necessities such as salt, sugar, soap, medicines, and ammunition, the *lagarto* is an ideal source of seasonal income. Nevertheless, the pattern of hunting *lagarto* primarily during the dry season, the period when most agricultural clearing is done, places this type of commercial hunting in direct conflict with farming.

In spite of offering a more stable supply of food, farming has never held much attraction for the Sirionó. In the precontact era, corn, manioc, and sweet potatoes were cultivated in small quantities by making use of the clearings provided by tree falls in the forest. Once the seeds or cuttings had been planted, the Sirionó abandoned the area in their continuing quest for food. Several months later they would return to these small gardens to harvest what remained after weeds and animals had taken their toll. Even with more than fifty years of contact and encouragement by missionaries, the Sirionó continue to farm on a small-scale and sporadic basis. In 1988, less than half of the village households put in crops, depending on wage labor, foraging, commercial hunting, surplus food commodities, and the goodwill of their farming relatives to support themselves. Hunting and fishing continue to occupy the time and interests of most of the male population in the village who acquire prestige and admiration from their peers as hunters, not farmers.

The Chimane

Location and History

The Chimane occupy a territory west of the Sirionó in a transitional zone between the humid tropical forest bordering the eastern reaches of the Andes Mountains and the westernmost edge of the Beni savannas. They are culturally related to the Yuracaré, Mosetene, and probably the Leco peoples who inhabited a large region of tropical forest along the eastern foot of the Andes (Metraux 1963a; 1963b). The Chimane living in the Beni Biosphere Reserve are located in the eastern limits of their traditional foraging territory (14° 38′S; 66°18′W.). The rainfall, altitude, and temperature ranges are all comparable to those discussed for the Sirionó area. However, the Chimane have larger expanses of forest to exploit, and there are several rivers in the region that are convenient for fishing.

The Chimane have had contact with Europeans since the seventeenth century and the establishment of Spanish missions in the region. (Steward and Faron 1959). Thus the Chimane have a long history of acculturation, although isolation and sporadic mission activity have contributed to the preservation of much of their traditional culture (Pérez Díez 1983; Riester 1976). Modern technologies such as firearms and hook and line are used for hunting and fishing, although unlike the Sirionó, most Chimane continue to bow hunt when they are unable to purchase ammunition. The Chimane have actively practiced horticulture since pre-Columbian times and combine this activity with hunting and fishing to subsist. In recent history, they have become a part of the social and economic network of the region involving patron-client relationships with large landholders, ranchers, and merchants. These ties have been strengthened through bonds of fictive kinship established through the ritual of *compadrazgo*, or co-godparenthood.

Settlement Patterns of Local Mestizo Peoples

The Beni Biosphere Reserve is located in an area of seasonally inundated and upland tropical forest and savanna in the Mamoré River drainage system. The government-managed reserve consists of 135,000 hectares and is surrounded by privately owned ranches or small farms held in usufruct by lowland campesinos, or small farmers. The center of the reserve is devoid of human inhabitants, but a few Chimane extended families live near this core area. The remainder of the inhabitants, consisting of Chimane, several families of other indigenous groups, and lowland campesinos are located around the periphery of the reserve, particularly along the Maniqui River, which forms the northern boundary of the protected area.

The Beni Biosphere Reserve is part of the Man and the Biosphere Reserve Program, whose goal is to incorporate local populations into the management of the conservation areas. This approach has not been tested fully as yet, and the management of human populations inhabiting reserves remains problematical (Hill 1988). One of the issues to be addressed inevitably will be Chimane use of game animals within the reserve, and in particular of certain endangered primates such as spider monkeys (*Ateles* spp.) which are an especially favored item in the Chimane diet. Preliminary research indicates that local mestizo people in the reserve hunt very little, depending on the availability of beef and crops for subsistence. The Chimane, therefore, have minimal competition from neighbors for available game. The Maniqui River and the smaller Curiraba River, which transects the reserve, both provide fish to the population, but again it is primarily the Chimane who take advantage of this resource. Lowlanders enjoy consuming fish, but most do not enjoy it as a steady diet. Unlike indigenous peoples who thrive on wild foods, campesinos in this area have developed dietary preferences in which game meat and fish are viewed as foods to be eaten as specialties and not as a regular source of nutrition.

The reserve is still relatively isolated, and large national markets are distant and difficult to reach. Consequently, mestizo settlement of the area has been sparse and often quite transitory. Most of the local people, including many of the Chimane, have been in the area less than five years. For mestizos, the area does not offer any particular economic advantage, and much of the better land has been tied up in large cattle ranches for several generations. As is customary in ranching zones, many ranch owners do not live on their property but hire overseers. Owners of small ranches, who do remain on their holdings, form the elite of the region. As with the Sirionó situation, population densities on ranches are low. Until recently, most mestizo settlement was confined to areas near rivers, since waterways were the primary conduit to the outside world. A new road through the area, forming a southern boundary of the reserve, has changed this pattern somewhat, and settlement along this route is appearing as well.

Most of the mestizos interviewed had come to the area at the request of ranchers in need of labor or to take over the homesteads of relatives or acquaintances who had abandoned their farms. Thus inmigration replaces rather than adds to the existing population. A relatively stable mestizo population consists of those who own cattle ranches, those who live in the town of San Borja to the west of the reserve, the residents of the village of Totaizal, a struggling community of small farmers on the southern boundary near the new road, and the itinerant merchants who travel the rivers of the region to buy and sell commodities.

Many of the mestizos in the area of the reserve are growing small patches of coca in an attempt to increase what appear to be otherwise meager earning from crops such as plantains and rice and employment on local ranches. When queried about hunting activities, most mestizos claimed that they seldom hunted, and if they consumed wild meat, it was usually purchased or traded from Chimane. Their preference was to purchase beef, often in the form of jerky. When queried as to what meat they consumed most often, 62 percent of the mestizos surveyed responded "beef."

Chimane Subsistence Patterns

The Chimane have a history of being seminomadic, widely ranging the territory that they have hunted and fished for generations. Many have relatives in older mission settlements near the mountains whom they visit on occasion, often for extended periods. Thus the reserve is only one of many areas in the region that is regularly exploited by this group. The Chimane, like the Sirionó, also work on ranches as day laborers and have established long-term labor relationships with local peoples both in the town of San Borja as well as among those who have large or small land holdings.

The Chimane hunt on average about once a week, and the presence of large mammals such as white-lipped peccaries is important to their continuing success. More than half the animals that appeared in the hunting survey were peccaries, and most of these were white-lipped. Although the Chimane consistently listed game animals such as peccaries as preferred meat species, when asked what they actually ate most often, 77 percent listed fish. Fishing, both within the reserve in forest lagoons and in rivers, contributes significantly to Chimane subsistence. The fish that are taken normally range from 1 to 2 kilograms in weight and are not among the larger, more lucrative commercialized species such as *surubí* (*Pseudoplatystoma* spp.) and *pacú* (*Colossoma* spp.). Thus while river systems farther downstream are of interest to commercial fishermen, the headwater rivers in the Chimane area are seldom exploited for this purpose.

Commercial hunting for skins and for meat has been practiced at times but does not appear to have major importance to the economic well-being of the Chimane. Most cash is secured through wage labor, the gathering and weaving of *jatata*, a roofing material that the Chimane have traditionally produced for the region, and by selling handicrafts such as weavings produced by Chimane women. Thus hunting among this group is used primarily as a form of direct subsistence.

The Chimane are excellent farmers and depend on gardens as a consistent and dependable source of nutrition. They cultivate upland rice, plantains, manioc, corn, squash, and a variety of lowland fruits. Most Chimane farms also have a supply of small domestic fowl and pigs. In addition to producing crops for subsistence, the Chimane sell their surplus to local ranchers as part of the long-standing relationship between ranchers and indigenous peasants who provide staple foods for ranch personnel.

The Yuquí

History and Location

The Yuquí are the least acculturated of the three groups discussed, having been contacted and "pacified" successfully by a group of Protestant missionaries in the mid-1960s. They are located on a mission station on the Chimoré River (64° 56′ W; 16° 47′ S). This region is only 60 kilometers from the eastern edge of the Andes and falls within the rain shadow of the mountains. Annual rainfall averages between 4,000 and 5,000 millimeters. The altitude, at 250 meters, is within the same range as the other areas described. Unlike the Sirionó and Chimane situations, however, the Yuquí habitat is entirely tropical forest, broken only by recent river-course changes causing the formation of oxbow lakes, swamps, and seasonally flooded forest.

The Yuquí are culturally and linguistically related to the Sirionó and are probably a subgroup of the original Guaraní populations that migrated into the area prior to the European conquest (Stearman 1984). Their level and style of material culture is equivalent to that of the Sirionó, with the exception that the Yuquí practiced no horticulture whatsoever prior to contact with the mission. Their previous pattern of total dependency on forest resources for subsistence makes them extremely vulnerable to changes in fish and game availability. The mission supplies the Yuquí with supplemental food in the form of the same surplus supplies (flour, dried milk, oil) that the Sirionó receive, and plantains have been established around the mission. The Yuquí, however, still derive the majority of their caloric intake and animal protein from game and fish.

Settlement Patterns of Local Mestizo Peoples

Until 1986, the present territory of the Yuquí was, with the exception of one Yuracaré Indian family, uninhabited by local peoples. The region was considered extremely remote and did not interest settlers because of the lack of access to markets and the availability of land in less difficult areas. In recent years, the boom in the cocaine trade has encouraged the opening and settlement of the area contiguous to the Yuquí for the purpose of growing coca. A new road has also made the area more appealing to settlers who have extended the Chapare coca zone into the Chimoré. Settlers are now located both above and below the Yuquí along the Chimoré River and in the region southwest of the mission. Land-clearing by local people in the 314-square-kilometer catchment zone used by the Yuquí for foraging has increased from 4 hectares in 1985 to 60 hectares in 1988. The population also now includes highland peoples who consider themselves temporary residents in the area only to take advantage of the market for coca.

Yuquí Subsistence Patterns

As previously noted, until very recently, the Yuquí had sole use of the forest surrounding the mission and freely exploited the Chimoré River and oxbow lakes. The introduction of new technologies such as hook and line, nets, and firearms have increased their opportunities for harvesting local faunal resources. In particular, new fishing techniques (formerly restricted to bow fishing and the use of *barbasco* poison) have enabled the Yuquí to make use of the Chimoré River. Prior to contact, the Yuquí seldom ventured near open water for fear of being detected and shot by mestizos. With settlement in the protected mission environment, the Yuquí were able to exploit the river, which became an important source of animal protein for them.

The pattern of production peculiar to coca cultivation has had particularly negative effects on the Yuquí's success in hunting and fishing. Unlike most settlers who grow food crops for subsistence, the new settlers in the Chimoré are interested only in growing coca. Colonists seldom remain on their farms year round, returning instead to highland settlements to continue other occupations. While in the lowlands, these colonists make extensive use of natural resources to feed themselves while they are planting or harvesting coca. Fishing is especially important to those who live along the Chimoré River, and who are therefore conveniently located to make daily use of this resource. Unfortunately, many of the colonists are ex-miners who use their skill in the handling of dynamite to quickly extract large quantities of fish. This has had a devastating effect on the river's fish populations. Additionally, with new market access, commercial fishermen have found the Chimoré River increasingly attractive. By late 1987, the river had been virtually depleted of fish as the result of fishermen repeatedly stretching gill nets across the entire width of the river.

Although the Yuquí consider themselves hunters, until recently fishing provided them with more than 50 percent of their animal-protein intake, and most of these fish came from the Chimoré River. The loss of the river as a food source has had severe repercussions. In 1985, the Yuquí were consuming 88 grams of animal protein per capita per day. By 1988, this amount had dropped to 40 grams.

In addition to exploiting fish resources, colonists also hunt, using dogs and pursuing game species such as peccaries, tapir, pacas, and deer favored by mestizos. According to the Yuquí, land clearing has interfered with the normal migratory patterns of white-lipped peccaries. As a consequence, this species has apparently disappeared from the area. As of mid-1988, the Yuquí had not seen a white-lipped peccary for three years. Tapir, deer, and capybara also are becoming scarce, forcing the Yuquí to embark upon hunting patterns that may lead to overhunting their catchment area. For example, the data indicate that they are now killing greater numbers of individuals from the smaller, less preferred species.

The Yuquí have periodically sold animal skins through the mission, which then markets them in major urban areas in Bolivia. The mission dislikes engaging in this activity, however, for several reasons. First, missionaries must keep track of which skins belong to which hunter, frequently resulting in confusion and disputes. Second, the skins must be stored safely until transportation is available and there are enough stockpiled to merit the marketing effort. Finally, the mission is scrupulous in avoiding any activity that may be illegal or appears so and would therefore threaten the church's ability to continue working in Bolivia. Since markets for skins vary, and certain species may be taken from or added to endangered lists, missionaries are uncomfortable about participating in this business. Consequently, although the Yuquí would like to have this additional source of income, especially for the purchase of ammunition, it is not often available to them.

The missionaries discourage wage labor away from the mission now that colonists are within a day's walk from the camp. Some Yuquí are beginning to avail themselves of this opportunity, however, often more from sheer curiosity to learn about colonists' lives than from the desire to earn cash. For the most part, however, the Yuquí still operate in a closed economic system at the mission, providing labor for missionaries in return for wages, which are then spent at the mission store.

Because of the Yuquí's continuing success until recently as foragers and the mission's willingness to provide additional nutrition in the form of surplus foodstuffs, there has been no concerted effort to teach the Yuquí to farm. Consequently, even after twenty-two years at the mission station, the Yuquí are not self-sufficient in growing their own staple foods such as rice, corn, or manioc. Their gardens remain too small even for minimal subsistence needs and are poorly tended. The Yuquí have an adequate supply of plantains, but these are generally recognized as a poor source of nutrition when not supplemented by other foods. With the decline in fish and game in the last several years, the Yuquí are now faced with a lowered meat intake, although the mission has managed to offset the nutritional loss to some extent by importing food supplies. However, this only increases the Yuquí's dependency on the mission and places them in greater jeopardy should they for any reason lose this type of assistance.

The presence of populations contiguous to indigenous groups does not inherently constitute a threat to foraging activities that emphasize the use of faunal resources. The nature of that contact must be considered as well. In the case of the Sirionó, the presence of large ranching interests on natural grassland has actually served to impede the continued alteration of the natural environment and has insulated the group from possible negative effects of the proliferation of small farms in the area. The Sirionó and ranchers have formed a symbiotic relationship that has benefitted both groups and has led to the preservation of hunting resources for the Sirionó. Cattle ranching is often described as the primary force in the destruction of rain forest, particularly in Brazil, and as such indirectly contributes to the demise of indigenous populations that depend on faunal resources to provide animal protein. However, when practiced in areas of natural pasture interspersed with forest, ranching in and of itself may not be harmful either to the environment or to the native peoples that depend on it to survive in traditional ways, as the experience of the Sirionó suggests.

If economic activities of local mestizo populations do not directly impinge on or compete with the foraging interests of indigenous peoples, then faunal use may not be immediately threatened. Overall human density patterns, of course, are critical to the maintenance of faunal reserves, but population distribution also must be considered. In the case of the Chimane, the territory that is hunted and fished is not heavily inhabited by local peoples. Instead, they have congregated along waterways and in areas with greater accessibility to markets and employment. Since this region has now been designated a Man and the Biosphere Reserve, there is great potential for the Chimane's continued productive use of faunal resources in the area. However, if conflicts arise over how these resources are to be exploited with regard to conservation interests, the Chimane may find that their access to fauna traditionally considered a source of food may be radically altered. In this case, protein substitution or subsidies, particularly in the form of beef, may become necessary to reach an equitable solution to these conflicting interests.

In the case of the Yuquí, both increased population densities as well as unusual patterns of natural-resource exploitation by settlers have conspired to pose a severe threat to the Yuquí's hunting and fishing success. Colonists and commercial fishermen are competing with the Yuquí for limited resources that cannot continue to provide for everyone at present levels of use.

Commercial hunting is of differing importance to each of the three groups addressed. Yet even the most active commercial hunters, the Sirionó, still combine most of their commercial hunting with subsistence needs. The most aggressively exploited species, the caiman, has been hunted successfully for many years because of a unique environmental situation. Had this not been the case, available markets and the need for cash income probably would have led to the extinction of these animals in the Sirionó region of the Beni. For the Sirionó, caiman hunting is important economically, but none felt that it was crucial to the group's survival. Many Sirionó commented that given the low prices paid by middlemen, intensive hunting of caiman often was not really worth the effort. Increasingly stringent international laws affecting marketability have made the pursuit of certain species such as otters no longer profitable, underscoring the efficacy of these laws when they are enforced. The Sirionó, like many indigenous peoples, simply respond to world demand for skins, legal or otherwise. When these markets are no longer viable, hunting of these species would be limited to subsistence needs.

A few key species of ungulates appear to contribute to the long-term success of hunting peoples in this region.* The white-lipped peccary in particular has been important in maintaining sufficient yields to enable continual use of catchment areas without game depletion becoming significant enough to adversely affect human well-being (although some species, particularly primates, may be absent from game inventories due to overhunting or movement out of the area). White-lipped peccaries offer the unique advantage of having high body weights along with congregating in herds of thirty to two hundred animals. With most indigenous hunting techniques, entire herds are rarely destroyed as the result of a single hunt. Because these animals move seasonally, they are available on a consistent but periodic basis; but they also are able to move out of hunting range as well, decreasing the frequency of harvesting. With the disappearance of white-lipped peccaries (as well as other important but less frequently killed large mammals) and fish, Yuquí hunting patterns changed profoundly, indicating that overexploitation of the Yuquí catchment area is now in progress.

Fishing is a significant, but often overlooked or discounted activity when the focus of research is on hunting. In each of the cases presented above, fishing provides an important seasonal or year-round source of animal protein and may relieve hunting pressure on certain prey species. For foragers who have become sedentary or semisedentary due to contact and circumscription, the presence of fish along with certain ungulates may be critical to long-term foraging success (Ross 1987: 9).

As settlement of the Amazon Basin continues at an unprecedented rate, new strategies are being developed and tested to ensure the protection of the environment and of native peoples, while at the same time ways are being sought in which the nations that desire to exploit this vast domain may do so in a rational manner. Concepts such as

*Vickers (1988) also notes that ungulates, particularly the peccaries, provide most of the meat for the Siona-Secoya, and that their variability in any year can have significant effects on yields.

farming-systems research that attempts to correlate local agricultural patterns with environmentally and economically sound farming practices offer hope in many sectors. Ultimately, people like the Yuquí, now sedentary and dependent on mission support, will need impetus to produce their own food. As a people without a history of farming, they are in an ideal position to develop a diversified system of agriculture that would also ensure the maintenance of the natural environment that they depend on for many of their nutritional needs.

The establishment of Man and the Biosphere Reserves, which incorporate conservation interests as well as those of traditional or indigenous human populations into their management plans, is another approach with great potential. Indigenous peoples must be included directly in the management of these areas and, in doing so, encouraged to consider agreements regarding the preservation of endangered or protected species. At the same time, conservationists involved with native peoples who depend on game animals as a major source of animal protein must understand that trade-offs may be necessary in order to assure equitable treatment of the human population.

The direction of these innovative strategies implies that achieving an understanding of the interaction between human and natural systems will require more than macrolevel decision-making. Natural resource use must be analyzed at the level of individual groups and local economies, with the recognition that each situation encountered presents special challenges and opportunities that should be considered in any type of long-range regional planning decision.

ACKNOWLEDGMENTS

I am grateful to the many organizations that have funded segments of my research in Bolivia since 1982: the University of Central Florida Division of Sponsored Research, the University of Florida Amazon Research and Training Program, Conservation International, the National Science Foundation (Research Grant no. BNS-8706958), the L. S. B. Leakey Foundation, the Charles A. Lindbergh Fund, and the Explorer's Club.

I would also like to thank Ida Cook, Joyce Lilie, Kent Redford, and Marianne Schmink for their critical commentary on the manuscript and suggestions for improvement. Parts of this paper were presented at the workshop, Traditional Resource Use in Neotropical Forests, held January 19–22, 1989, at the University of Florida.

Many individuals in Bolivia have contributed greatly to my interests in pursuing research projects there over the years. I would especially like to acknowledge the assistance of Jack and Darlene Anderson, Macri Bastos, Phil and Jackie Burns, Ray Cowell, Chiro and Nancy Cuellar, Bob and Mary Garland, Sister Mary Gerald McClosky, Wilson and Doroty Melgar, Carmen Miranda, José Mirtenbaum, María Teresa Ortiz, Jürgen Riester, and Hugo Daniel Ruíz. I am also indebted to my research assistant, Linda Moore, who accompanied me to the field from 1987 to 1988.

Without the help of the Chimane, Sirionó, and Yuquí peoples, most of this work would not have been possible. It is my hope that the results of this research may in some way contribute to their survival as unique ethnic groups in a world that would be much poorer without them.

LITERATURE CITED

Clad, J. 1984. Conservation and indigenous peoples: A study of convergent interests. *Cultural Survival Quarterly* 8(4): 68–73.

Gross, D., G. Eiten, N. M. Flowers, F. M. Leoi, M. L. Ritter, D. W. Werner. 1979. Ecology and adaptation among native peoples of central Brazil. *Science* 206:1043–1050.

Henley, P. 1982. *The Panare: Tradition and Change on the Amazonian Frontier.* New Haven, Connecticut: Yale University Press.

—— 1978. Review of 'Os Indios e a Civilização' by Darcy Ribeiro. *Cambridge Anthropology* 4: 88–111.

Hill, K. 1988. International conservation and indigenous peoples: General principles and three case studies from lowland South America. Washington, D.C.: World Bank Contract L1049. ———. Typescript.

Holmberg, A. 1950. *Nomads of the Long Bow.* Washington, D.C.: Smithsonian Institution.

Jones, J. 1980. Conflict between whites and Indians on the Llanos de Moxos, Beni Department: A case study in development from cattle regions of the Bolivian Oriente. Ph.D. dissertation, University of Florida, Gainesville.

Llewelyn-Davies, Melissa, prod. 1982. *The Panare—Scenes from the Frontier.* London: British Broadcasting Corporation.

Métraux, A. 1963a. The tribes of Mato Grosso and eastern Bolivia. In Julian Steward, ed., *Handbook of South American Indians,* Vol. 3, 349–360. New York: Cooper Square.

—— 1963b. Tribes of the eastern slopes of the Bolivian Andes. In Julian Steward, ed., *Handbook of South American Indians,* Vol. 3, 465–506. New York: Cooper Square.

Ministerio de Transportes y Communicación. 1975. *Anuario Meterológico.* Ministerio de Transportes y Comunicación, La Paz, Bolivia.

Paolisso, M. and R. Sackett. 1985. Traditional meat procurement strategies among the Irapa-Yukpa of the Venezuela-Colombia border area. *Research in Economic Anthropology* 7: 177–199.

Pérez Díez, A. A. 1983. Etnografía de los Chimane del Oriente Boliviano. Ph.D. dissertation, University of Buenos Aires, Argentina.

Posey, D. A., J. Frechione, J. Eddins, L. Francelino da Silva, with D. Myers, D. Case, and P. MacBeath. 1984. Ethnoecology as applied anthropology in Amazonian development. *Human Organization* 43(2): 95–107.

Redford, K. H. and J. Robinson. 1987. The game of choice: Patterns of Indian and colonist hunting in the Neotropics. *American Anthropologist* 89(3): 650–667.

Redford, K. H. and A. M. Stearman. 1989. Local peoples and the Beni Biosphere Reserve, Bolivia. Vida Silvestre. *Neotropial* 2(1):49–56.

Ribeiro, D. 1970. *Os Indios e a civilização: A integração das populacoes indigenas no Brasil moderno.* Rio de Janeiro, Brazil: Editora Civilização Brasileira.

Riester, J. 1976. *En Busca de la Loma Santa.* La Paz, Bolivia: Los Amigos del Libro.

Ross, E. 1987. An overview of trends in dietary variation from hunter-gatherer to modern capitalist societies. In Marvin Harris and Eric Ross, eds., *Food and Evolution: Toward a Theory of Human Food Habits,* 7–55. Philadelphia: Temple University Press.

Saffirio, J. and R. Hames. 1983. The forest and the highway. In Kenneth M. Kennsinger, ed., *Working Papers on South American Indians.* Bennington, Vermont: Bennington College. 1–52.

Saffirio, J. and R. Scaglion. 1982. Hunting efficiency in acculturated and unacculturated Yanomama villages. *Journal of Anthropological Research* 38: 315–327.

Sanabria Fernández, H. 1973. *Breve historia de Santa Cruz.* Colección Ayer y Hoy. La Paz, Bolivia: Librería-Editorial "La Juventud."

Stearman, A. M. 1987. *No Longer Nomads: The Sirionó Revisited.* Lanham, Maryland: Hamilton Press.

—— 1984. The Yuquí connection: Another look at Sirionó deculturation. *American Anthropologist* 86(3): 630–650.

Steward, J. and L. C. Faron. 1959. *Native Peoples of South America.* New York: McGraw-Hill.

Vickers, W. 1991. Hunting yields and game composition over ten years in an Amazonian Indian territory. In J. G. Robinson and K. H. Redford, eds., *Subsistence and Commercial Uses of Neotropical Wildlife,* 53–81. Chicago: University of Chicago Press.

—— 1988. Game depletion hypothesis of Amazonian adaptation: Data from a native community. *Science* 239: 1521–1522.

PART II

Folk Societies

Introduction

KENT H. REDFORD AND
CHRISTINE PADOCH

Residents of Neotropical forests have often been divided, erroneously, into only two groups: tribal indigenous peoples and recently arrived colonists. This distinction has ignored many traditional, indigenous, but nontribal peoples who frequently are numerically in the majority in forested regions. These are the "folk societies" discussed in part 2.

Because of their heterogeneous cultural and biological heritage, their frequent lack of strong political organization, and their "nonexotic" appearance, little attention has been devoted to the study of these peoples, including their resource-use patterns. In recent years, however, considerable interest in the societies that inhabit the margins of the Amazon River and its tributaries has been awakened among scientists and other researchers. The researchers who study these *caboclo* or *riberinho* groups in Brazil and the *ribereños* of Peru argue that two characteristics of their resource management strategies are particularly important. First, *caboclos/ribereños* are almost the only traditional peoples still found in the Amazon floodplain, an area that is environmentally very diverse and highly productive but poorly understood by scientists and ill-used by modern technologies. Second, *caboclos/ribereños,* as inheritors and changers of tribal traditions, have adapted some age-old subsistence practices to the demands and opportunities offered by markets. Thus, some have argued, they already have done a job that development experts had set for themselves.

Part 2 begins with an invaluable review, by Mário Hiraoka, of recent research done on *caboclos* and *ribereños*, with particular emphasis on resource management systems. Hiraoka finds that although the last decade has seen a considerable increase in *caboclo/ribereño* studies, the research and its effect on policy remain quite limited.

Hiraoka points out that most of the studies have been done by a small cadre of individuals and that their work is geographically limited to areas close to major cities and to a handful of river systems. Furthermore, he finds that despite the recent spate of published works, many regional researchers and policymakers remain unaware of the findings and "refuse to recognize the value of this cultural group . . . [as] one of the best-adapted human groups of the region." Hiraoka goes on to challenge future researchers by providing an extensive agenda for research and for the dissemination of findings to those involved in policy-making for development and conservation.

The next essay, by Christine Padoch and Wil de Jong argues for the use of larger samples and more in-depth work in research on Amazonian resource management. Limiting their study to just one *ribereño* village in the Peruvian Amazon, these researchers uncover not only a large number of distinct agricultural types but surprising variation among households and significant change in agricultural activities from year to year. They point out that many of the locally important production systems they briefly describe have yet to be noted and appreciated by researchers. Padoch and de Jong suggest that investigators have greatly underestimated the diversity, variation, and change that they believe characterizes traditional resource use. They argue that improvements in the way research is done will uncover similar complexity and dynamism in other traditional villages.

The last paper in part 2 presents a case study of a floodplain community near the opposite end of the Amazon Basin. Anthony Anderson and Edviges Ioris describe the forest extraction and resource management activities of several *caboclo* households on Combu Island near the city of Belém, Brazil. Managed forests rich in *açai* palms, cacao, and rubber supply Combu Islanders with part of their subsistence needs, as well as with valuable goods for sale in the city. Carefully examining household incomes and budgets, Anderson and Ioris show that the management, extraction, and marketing of forest products is not only a sustainable form of resource use on Combu but also a low-risk, relatively lucrative activity.

Echoing Hiraoka's concern over the geographic limitation of recent research, the authors warn the reader that Combu Island, located in the Amazon estuary and outside the largest city in the basin, presents a highly specific situation for commercial extraction. Unlike many other sites, on Combu agricultural activities are severely limited and perishable forest fruits are easily marketed. Despite the unusual characteristics of Combu, the authors suggest that their principal findings can be generalized to other Amazonian regions. They end by pointing out, like the other authors in part 2, that more research is needed, particularly "holistic, cost-benefit analysis of [the economic] benefits" of extraction to rural households.

SEVEN

Caboclo and *Ribereño* Resource Management in Amazonia: A Review

MARIO HIRAOKA

Debates concerning the uncontrolled clearing of Amazonian tropical forests for agricultural purposes with possible negative effects on humans and the global environment are no longer limited to exchanges among specialists. As the increasing media coverage indicates, the rapid conversion of tropical rain forests and consequent ecological changes, have begun to influence popular opinion. Much of the blame is placed on inappropriate resource management strategies. Although experiments in systems appropriate for the region are being carried out, none thus far have proved to be satisfactory in terms of environmental and economic costs. As an alternative to such development, anthropologists, ethnobotanists, and geographers, among othrs, have been seeking ecological and resource-use knowledge from indigenous people (Denevan and Padoch 1988; Posey 1982; Posey 1983; Posey et al. 1984). A recent report by the National Academy of Sciences (1982), calling for more research on traditional knowledge systems, lends additional support to this trend.

The Amerindians' familiarity with the structure, composition, and dynamics of Neotropical ecosystems, resulting from their long association with these local environments, is believed able to contribute to a better understanding of, as well as formulation of sound guidelines for, the sustained management of Amazonian flora and fauna. The resource management systems devised by these people seem to offer economically and ecologically viable alternatives to the ones currently practiced by newcomers. Occupying a great variety of environments within Amazonia, Amerindians have developed patterns appropriate for each of the specific ecological zones they occupy. However, present-day indigenous groups are found almost exclusively on the interfluvial terrain. They are conspicuously absent from the riverine zones of major rivers.

The Amazonian inhabitants known as *caboclos* in Brazil and *ribereños* in Peru have been consistently overlooked both by students of the region and by government planners. They are detribalized Amerindians, the offspring of European-Amerindian marriages, and descendants of early immigrants of varying ethnic and national origins. The *caboclos/ribereños* are numerically the dominant group in lowland riparian Amazonia, yet they are largely invisible to the scholarly community and the regional development agencies. The lack of attention given to *caboclos/ribereños* results from several factors, including the dispersed nature of their settlements, the absence of political and economical organizations that represent them in regional, national, and international forums, and their "nonexotic" life-styles. Yet, along with tribal peoples, they are potentially the most important group to assist with the development of economically and socially sound management practices. Through their market activities this demographic group is linked with national and international economies, and they are in varying degrees integrated with the dominant Ibero-American societies. Thus *caboclo-ribereño* cultural patterns are easier to access than those of tribal groups.

Amazonians are generally categorized into three broad groups: tribal Amerindians, recently arrived colonists, and *caboclos/ribereños*. The culturally diverse tribal Indians and their varied solutions for rain forest survival have been the focus of extended studies by anthropologists and ethnobotanists. Similarly, the large-scale environmental modifications brought about by family farmers, cattle ranchers, and lumbermen of the region, although recent in time, continue to attract a large number of investigators (Moran 1982). But the *caboclos/ribereños*, despite their ubiquitousness and importance in the human and economic drama of Amazonia within the last century, have largely escaped the attention of scientists.

The following brief review of *caboclo/ribereño* research particularly emphasizes their resource management systems. It is not intended as a comprehensive review of the literature, but rather takes stock of recent investigations (including basic literature, agriculture, fishing and hunting, and forest product extraction, as well as the location of existing research) and offers some directions for future studies on the subject.

Basic Literature on the Ribereños

The first studies to focus on the *caboclos/ribereños* began to appear in the 1950s. The basic works include those by Wagley (1953), Galvão (1952), and Sternberg (1956). The classic essay about a *caboclo* community is Wagley's study of Ita (Gurupá), in the estuarine zone near Belém, Brazil. The first major account of the religious syncretism of the *caboclos,* is that of Galvão, who also collected information at Gurupá. Sternberg offers a historical geographical interpretation of land-use changes at Careiro, a Brazilian floodplain agricultural village that was being integrated into the market economy in the 1950s. Located a short distance from Manaus, across the Amazon, the area around Careiro was rapidly becoming a truck farming belt. Although not entirely devoted to the *ribereños* of Peru, the historical synthesis by San Román (1975) offers valuable insights into the evolution of this fluviatile population. Another useful compendium of people and resources in northeastern Peru is that of Villarejo (1943). Each of the foregoing volumes offers a brief section on the resource-use patterns of the *caboclos/ribereños.*

The period after 1970, witnessed a rapid expansion of studies on the region (Moran 1982), an attempt to assess the environmental and socioeconomic effects of large-scale Amazonian development. An outgrowth of these activities was the diversion of interest toward the fluviatile zone. Goulding (1979; 1980; 1981) and Smith (1981a) provided the first updated treatises on the nature of *caboclo* fishing since the work of Veríssimo (1970). The rubber boom, an economic phase so important in understanding the riverine people, is detailed by Weinstein (1983). The relationships between production and exchange, the influences of the environment on economic activity, and the role of the state in defining the outcome of the boom and its effects on the regional population are her basic themes. The cultural ecology of a *caboclo* community on the estuary of the Tocantins River, to the southwest of Belém, is presented by Parker (1981). In it, he describes the socioeconomic organization of a small community and its inhabitants' adaptation to the flood zone. The first volume to focus on the diverse aspects of change among the riparian inhabitants of both Brazil and Peru was *The Amazon Caboclo: Historical and Contemporary Perspectives* (Parker 1985a).

Caboclo/Ribereño Farming

The riparian habitats of Amazonia, reflecting a long and complex tectonic history, are diverse (Bigarella and Ferreira 1985:49–71). Until recently, it was common to portray the basin's environment in terms of two contrasting ecosystems: the interfluvial *terra firme* and the floodplain *várzea.* Accordingly, two farming systems, swidden and floodplain agriculture, were characterized as having developed in adaptation to the differing ecosystems (Meggers 1971; Meggers 1985; Fearnside 1985; Sternberg 1956). Although it serves as a first-order generalization, the upland-lowland dichotomy is increasingly insufficient to account for the diverse ecological zones and resource uses found in the region. In the fluviatile zone alone, where the majority of the *caboclos/ribereños* live, four ecological zones are recognized and differentiated according to floodwater sources: *várzea do rio, várzea da maré, várzea do mar,* and *várzea da chuva* (Moreira 1981; Parker 1981:59–60; Sombroek 1966:38). In a similar manner, the Tertiary and older uplands that often line the stream courses can be subdivided into multiple ecological zones according to variables such as climate, flora, soils, and terrain. Denevan (1971), Gross (1975), and Lathrap (1970), for example, attempt to demonstrate Amerindian sociocultural and economical variations according to ecological differences. It is logical to surmise that these peoples devised a number of site-specific resource-use systems to adapt to environmental variations.

Certainly, the riverine zone of Amazonia supported a large and dense population at the time of European arrival (Denevan 1976; Meggers 1971; Roosevelt 1980). Historical records, however, offer only fragmentary evidence about agricultural practices in the riverine sector (Meggers 1971:121). Descriptions indicate that the diverse biotopes extending from the annually submerged mud- and sandbars to the uplands were employed for crop cultivation (Acuña 1859; Fritz 1922; Figueroa 1986). Without the accounts of cultivation techniques and spatial organization, one can only conjecture about the continuity of agricultural systems currently carried out by the *caboclos/riber-*

eños. Uncertainty regarding the origins of contemporary farming is especially true for the flood zone of the Middle and Lower Amazon, whose inhabitants disappeared early in the colonial period. In the Upper Amazon, where the native population survived along the floodplains of rivers including the Ucayali and Huallaga, the present *ribereño* practices may show a closer affinity to those of the past than farther downstream (Bergman 1980; Stocks 1981). However, considerable uncertainty exists regarding the representativeness of existing patterns for an extensive area.

The structure of *caboclo/ribereño* flood zone farming in Peru is beginning to be understood. Basin-wide generalizations are still far from possible because of ecological variations, as pointed out above, and differing socioeconomic factors, including local histories, accessibility to markets, and cultural traditions. The floodplains of major Amazonian rivers are characterized by a variety of aquatic and land biotopes resulting from river dynamics (i.e., the annual rise and fall of water) and the meanders caused by the water flowing over a terrain of low profile (Lathrap 1970). Deposition of nutrient-bearing silt is heaviest near the river, and the length of the flood season is dependent on elevation. Within this zone, farming is practiced on the flood-free *restingas* or natural levees, the seasonally-appearing low levees, as well as mud- and sandbars. General descriptions of farming on the floodplains of the Peruvian Amazon was first presented by Higbee (1945). Denevan (1984), basing his discussion on data from the Shipibos of the Ucayali, argues that floodplain farmers use a principle analogous to the altitudinal zonation of Andean agriculture. Agricultural differentiation is combined with ecological zonation to create the diverse biotopes that in turn are integrated to satisfy subsistence needs. Information from the Amazonian floodplain around Iquitos suggests that *ribereños* use similar strategies (Hiraoka 1985a; Hiraoka 1985b).

In recent years, with the increasing emphasis on such cash crops as rice and jute that receive government encouragement and subsidies, this complex traditional system is being simplified (Chibnik 1990; Chibnik and de Jong 1989; Padoch 1988b; Hiraoka 1989a; Hiraoka 1989b; Pinedo-Vasquez 1988). The streamlining of agricultural land use, analogous to such processes elsewhere, is affecting *ribereño* life in a number of ways, including increasing disparities in access to land and wealth, stimulating rural emigrations, and changing flora and fauna (Hiraoka 1985c; Mitschein, Miranda, and Paraense 1989;

Santos 1982). Increasing urbanization and improved accessibility to markets also are changing the farming structure of the *caboclos/ribereños* elsewhere. For example, Schmink (1985) reports the changes at São Felix, a community along the Xingu River. Barrow (1982), Brabo (1979b), Faulhaber (1987), Miller (1985), Petrick (1978), and Wesche (1985) describe the socioeconomic and cultural concomitants of increasingly market-oriented farming in riparian *caboclo* communities in Brazil.

Although the outlines of riverine floodplain farming are beginning to be known, studies of *caboclos/ribereños* inhabiting other flood zones have lagged. Wagley's work on Gurupa (1953) sheds almost no light on estuarine floodplain cropping. Parker's (1981:152–53) findings on Limoeiro do Ajuru, another estuarine site, indicate virtually no utilization of the *várzea,* arguing that the location of Limoeiro in a tidal zone precludes the use of estuarine floodplains. A lack of uses for the tidal flood zone does not seem to be universal. Lima (1956:11–13), basing his comments on historical sources, indicates that missionaries, and later the *caboclos,* developed successful farming patterns on diverse sites along the estuarine *várzea.* Such conflicting findings suggest that knowledge of farming in this area is still very limited.

The lowland savannas of Amazonia are also poorly known. During the pre-Columbian period, these seasonally flooded lands supported a considerable population. The Llanos de Mojos, for example, with an area of approximately 180,000 square kilometers, were estimated to be inhabited by 350,000 people at the time of European arrival, giving an average density of 2 people/square kilometer, (Denevan 1966:112–120). The savannas of eastern Marajó appear to have supported a large population as well (Meggers and Evans 1957).

The variety of ecological zones on the interfluves is also skillfully utilized by the *caboclos/ribereños.* Farming on these diverse forest-covered sites has sometimes served as a complement to floodplain cropping. Without the seasonal rejuvenation of the soil through the deposition of mineral-bearing floodwaters, these higher lands show great spatial variations in agricultural potentials due to variables including climate, flora, parent materials, length of weathering processes, and past human actions. Generally speaking, however, about 75 percent of upland soils are classed as nutrient-poor oxisols and ultisols (Nicholaides et al. 1983:110; Sanchez et al. 1982).

The *caboclos/ribereños,* like their forebears, use the time-tested swidden system to produce their subsistence needs. The literature on the *caboclos* suggests that the techniques involved are similar throughout the basin (Wagley 1953; Moran 1974). The plant nutrients locked in the biomass and organic litter at the surface are released through forest cutting and burning (Nye and Greenland 1960; Moran 1981; Parker 1981). Then a short cropping cycle, with carbohydrate-rich items like manioc, taro, other tubers, and plantains/bananas as the principal crops, follows. Most investigators, relying on their own investigations and on research done elsewhere in the tropics, believe the fields are abandoned to fallow following declining yields after one or two years of cultivation. A new forested patch, either a formerly abandoned site or climax flora, is then cleared, and a new cropping cycle ensues. Shifting cultivation has been viewed as an agronomically sustainable system and environmentally sound forest management technique requiring few external inputs in situations of low population density (Harris 1971; Posey 1982; Posey 1983; Lathrap 1970; Moran 1981). Recent findings, first among Amerindian groups, and later among *ribereños*, suggest that much indigenous farming more appropriately should be called swidden-fallow agroforestry (Denevan et al. 1984; Denevan and Padoch 1988). Researchers have reported that swidden plots include, in addition to short-term staples, various tree species that are harvested for a number of years after the sites are left to fallow (Harris 1971; Hecht 1982; Hiraoka 1986; Parker 1981; Posey 1983). An increasing number of studies points out that "the use of swidden fallows is a widespread . . . practice in the tropics," and that it is "a form of agroforestry involving a combination of annual crops, perennial tree crops, and natural forest regrowth" (Denevan and Padoch 1988:1). Among the *ribereños* of Peru the system has been adapted to produce for the market (Hiraoka 1986; Padoch 1988c; Padoch et al. 1985). Access to market, personal preferences, ecological variations, and other elements are responsible for minor variations in the plant combinations and the length the gardens are left in agroforestry. Although the long-term environmental effects of swidden-fallow agroforestry remain to be demonstrated, preliminary results from a *ribereño* community in northeastern Peru indicate that the economic potential of agroforestry products in the *ribereños'* well-being is considerable (Padoch et al. 1985).

The riparian area of lesser tributaries of the Amazon are also inhabited by *caboclos* and *ribereños*. Without access to the biologically productive floodplains, the people living along black and clearwater rivers have had to depend essentially on the forested interfluves for their subsistence. Scattered and brief studies provide only a preliminary understanding of farming along these watercourses (Hiraoka 1989a; Oliveira 1975).

Agricultural work among the *caboclos/ribereños* normally is done by the nuclear family, which is the basic unit of production. On occasions when extra-familial labor is required (e.g., forest felling, weeding, and harvesting of grains), farmers resort to hired and/or exchange labor. The former is common among individuals with large and diversified operations or among villagers who are facing a labor shortage because everyone is engaged in a task such as rice harvesting. Informal labor exchange involving ten to thirty people is still a common institution in Amazonia; it is called *minga* in Peru and *puxirão* or *convite* in Brazil (Wagley 1953:68). Minor differences do exist between communities; however, such work parties generally are called to perform a specific task and last no longer than a day. The recipient of the labor serves food and beverages to the guests and is obligated to reciprocate with labor at a future date (de Jong 1988; Chibnik and de Jong 1988; Wagley 1953; San Román n.d.; vol. 2, 60–68). Despite the increasing participation of the fluviatile groups in the monetary economy, these *mingas/puxirões* continue to exist; they do not entail expenditures of money and allow for socializing among members.

Among the most important contributions of research into *caboclo-ribereño* agriculture to date has been the "discovery" of the practice of swidden-fallow agroforestry. It is true that soil nutrient dynamics under the agroforest, ideal crop associations and reconversion periods, the appropriate scale for small holders, and other questions remain to be answered. However, the system appears to be a promising alternative to the swidden-monoculture-ranching land use practiced on interfluvial lands. Current investigations reveal only site-specific findings. In a region as ecologically varied as the Amazon, planning will have to be based on site-specific studies. This implies that future land-use models will have to allow for great flexibility.

Fishing and Hunting

The *caboclo/ribereño* is a jack-of-all-trades. Along with farming, collecting of forest products, and manufacturing of handicraft items, fishing and hunting contribute significantly toward his/her subsistence and cash generation. The role of land and aquatic faunae in the economy depends on the characteristics of the habitats available to the *caboclo/ribereño*. The number of faunal species is large, reflecting the physical, biological, and chemical diversities of the region. Identified fish species alone number between thirteen hundred and two thousand in the basin (Junk 1975:109).

Within the basin, productivity and available catch vary enormously with hydrochemical properties and the biophysical habitats bordering the rivers (Goulding 1983; Goulding 1985). The whitewater rivers with their floodplains are considered to be the most productive. Along such watercourses the population has become dependent on aquatic fauna for animal protein, since it is plentiful and easily available year round (Acuña 1859; Fritz 1922). However, uncontrolled harvests for shipment outside the watershed, begun in the colonial period, have seriously depleted many of the aquatic animals. Turtles belonging to the genus *Podocnemis*, manatees (*Trichechus manatus*), caimans (*Melanosuchus niger* and *Caiman crocodilus*), capybaras (*Hydrochoaeris hydrochaeris*), and otters (*Pteronura brasiliensis* and *Lutra* sp.) face serious survival problems (Smith 1974; Smith 1979; Smith 1981b; Ojasti 1967). Many of the fishes such as the *pirarucu/paiche*) (*Arapaima gigas*), *tambaquí/gamitana* (*Colossoma macropomum*), *pirapitinga/paco* (*Colossoma bidens*), *piraíba/saltón* (*Brachyplatystoma filamentosum*), and *piramutaba/dorado* (*Brachyplatystoma vaillanti*) have become scarce as a result of increasing rural and urban population pressures on resources, extraregional markets, technological innovations, and ecological changes brought about by activities like agricultural expansion in the floodplains and construction of hydroelectric dams (Goulding 1981; Goulding 1983; Hiraoka 1988; Smith 1985).

From the historical record, it is hard to ascertain whether fish or other aquatic fauna were more important to the *caboclos/ribereños* in earlier times, since both were plentiful (Goulding 1983:189). Today, fish are the undisputed animal protein source for the rural inhabitants and lower-income groups in the basin's cities (Junk 1983; Goulding 1981). Demand is met by commercial fishermen based in Iquitos, Itacoatiara, Manaus, and other major regional centers, and by the *caboclos/ribereños* (Petrere 1978; Smith 1981a; Hiraoka 1985; Silveira 1979; Simões 1981; Brabo 1981; Furtado 1987).

Fishing for food and sale is done by virtually every riparian rural dweller, regardless of sex or age. Familiarity with fish behavior, life cycles, and environments enable the *caboclos/ribereños* to deliver fish to the family on a continuous basis. Surpluses, which occur seasonally, are sent to the market. Fishing efficiency continues to improve as new equipment is incorporated. Traditional tools including harpoons, bow and arrow, and weirs are still employed, but the introduction of gill nets, cast nets, and outboard motors are increasing catches through exploitation of larger areas. But competition for the same resources and the inability of the *caboclos/ribereños* to control resource sites because of ambiguous legislation or lack of enforcement are causing conflicts between commercial fishermen and the riparian people. These issues are briefly reported by Hiraoka (1988), Pinedo-Vásquez (1986), and Smith (1981a).

The role of magic, folklore, and taboos in fishing, although not as important as in hunting, influence *caboclo/ribereño* fishing to this day. Research on the subject is inadequate, but the relationships between the various beliefs and aquatic faunal exploitation should be given more attention (Wagley 1953: 78; Smith 1981a).

In spite of the importance of fish in the local diet, much remains to be learned about productivity, reproductive behavior, and consumption. The ichthyofaunal population is not as large as suggested by the *piracemas/mijanos*, the large seasonal migrations of schools of Characins. Junk (1983) estimates that up to 350,000 metric tons of fish can be harvested annually from the Amazon waters on a sustained basis. In 1980, between 100,000 and 130,000 metric tons are believed to have been taken from the basin (Goulding 1983:193). The real figure is probably much higher because of underreporting and inaccuracies in accounting for the riparian peoples' harvests. Fishing data, including efficiency and total catch, exist for several points within the basin, but they offer only a general picture. Reflecting the great variability in techniques and environment that exists, as well as differences in survey techniques, these data show great variation by area (Bergman 1980; Smith 1981a; Goulding 1981).

The capture of game for meat and cash constitutes another important activity among the *caboclos/ribereños*. For villages without access to aquatic fauna, land fauna become the primary animal protein source. Although hunting among Amerindians and colonists has been treated extensively, no analogous literature exists for the *caboclos/ribereños* (Redford and Robinson 1987). Brief references on game capture usually are included in reports. Fish is relatively abundant along the whitewater rivers, but even in these areas, game is an important element in the diet. Although game populations depend on variables like hunting pressures, species preferences, and environmental variations, studies of aboriginal communities seem to suggest that animal protein has not been scarce in absolute terms (Vickers 1980). Limited data, however, indicate that a combination of techniques provides most of the meat needed. The most common technique used seems to be the combination of hunting in high and secondary forests and limited raising of domesticates such as fowl and hogs. Recent findings point out the significance of fallow-agroforest sites as a major producer of game. Animals like armadillos (*Dasypus novemcintus*), brocket deer (*Mazama* spp.), monkeys, birds, and a number of caviomorphs including agoutis (*Dasyprocta variegata*), pacas (*Myoprocta* spp.), and *sacha cuy* (*Proechimys brevicauda*) are harvested in a semimanaged fashion (Hiraoka 1988). Linares (1976) indicates from her archeological findings in Panama that the association of man and animals in the fallow-agroforest may have obviated the need for domestication in the neotropics. A better understanding of the *purmeros,* or fallow-agroforest animals, may enhance still more the value of the swidden-fallow agroforestry systems as sites for food production.

Forest Product Extraction and Forest Management

Caboclos/ribereños possess a knowledge of the forest that rivals that of contemporary Amerindians of the interfluvial zone. They collect items ranging from food, fiber, dyes, resins, and medicines to fish poison and construction materials from the forest. Since the early colonial period, many of these goods acquired commercial value and were extracted for sale. The most dramatic example of such an economy was the Rubber Boom. At times, the rapacious extraction of products like sarsaparilla and cabinet woods led to their serious overexploitation in many places. With such a history, extractivism in the region came to be equated in the public's mind with environmental destruction, and the *caboclos/ribereños* labeled as the major culprits. The rural inhabitants of fluviatile Amazonia continue to collect a wide variety of products both for subsistence and for sale, but the new findings increasingly typecast them as rational stewards of the forest.

Floral products are obtained from several forest biomes. The diverse floral associations are grouped broadly into two forest types: managed and unmanaged. Sites receiving some input of labor (e.g., fallowed fields and agroforests where economically valuable species are protected or cared for) fall under the former category. Flora including vines, palms, and trees and cultivated species such as plantains, cacao, and *guaba* (*Inga edulis*) are pruned, cleared of competing species, propped up with stakes, kept free of insect pests, and mulched in some cases. These practices are reported from both the riparian and estuarine floodplains, suggesting a common practice among the *caboclos/ribereños* (Anderson 1988; Anderson et al. 1985; Hiraoka 1985a; Hiraoka 1985b; Padoch 1988b; Padoch and de Jong 1990). These managed sites are noted for their impermanence. They are rarely viewed by the people as places of permanent production. As several successional species are included, the floral composition undergoes frequent change, and after a period of twenty-five to forty years, management ceases or the site is reconverted to a swidden plot. As Anderson (1988), and Padoch et al. (1985) point out, the income from these forests often surpasses the combined total of other activities.

Diverse unmanaged floral communities in the neighborhood of *caboclo/ribereño* settlements are also places of extraction. The flood zone often contains relatively few species. Variations in edaphic conditions within short distance and unaerobic conditions contribute to the restricted floral speciation in the flood zone (Denevan 1984; Irion 1984). Backswamps with stagnant water, known as *aguajales* in Peru, are often found with quasi-monospecific stands of *Mauritia flexuosa* palms. Somewhat analogous characteristics are found in the estuarine flood zone. The products of these environments, especially palms, are prominent sources of food and income for the regional population (Anderson 1988; Hiraoka 1985a; Padoch 1988a). The results of investigations of *caboclo/ribereño* use of flora in the waterlogged biotopes point out an important direction for future inquiries, i.e., the sustainable uses of other agriculturally marginal biotopes for income generation such as the *igapó/tahuampa* and the boggy terrain in the interface between floodplain lakes and natural levees (Moreira 1970). As is pointed out by Frechione, Posey, and da Silva (1989), the ethnoecological knowledge of the *caboclos* can contribute toward a fuller use and management of local environments.

The unmanaged forests of the interfluvial terrain, where the flora attains a great diversity of species, are also source areas of valuable products. Even here, concentration of some species like the Brazil nut tree (*Bertholletia excelsa*) and *babassu* palm (*Orbignya* spp.) may occur under appropriate ecological conditions (Laraia and Matta 1967; Balick 1985; Balick 1987). Many writers refer to the numerous uses *caboclos* and *ribereños* have for forest products (Anderson 1988; Padoch 1988c; Hiraoka 1986; San Román 1975; Villarejo 1943), but detailed studies of issues such as their specific economic contribution, the proportion of labor inputs, and the spatial structure of the exploited area remain to be conducted (Peters, Gentry, and Mendelsohn 1989). Among the diverse plants, palms have received considerable attention. The manifold uses of palms by indigenous peoples are well known. *Caboclos/ribereños,* as the cultural representatives of the aborigines in the riparian zone, possess a vast familiarity with the botanical uses of these plants. As the recent spate of publications indicates, interest in documenting people's knowledge on palms has been aroused (Anderson 1988; Brabo 1979a; López-Parodi 1988; Mejía 1988; Padoch 1988a).

Palms and other flora are frequently employed as diagnostic features of the underlying terrain; they are particularly useful for agricultural land-use decision making. Plants growing in secondary and high forest offer the *caboclos* a first-order index of some edaphic properties including litter accumulation, plant nutrient content, and soil moisture. For example, the decision to convert a plot in successional growth into a swidden is based on the height and composition of the standing flora. The choice of crop combinations in a plot is influenced by the presence or absence of key plants. On the other hand, folk beliefs can lead to the preservation of floral stands. In northeastern Peru, sites overgrown with a tree, *supay casha,* are not used for cultivation. The role of folklore in forest management is an area requiring further investigation.

Concepts of resource use and conservation ethics among the *caboclos/ribereños,* as among their aboriginal counterparts, appear to be either empirically derived or rooted in beliefs in spirits and supernatural beings. Wagley (1953:76–81) suggests that the belief in folklore has a greater influence on hunting and fishing, and, one might also add, collecting than on farming. "The reason for this difference . . . lies in the nature of agricultural pursuit. In the Amazon, agriculture centers upon manioc, a plant that is exceedingly hardy. A failure of a manioc crop is almost unheard of" (Wagley 1953:81). In a brief article, Smith (1983) details the link between the supernatural and its roles in environmental management among the *caboclos*. A compendium of such myths and spirits for Brazil was assembled by Orico (1975), while the Peruvian counterpart can be found in Huamán Ramírez (1983). Galvão (1951; 1952) and Silva (1959) offer additional insights. Continued site-specific research on this topic may serve to better evaluate the roles of folklore in environmental management.

Location of Existing Research

A review of extant study sites shows an irregular distribution. Important ground has been broken in comprehending *caboclo/ribereño* resource-use systems, but studies to date remain spatially skewed (see figure 7.1). The locational biases are evident. Findings occur in two types of places: near major urban centers like Belém, Manaus, Pôrto Velho, and Iquitos, or along the Amazon and its major affluents such as the Ucayali, the Madeira, and the Tocantins. Such a pattern suggests strong market-influenced land uses, and life-styles close to those of urban folk. Generalizations formulated from such bases are bound to illustrate only a segment of reality. The research sites also indicate an overwhelming concentration on floodlands of whitewater rivers. These lands comprise only a minute portion of the basin and possess unique ecological features. If *caboclo/ribereño* inquiries, in addition to contributing to academic understanding, are intended to contribute toward defining economically and ecologically compatible resource management systems, information from more varied ecological zones is desirable. Within this deceptively uniform-looking region, subtle environmental differences exist. Local inhabitants, recognizing these microecological variations, have developed livelihood systems adapted to them. It remains for the students of the region to recognize these differences and uncover the appropriate practices developed by the inhabitants.

Research on *caboclo/ribereño* resource use began to receive a major impetus within the last decade. Most of the findings came from a small core of individuals. The current foci are on the themes of agricultural systems, forest management, and local environmental conservation concepts. Research increasingly is directed toward evaluating the practicality of incorporating indigenous knowledge in re-

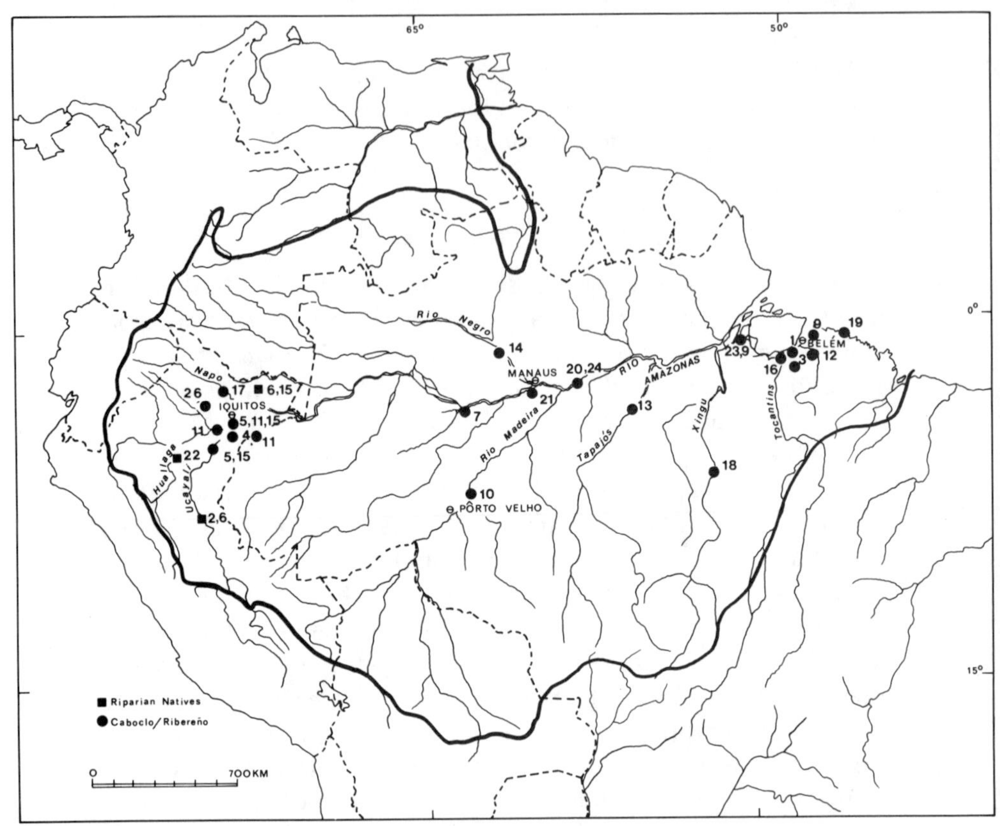

Figure 7.1
Location of *Caboclo/Ribereño* Studies.
1. Anderson 2. Bergman 3. Brabo 4. Chibnik 5. De Jong 6. Denevan 7. Frechione et al. 8. Furtado 9. Galvão 10. Goulding 11. Hiraoka 12. Lima 13. Miller 14. Oliveira 15. Padoch 16. Parker 17. Regan 18. Schmink 19. Silveira 20. Smith 21. Sternberg 22. Stocks 23. Wagley 24. Wesche 25. Faulhaber 26. Peters, Gentry, and Mendelsohn 27. Silva 28. Simões

source utilization and environmental planning. These objectives offer new directions in the study of Amazon peasantry. The themes are also in line with those of other tropical forest lands of the world, where ecologically and economically balanced ways of development are being sought. The new investigations, however, go beyond the novel themes; they are interdisciplinary and ecological in approach.

The expanding literature is making the *caboclos/ribereños* known to the scholarly world. The riparian inhabitants are seen as the most easily accessible repository of Amazonian environmental and environmental-management knowledge. Yet many students of the region and individuals in policy-making positions refuse to recognize the value of this cultural group. Several factors seem to be responsible for the slow acceptance. First, even specialists on the region, e.g., Hemming (1987), Bunker (1985), and Foweraker (1981), do not differentiate the group from the newly arrived colonists and tribal Indians. There is also a lack of recognition that the *caboclos/ribereños* is one of the best-adapted human groups in the region (Moran 1974), thus making them valuable intermediaries for technological change and transfer. The common perception is to view them as cultural mongrels and therefore of peripheral value for academic inquiries. Furthermore, the regional society tends to downgrade the *caboclos/ribereños.* The terms *caboclo* and *ribereño* in themselves carry a pejorative connotation analogous to the use of the term *cholo* in the Andean uplands. The stigma is based on a combination of attributes, including family name, rural location, socioeconomic standing, and a number of cultural traits associated with Amerindians. Finally, they are too ubiquitous: *caboclos/ribereños* are found all over the riparian zone and their outward appearances are not sufficiently exotic.

As in any new field of knowledge, the initial phase consists essentially of exploratory work. The general outlines of *caboclo/ribereño* resource-use systems are beginning to emerge. Among the several tasks for the next phase are:

1. Collection of data from communities located in diverse ecological zones. Information from areas other than the estuarine and whitewater river floodplains is needed.
2. Comparative studies are desirable. Findings from different points within the basin or with analogous tropical forest areas, may be compared to identify distinctive solutions for similar problems in resource use.

3. Adoption of an ecosystemic approach in data interpretation. Since a large number of microecological units comprise the resource sphere of the *caboclo/ribereño,* an understanding of their economy or environment has to take into account how distinctive spaces are integrated and managed as a whole.
4. Application of findings to colonists arriving from outside the region.
5. Diffusion of research findings to regional policymakers.

ACKNOWLEDGMENTS

I wish to thank C. Padoch and M. Pinedo-Vasquez for their valuable comments during the preparation of this paper. Funds and released time for this study were provided by the Millersville University Grants Committee.

LITERATURE CITED

Acuña, C. de. 1859. New discovery of the great river of the Amazons. In C. R. Markham, ed., *Expeditions into the Valley of the Amazons, 1539, 1540, 1639,* 46–134. London: Hakluyt Society.

Anderson, A. B. 1988. Use and management of native forests dominated by *acai* palm (*Euterpe oleracea* Mart.) in the Amazon estuary. In M. J. Balick, ed. *The Palm-Tree of Life: Biology, Utilization, and Conservation. Advances in Economic Botany* 6:144–154. New York: New York Botanical Garden.

Anderson, A. B., A. Gély, J. Strudwick, G. L. Sobel, and M. G. C. Pinto. 1985. Um sistema agroflorestal na várzea do estuário amazônico (Ilha das Onças, Município de Barcarena, estado do Pará). *Acta Amazônica,* Supplement 15 (1–2):195–224.

Balick, M. 1985. Useful plants of Amazonia: A resource of global importance. 339–368. In G. T. Prance and T. E. Lovejoy, eds., *Key Environments: Amazonia.* Oxford, England: Pergamon Press.

—— 1987. The economic utilization of the babassu palm: A conservation strategy for sustaining tropical forest resources. *Journal of the Washington Academy of Sciences* 77(4):215–223.

Barrow, C. J. 1982. The development of the *várzeas* (floodlands) of Brazilian Amazonia. Paper presented at the 44th International Congress of Americanists, Manchester, England, September 1982.

Bergman, R. W. 1980. *Amazon Economics: The Simplicity of Shipibo Indian Wealth.* Dellplain Latin American Studies, no. 6. Ann Arbor, Mich.: University Microfilms.

Bigarella, J. J. and A. M. M. Ferreira. 1985. Amazonian geology and the Pleistocene and the Cenozoic environments and paleoclimates. In G. T. Prance and T. E. Lovejoy, eds., *Key Environments: Amazonia,* 49–71. Oxford, England: Pergamon Press.

Brabo, M. J. C. 1979a. Palmiteiros de Muaná: Estudo sobre o processo de produção no beneficiamento do açaizeiro. *Antropología,* (Boletim Museu Paraense Emílio Goeldi, nova série) no. 73:1–22.

—— 1979b. Os roceiros de Muaná. *Publicações Avulsas* (Museu Paraense Emílio Goeldi, Belém, Brazil) no. 32.

—— 1981. Pescadores, geleiros, fazendeiros: Os conflitos da pesca em Cachoeira do Ararí. *Antropología,* (Boletim Museu Paraense Emílio Goeldi, nova série), no. 77.

Bunker, S. 1985. *Underdeveloping the Amazon.* Urbana, Ill.: University of Illinois Press.

Chibnik, M. 1990. Double-edged risks and uncertainties: Choices about rice loans in the Peruvian Amazon. In E. Cashdan, ed., *Risk and Uncertainty in Tribal and Peasant Economies.* Boulder, Colo.: Westview Press.

Chibnik, M. and W. de Jong. 1989. Agricultural labor organization in the *ribereño* communities of the Peruvian Amazon. Paper presented at the 46th International Congress of Americanists, July 1988, Amsterdam.

De Jong, W. A. 1988. Organización del trabajo en la Amazonía peruana: el caso de las sociedades agrícolas de Tamshiyacu. *Amazonía Indígena* 7(13):11–17.

Denevan, W. M. 1966. The aboriginal cultural geography of the Llanos de Mojos of Bolivia. Ibero-Americana, no. 48. Berkeley, Calif.: University of California Press.

—— 1971. Campa subsistence in the Gran Pajonal, eastern Peru. *Geographical Review* 61:496–518.

—— 1976. The aboriginal population of Amazonia. In W. M. Denevan, ed., *The Native Population of the Americas in 1492,* 205–234. Madison, Wis.: University of Wisconsin Press.

—— 1984. Ecological heterogeneity and horizontal zonation of agriculture in the Amazon floodplain. In M. Schmink and C. H. Wood, eds., *Frontier Expansion in Amazonia,* 311–336. Gainesville, Fla.: University of Florida Press.

Denevan, W. M. and C. Padoch, eds. 1988. *Swidden-fallow agroforestry in the Peruvian Amazon. Advances in Economic Botany.* New York: New York Botanical Garden.

Denevan, W. M., J. M. Treacy, J. B. Alcorn, C. Padoch, J. S. Denslow, S. Flores Paitan. 1984. Indigenous agroforestry in the Peruvian Amazon: Bora Indian management of swidden fallows. *Interciencia* 9:346–357.

Faulhaber, P. 1987. *O navio encantado: Etnías e alianças em Tefé.* Belém, Brazil: Museu Paraense Emílio Goeldi.

Fearnside, P. M. 1985. Agriculture in Amazonia. In G. T. Prance and T. E. Lovejoy, eds., *Key environments: Amazonia,* 393–418. Oxford, England: Pergamon.

Figueroa, F. de. 1986. Informe de las misiones de el Marañon, Gran Pará ó Rio de las Amazonas. 143–309. In A. Chirif, ed., *Informes de jesuitas en el Amazonas, 1660–1684*. Monumenta Amazónica, Vol. B1. Iquitos, Peru: CETA.

Foweraker, J. 1981. *The Struggle for Land: A Political Economy of the Pioneer Frontier in Brasil from 1930 to the Present Day.* Cambridge, England: Cambridge University Press.

Frechione, J., D. A. Posey, and L. F. da Silva. 1989. The perception of ecological zones and natural resources in the Brazilian Amazon: An ethnoecology of Lake Coari. In D. A. Posey and W. Balée, eds. *Resource Management in Amazonia, Advances in Economic Botany* 7:260–282. New York: New York Botanical Garden.

Fritz, S. 1922. *Journal of the Travels and Labors of Father Samuel Fritz in the River of the Amazon between 1686 and 1723*. London: Hakluyt Society.

Furtado, L. F. 1987. *Curralistas e redeiros de Marudá: pescadores do litoral do Pará.* Belém, Brazil: Museu Paraense Emílio Goeldi.

Furtado, L. F., R. C. Souza, and M. L. van den Berg. 1978. Notas sobre uso terapêutico de plantas pela população cabocla de Marapanim, Pará. *Antropología,* (Boletim Museu Paraense Emílio Goeldi, n.s.) no. 70.

Galvão, E. 1951. Panema: Uma crença do caboclo amazônico. *Revista do Museu Paulista* 5:221–225.

—— 1952. The religion of an Amazon community: A study in culture change. Ph.D. dissertation, Columbia University.

Goulding, M. 1979. *Ecología de pesca do Rio Madeira.* Manaus, Brazil: Centro Nacioncal de Pesqiusa-Instituto Nacional de Pesquisa Amazonica.

—— 1980. *The Fishes and the Forest: Explorations in Amazonian Natural History.* Berkeley, Calif.: University of California Press.

—— 1981. *Man and Fisheries on an Amazonian Frontier.* The Hague: Dr. W. Junk.

—— 1983. Amazon fisheries. In E. F. Moran, ed., *The Dilemma of Amazonian Development*, 189–210. Boulder, Colo.: Westview.

—— 1985. Forest fishes of the Amazon. In G. T. Prance and T. E. Lovejoy, eds., *Key Environments: Amazonia,* 267–276. Oxford, England: Pergamon Press.

Gross, D. 1975. Protein capture and cultural development in the Amazon basin. *American Anthropologist* 77:526–549.

Harris, D. R. 1971. The ecology of swidden agriculture in the upper Orinoco rainforest, Venezuela. *Geographical Review* 61:475–495.

Hecht, S. B. 1982. Agroforestry systems in the Amazon Basin. In S. B. Hecht, ed., *Amazonia: Land Use and Agricultural Research*, 331–372. Cali, Colombia: CIAT.

Hemming, J. 1987. *Amazon Frontier: The Defeat of the Brazilian Indians.* Cambridge, Mass.: Harvard University Press.

Higbee, E. C. 1945. The river is the plow. *Scientific Monthly* 60:405–416.

Hiraoka, M. 1985a. Floodplain farming in the Peruvian Amazon. *Geographical Review of Japan* 58 (series B, no. 1):1–23.

—— 1985b. Mestizo subsistence in riparian Amazonia. *National Geographic Research* 1(2):236–246.

—— 1985c. Cash cropping, wage labor, and urbanward migrations: Changing floodplain subsistence in the Peruvian Amazon. In E. P. Parker, ed., 199–

242. *The Amazon Caboclo: Historical and Contemporary Perspectives.* Studies in Third World Societies, Vol. 32. Williamsburg, Va.: Department of Anthropology, College of William and Mary.

—— 1986. Zonation of mestizo farming systems in Northeast Peru. *National Geographic Research* 2(3):354–371.

—— 1988. Aquatic and land fauna exploitation among the floodplain *ribereños* of the Peruvian Amazon. Paper presented at the 46th International Congress of Americanists, July 1988, Amsterdam.

—— 1989a. Agricultural systems on the floodplains of the Peruvian Amazon. In J. O. Browder, ed., *Fragile Lands of Latin America: Strategies for Sustainable Development,* 75–99. Boulder, Colo.: Westview Press.

—— 1989b. Riverine subsistence patterns along a blackwater river in the northeast Peruvian Amazon. *Journal of Cultural Geography* 9(2):103–119.

Huamán Ramírez, C. 1983. *Los mistérios de la selva.* Lima, Peru: Imprenta Editores Tipo-Offset.

Irion, G. 1984. Sedimentation and sediments of Amazonian rivers and evolution of the Amazonian landscape since Pliocene times. In H. Sioli, ed., *The Amazon: Limnology and Landscape Ecology of a Mighty Tropical River and Its Basin,* 210–214. The Hague: Dr. W. Junk.

Junk, W. J. 1975. Aquatic wildlife and fisheries. In *The Use of Ecological Guidelines for Development in the American Humid Tropics,* 109–125. Morges, Switzerland: International Union for Conservation of Nature and Natural Resources.

—— 1983. As águas da região amazônica. In E. Salati, W. J. Junk, H. O. R. Shubart, and A. E. Oliveira, eds., *Amazônia: desenvolvimento, integração e ecología,* 45–100. Brazil: São Paulo, Brasiliense-CNPq.

Kahn, F. 1988. Ecology of economically important palms in Peruvian Amazonia. In M. J. Baleck, ed. *The Palm-Tree of Life: Biology, Utilization,* and *Conservation. Advances in Economic Botany* 6:42–49. New York: New York Botanical Garden.

Kahn, F. and K. Mejía. 1987. Notes on the biology, ecology and use of a small Amazonian palm: *Lepidocaryum tessmanii. Principes* 31(1):14–19.

Laraia, R. B. and R. da Mata. 1967. *Indios e castanheiros.* São Paulo, Brazil: Difusão Européia do Livro.

Lathrap, D. W. 1970. *The Upper Amazon.* New York: Praeger Publishers.

Lima, R. R. 1956. A agricultura nas várzeas do estuário do Amazonas. *Boletim Technico de Instituto Agronomia Norte* (Belém) no. 33.

Linares, O. 1976. "Garden hunting" in the American tropics. *Human Ecology* 4:331–349.

López Parodi, J. 1988. The use of palms and other native plants in non-conventional, low cost rural housing in the Peruvian Amazon. In M. J. Balick, ed. *The Palm-Tree of Life: Biology, Utilization,* and *Conservation. Advances in Economic Botany* 6:119–129. New York: New York Botanical Garden.

Meggers, B. 1971. *Amazonia: Man and Culture in a Counterfeit Paradise.* Chicago: Aldine.

—— 1985. Aboriginal adaptation to Amazonia. In G. T. Prance and T. E.

Lovejoy, eds., *Key Environments: Amazonia*, 307–327. Oxford, England: Pergamon Press.

Meggers, B. and C. Evans. 1957. *Archeological investigations at the mouth of the Amazon*. Bureau of American Ethnology Bulletin 167. Washington, D.C.: Smithsonian Institution.

Mejía, K. 1988. Utilization of palms in eleven mestizo villages of the Peruvian Amazon (Ucayali River, Department of Loreto). In M. J. Balick, ed., *The Palm-Tree of Life: Biology, Utilization, and Conservation Advances in Economic Botany* 6:130–136.

Miller, D. 1985. Highways and gold: Change in a caboclo community. In E. P. Parker, ed., *The Amazon Caboclo: Historical and Contemporary Perspectives*, 167–198. Studies in Third World Societies, Vol. 32. Williamsburg, Va.: Department of Anthropology, College of William and Mary.

Mitschein, T. A., H. R. Miranda, and M. C. Paraense. 1989. Urbanização selvagem e proletarização passiva na Amazônia, o caso de Belém. Belém, Brazil: Centro de Estudos Juridicos do Para—Nucleo de Altos Estudos de Amazonía/Universidade Federal do Para.

Moran, E. F. 1974. The adaptive system of the Amazonian caboclo. In C. Wagley, ed., *Man in the Amazon*, 136–159. Gainesville, Fla.: University Presses of Florida.

—— 1981. *Developing the Amazon*. Bloomington, Ind.: Indiana University Press.

—— 1982. Ecological, anthropological, and agronomic research in the Amazon basin. *Latin American Research Review* 17(1):3–41.

Moreira, E. 1970. *Os igapós e seu aproveitamento*. Belém, Brazil: Imprensa Universitária.

—— 1981. Os igapos. Belem, Brazil: Universidade Federal do Para.

National Academy of Sciences. 1982. *Ecological Aspects of Development in the Humid Tropics*. Washington, D.C.: National Academy Press.

Nicholaides J. J. III, P. A. Sanchez, D. E. Bandy, J. H. Villachiea, A. J. Coutu, C. S. Valverde. 1983. Crop production systems in the Amazon basin. In E. F. Moran, ed., *The Dilemma of Amazonian Development*, 101–154. Boulder, Colo.: Westview Press.

Nye, P. and D. Greenland. 1960. *The Soil under Shifting Cultivation*. Harpenden, U.K.: Technical Communication No. 51. Commonwealth Bureau of Soils.

Ojasti, J. 1967. Consideraciones sobre la ecología y conservación de la tortuga *Podocnemis expansa*. *Atas do Simpósio sobre a Biota Amazônica* 7:201–206.

Oliveira, A. E. 1975. São João: Povoado do Rio Negro (1972). *Antropología* (Boletim Museu Paraense Emílio Goeldi, n.s.), no. 58:1–56.

Orico, O. 1975. *Mitos ameríndios e crendices amazônicas*. Rio de Janeiro, Brazil: Civilização Brasileira.

Padoch, C. 1988a. Aguaje (*Mauritia flexuosa* L. f.) in the economy of Iquitos, Peru. In M. J. Balick, eds., *The Palm-Tree of Life: Biology, Utilization, and Conservation, Advances in Economic Botany* 6:241–224, New York: New York Botanical Garden.

—— 1988b. People of the floodplain and forest. In J. S. Denslow and C. Padoch, eds., 127–140. *People of the Tropical Rain Forest.* Berkeley, Calif.: University of California Press.

—— 1988c. The economic importance and marketing of forest and fallow products in the Iquitos region. In W. M. Denevan and C. Padoch, eds. *Swidden-Fallow Agroforestry in the Peruvian Amazon. Advances in Economic Botany* 5: 74–89. New York: New York Botanical Garden.

Padoch, C. and W. De Jong. 1990. Santa Rosa: The impact of the forest products trade on an Amazonian village. In G. T. Prance and M. Balick, eds., *New Directions in the Study of Plants and People. Advances in Economic Botany 8:*151–158.

Padoch, C., J. Chota Inuma, W. de Jong, and J. Unruh. 1985. Amazonian agroforestry: A market-oriented system in Peru. *Agroforestry Systems* 3:47–58.

Parker, E. P. 1981. Cultural ecology and change: A caboclo várzea community in the Brazilian Amazon. Ph.D. dissertation, University of Colorado.

—— 1985a. The Amazon caboclo: An introduction and overview. In E. P. Parker, ed., *The Amazon Caboclo: Historical and Contemporary Perspectives,* xvii–il. Studies in Third World Societies, Vol. 32. Williamsburg, Va.: Department of Anthropology, College of William and Mary.

—— 1985b. Caboclization: The transformation of the Amerindian in Amazonia 1615–1800. In E. P. Parker, ed., *The Amazon Caboclo: Historical and Contemporary Perspectives,* 1–49. Studies in Third World Societies, Vol. 32. Williamsburg, Va.: Department of Anthropology, College of William and Mary.

Peters, C. M., A. H. Gentry, and R. O. Mendelsohn. 1989. Valuation of an Amazonian forest. *Nature* 339:655–656.

Petrere, M. Jr. 1978. Pesca e esforço de pesca no estado do Amazonas. *Acta Amazonica* (supplement 2), 8(3): 1–54.

Petrick, C. 1978. The complementary function of floodlands for agricultural utilization. *Applied Science Development* 12:26–46.

Pinedo-Vásquez, M. 1986. Conservation and local control of resources in the Peruvian Amazon. Paper presented at the Annual Meeting of the American Anthropological Association, December 1986, Philadelphia.

—— 1988. The river people of Maynas. In J. S. Denslow and C. Padoch, eds., *People of the Tropical Rain Forest,* 141–142. Berkeley, Calif.: University of California Press.

Posey, D. A. 1982. Nomadic agriculture of the Amazon. *Garden* 6(1):18–24.

—— 1983. Indigenous ecological knowledge and development. In E. F. Moran, ed., *The Dilemma of Amazonian Development,* 225–257. Boulder, Colo.: Westview Press.

Posey, D. A., J. Frechione, J. Eddins, L. F. da Silva, 1984. Ethnoecology as applied anthropology in Amazonian development. *Human Organization* 43(2):95–107.

Redford, K. and J. G. Robinson. 1987. The game of choice: patterns of Indian and colonist hunting in the neotropics. *American Anthropologist* 89(3):650–667.

Regan, J. 1983. *Hacia la tierra sin mal: estudio sobre la religiosidad del pueblo en la amazonía*. Iquitos, Peru: Centro de Estudios Teologicos de la Amazonía.

Roosevelt, A. C. 1980. *Parmana: Prehistoric Maize and Manioc Subsistence along the Amazon and Orinoco*. New York: Academic Press.

Ross, E. 1978. The evolution of the Amazon peasantry. *Journal Latin American Studies* 10:193–218.

San Román, J. 1975. *Perfiles históricos de la amazonía peruana*. Lima, Peru: Paulinas-Centro de Estudios Teologicos de la Amazonía.

—— *Estudio sócio-econômico de los rios Amazonas y Napo*. 2 vols. Iquitos, Peru: Centro de Estudios Teologicos de la Amazonía.

Sanchez, P. A., D. E. Bandy, J. H. Villachica, and J. J. Nicholaides III. 1982. Soils of the Amazon basin and their management for continuous crop production. *Science* 216:821–827.

Santos, A. M. S. 1982. Aritapera: Uma comunidade de pequenos produtores na várzea amazônica, Santarém, Pa. *Antropología*, (Boletim Museu Paraense Emílio Goeldi, n.s.) no. 83: 1–58.

Schmink, M. 1985. São Felix do Xingu: A caboclo community in transition. In E. P. Parker, ed., *The Amazon Caboclo: Historical and Contemporary Perspectives*, 143–166. Studies in Third World Societies, Vol. 32. Williamsburg, Va.: Department of Anthropology, College of William and Mary.

Silva, A. B. 1959. Contribuição ao estudo do folklore amazônico na Zona Bragantina. *Antropologia*, (Boletim Museu Paraense Emílio Goeldi, n.s.) no. 5:1–66.

Silveira, I. M. 1979a. Formas de aviamento num povoado pesqueiro da Amazônia. *Antropología* (Boletim Museu Paraense Emílio Goeldi, n.s.) no. 74:1–24.

—— 1979b. Quatipurú: Agricultores, pescadores e coletores numa vila amazônica. *Publicações Avulsas* (Museu Paraense Emílio Goeldi, Belém, Brazil) no. 34: 1–96.

Simões, M. F. 1981. Coletores-pescadores ceramistas do litoral do Salgado (Pará). *Antropología* (Boletim Museu Paraense Emílio Goeldi, n.s.) no. 78: 1–26.

Smith, N. J. H. 1974. Destructive exploitation of the South American river turtle. *Yearbook of the Association of Pacific Coast Geographers* 35:86–102.

—— 1979. Aquatic turtles of Amazonia: An endangered resource. *Biological Conservation* 16:165–176.

—— 1981a. *Man, Fishes, and the Amazon*. New York: Columbia University Press.

—— 1981b. Caimans, capybaras, otters, manatees, and man in Amazonia. *Biological Conservation* 19:177–187.

—— 1983. Enchanted forest. *Natural History* 92(8):14–20.

—— 1985. The impact of cultural and ecological change on Amazonian fisheries. *Biological Conservation* 32:355–373.

Sombroek, W. 1966. *Amazon Soils: A Reconnaissance of the Soils of the Brazilian Amazon Region*. Wageningen, The Netherlands: Center for Agricultural Publications and Documentation.

Sternberg, H. O. 1956. *A água e o homem na várzea do Careiro*. Rio de Janeiro, Brazil: Universidade do Brasil.

Stocks, A. 1981. *Los nativos invisibles*. Lima, Peru: Centro Amazonico de Antropología y Aplieación Práctica.

Veríssimo, J. [1895] 1970. *A pesca na Amazônia*. Belém, Brazil: Universidade Federal do Pará.

Vickers, W. 1980. An analysis of Amazonian hunting yields as a function of settlement age. *Working Papers on South American Indians* 2:7–29.

Villarejo, Avencio. 1943. *Así es la selva*. Iquitos, Peru: CETA.

Wagley, C. 1953. *Amazon Town: A Study of Man in the Tropics*. New York: Macmillan.

Weinstein, B. 1983. *The Amazon Rubber Boom, 1850–1920*. Stanford, Calif.: Stanford University Press.

Wesche, R. 1985. The transformation of rural caboclo society upon integration into Brazil's Amazonian frontier: A study of Itacoatiara. In E. P. Parker, ed., *The Amazon Caboclo: Historical and Contemporary Perspectives*, 115–142. Study in Third World Societies, Vol. 32. Williamsburg, Va.: Dept. of Anthropology, College of William and Mary.

EIGHT

Diversity, Variation, and Change in Ribereño Agriculture

CHRISTINE PADOCH AND
WIL DE JONG

Scholars have long pointed out that the Amazon Basin is not the homogeneous, featureless expanse of forested but infertile lands that popular generalizations suggest (Denevan 1984; Sioli 1975; and many others). From one end to the other of the continent-sized area, rivers, soils, forests, climates, and flood regimes differ dramatically. The peoples of the Amazon, of course, have also differed for millenia: their various cultures, histories, preferences, and fortunes, as well as the unlike environments they exploit, are reflected in a great assortment of resource-use patterns.

Fundamental dissimilarities in resource availability and resource use techniques exist not only on a basin-wide scale, but also within very limited areas. The distinction that has been made in terms of resource availability and subsistence potential between *terra firme* and *várzea* environments is probably the most familiar contrast (Meggers 1971). Researchers have recently pointed out, however, that much finer distinctions are used by Amazonians and need to be recognized by researchers who hope to understand indigenous resource-use patterns (Frechione, Posey, and da Silva 1989; Denevan 1984; Hiraoka 1985a; Hiraoka 1985b; Bergman 1980).

That traditional resource-use practices are not only highly diverse, but also frequently changing, may be more difficult to glean from most of the available literature. With some notable exceptions (Bergman 1980; Stocks 1981), discussions of Amazonian resource uses appear to assume a virtually unchanging traditional past, disrupted only by catastrophic events such as the invasion of the basin by Europeans or, more recently, by the abrupt involvement of some traditional groups in markets.

Santa Rosa, Peru, presents a striking example of the diversity, variation, and change that characterize traditional resource use in the Amazon. While the idea that resource-use patterns in the Amazon—as in many humid tropical areas—are diverse, variable, and changing is not new, Santa Rosa's patterns are original in both degree and scale. Twelve distinct types of agriculture and thirty-nine variations in resource-use strategies among the forty-six households are found within territory of this single village on the lower Ucayali River. Moreover, these strategies underwent a high degree of change over a period of twelve months. Although the data presented on Santa Rosa are derived from only one community, the level of variation and change described probably is not atypical of settlements in the Amazon. Furthermore, an appreciation of this tremendous diversity and change should have an impact upon how research on traditional resource use is done, both in the floodplain of the Amazon Basin and elsewhere in the tropics.

Our data on Santa Rosa were gathered between June 1985 and August 1987 using a combination of techniques including interviews; participant observation; measurement, mapping, and inventory of fields; and agricultural activity diaries kept by a selected group of informants.

The Community of Santa Rosa

Most of the households that compose the *caserio* (village) of Santa Rosa have their main residences on the true right bank of the Ucayali River. Several families, however, continue to maintain their principal houses across the river, where most of the community was previously located. In May of 1986 the population of the settlement was 335 people, who were divided among a deceptively homogeneous-appearing forty-six households. All Santa Rosinos consider themselves *ribereños*, a term that denotes nontribal, rural residents of the Peruvian Amazon. They all speak Spanish, all engage in a combination of agricultural and foraging activities, all are descended from families that have lived along Amazonian rivers for many generations, and all are, by any standard measure, poor.

Upon closer investigation, however, very significant differences can be found among the village's inhabitants. Santa Rosinos trace their heritage to at least five different ethnic groups (Padoch and de Jong 1990), although few readily reveal or discuss their tribal roots. They also differ greatly in their life experiences: some have spent much of their lives as urban dwellers and river travelers, some have seen the sights of distant Lima, and others have stayed very close to home. The amount of time that any household has been part of the community is of particular importance in accounting for variable access to resources and differences in farming patterns. Customary law supports farmers' claims of ownership to particular farming sites, although no household in Santa Rosa has obtained formal, permanent title to the lands it uses. The state normally grants certificates of temporary usufruct to mudflat plots, and all Santa Rosinos who claim such areas do hold these certificates. Some people who farm the levees have also obtained certificates to those lands. Other rights to land, however, tend to be recognized by the community as a whole but have no formal recognition in Peruvian law.

The Floodplain and Forest

Much of the diversity of Ucayali resource use can be seen as a reflection of the highly diverse environment to which village households have access. While the nucleated village of Santa Rosa is located on *terra firme,* farmers use a variety of *várzea* and *terra firme* zones. Five distinct landforms are agriculturally important. The relative location of most of these is indicated in figure 8.1, an idealized view of the Ucayali floodplain in cross-section. Four of these landforms are located within the *várzea,* or floodplain: *barreales* (mudflats), *playas* (sandbars), *restingas* (natural levees), and *aguajales* (backswamps, dominated by the palm *Mauritia flexuosa).* The *altura* (*terra firme*), or area above the floodplain, is classified as only one zone, although it does include ecologically different sites and microgradients (Denevan 1984).

Across the Ucayali from the *caserio,* as well as upriver from it, are extensive areas of *barreales*—mudflats that appear yearly along the

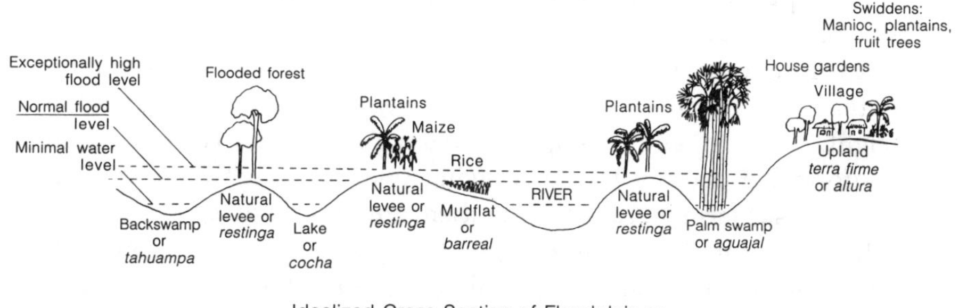

Figure 8.1

river channel when floods subside and are renewed annually by the flood. Optimal *barreales* have heavy, fertile, water-retaining soils and are high enough in elevation to stay above the flood for five to six months of the year. Coveted for their cash-cropping potential, *barreales* are the most desired and most limited agricultural landform in Santa Rosa.

Where the flood deposits coarse sand rather than fine silt, *playas* are formed. Because of their coarse texture and limited moisture-holding capacity, few crops can be grown on these beaches. Since farmers have little interest in using *playas*, they tend to be available to any farmer who desires them.

Much of the riverfront at Santa Rosa consists of a mature, permanently flooded *Mauritia flexuosa* swamp. Thick sediments deposited here in recent years, however, have caused limited areas to stay above water level for several months each year, making agriculture possible. The *aguajal* has been used for agriculture by Santa Rosinos only since 1983.

During much of the 1970s the majority of Santa Rosa's households lived across the river on a natural levee, or *restinga*. Several high floods, however, persuaded most families to locate their principal houses where flood danger was reduced. However, that large levee (as well as others downriver from today's village) continues to be used for agriculture. Levee areas differ greatly in elevation (and, consequently, in their susceptibility to flooding), as well as in soil texture and quality. Most levees, however, are considered desirable farming sites, and available space is usually in short supply.

The appearance and use of all the *várzea* landforms not only changes dramatically with the stages of the annual flood, but differences in landscape as well as in agricultural productivity, labor requirements, and risk can also be detected from year to year as the height, duration, and timing of annual floods change. The floodplain landforms, especially the lower-lying ones, are highly ephemeral, and the risks farmers run in using them are considerable.

In contrast to the limited but risky *várzea* sites, *altura* areas are very abundant in Santa Rosa. Many kilometers of old forest stretch eastward from the village to the Brazilian border. More limited areas of secondary vegetation are located closer to the village. Santa Rosa itself is located on a slight rise, and much of the community's upland territory is hilly. The dissected topography signals differences in drainage characteristics and soil fertility, and parts of Santa Rosa's territory have very different agricultural-use histories. (We have not, however, chosen to distinguish between *altura* areas here.)

A Diversity of Agricultural Types

No agricultural field in Santa Rosa is identical to any other field. Not only do sites differ ecologically, but each household makes different choices in exactly what to plant, how to manage the vegetation, and how long to keep up active management. Recognizing that extremely high variability exists, we have nevertheless classified the agricultural endeavors of Santa Rosinos into only twelve types. In deciding on types we chose to distinguish agricultural fields that shared similar land, labor, and capital requirements, as well as those similar in subsistence or market orientation, in levels of crop diversity (although not in specific mixes of crops), and in risk factors. None of the types discussed here is a homogeneous set. Some of the agricultural types could be considered "successional stages" of other types. That is, some of the types are apt to follow others. However, none *must* follow another type, and therefore no two types can be considered as a single production system.

House gardens, an important type of agriculture in Santa Rosa, are not included among the twelve types. Although all house gardens do share some characteristics and every household in the village had a plot that could be considered a house garden, we chose to exclude this type precisely because of its universality and the extreme variation we found within the class.

Várzea Agriculture

Reflecting the great ecological variability of the lower Ucayali floodplain, as well as the opportunities presented by its rather rather fertile soils, Santa Rosinos locate eight of their twelve agricultural types in the *várzea*. Many of these agricultural types have yet to be described in the scientific literature. Very brief sketches of these eight types follow:

The lowest-lying sites and those most prone to flooding are the *barreal* and the *playa*. One kind of agriculture is located on each of these formations. Both landforms begin to appear in Santa Rosa in early to mid-June as river levels fall. The fine-textured sites are used for the production of rice. Rice is broadcast on the mudflats as soon as the water level allows; no land preparation takes place, and little further labor is involved until the rice is ready for reaping. Harvests occur in October or November as the river is rising and are often completed in standing water. Following the harvest, slash weeding of the *barreal* ensures that when the waters again recede, a clean surface will be available for planting. Even though some part of the harvest may be lost to the flood, in good years returns of 3 to 4 metric tons of rice per hectare are obtained. Because of its productivity and the availability of credit for rice production, every available plot of *barreal* land in Santa Rosa is farmed.

Planting rice on the *barreal* is, however, a risky business. Small, unpredictable rises in the river level (*repiquetes*) can have disastrous effects especially if they occur soon after planting. Seeds must often be broadcast several times, and occasionally entire crops are lost. *Barreal* rice is additionally risky because many farmers obtain bank credit for its planting and management. If the crop fails, farmers not only lose one year's production but are also saddled with large debts.

Because of their coarse texture and consequent dry soils, sandbars or *playas* are used only for the planting of drought-resistant cowpeas and, occasionally, peanuts. Again, the labor needed is not large, but the risk of floods and the lack of credit for the growing of these crops makes *playa* agriculture a far less popular option.

Sites located in the *aguajal* palm swamp in Santa Rosa dry out earlier and flood later than *barreales*. Rice is the only crop now grown in the Santa Rosa *aguajal*, although techniques and labor requirements differ from *barreal* technologies. Rice is dibbled in rather than broadcast, and weeding is somewhat more onerous. Yields can be quite high, reaching about 3 metric tons per hectare in exceptional years. However, the success of *aguajal* farming depends on sufficiently high and well-timed floods that allow for easy site clearance and that destroy noxious weeds. Locally important varieties of jute (*Urena lobata*) also can be grown in *aguajales*, although the labor involved in farming and processing the product have made it an unpopular crop recently.

Santa Rosinos use natural levees, or *restingas*, for five different production types: monocultures of rice, maize, and plantains; diverse, mixed fields known as *chacras*, or swiddens; and agroforestry fields (*purmas*). Although we have chosen to distinguish these types largely by crop, each also involves the use of different techniques, has different ecological and labor requirements, and poses different risks.

Rice can be grown only on *restingas* with heavy soils and only in the first year after clearing from forest or for a single year after an exceptionally high flood. After the annual flood subsides and the *restinga* dries, vegetation is slashed, left to dry, and burned and often reburned again to remove the tangled, thorny brush. Alternatively, *restinga* vegetation may be cleared somewhat less thoroughly prior to the flood, with the hope that a sufficiently high flood will remove any remaining brush. Rice is dibbled in, and subsequent labor is confined to some limited weeding. Rice cultivation on the *restinga* can be followed with another rice crop only if an unusually high flood effectively covers the previous year's field with a good cap of sediment, reducing the problem of weed invasion. *Restinga* rice fields differ from those on the *barreal* also in the labor that is employed. While *barreales* often are farmed with paid labor, *restinga* farmers normally rely on the work of family members or of reciprocal labor groups (Chibnik and de Jong 1989).

Maize grown on the levees demands somewhat different ecological conditions than does rice. *Restinga* soils must be drier for corn planting, soils may be sandier, and weeds do not pose as much of a problem. Maize production can follow rice, or several years of maize farming can follow each other. Maize can be grown almost indefinitely, as long as a major flood occurs every few years. During the years of research in Santa Rosa, high floods were the rule, and continuous cultivation of maize on lower *restingas* was possible.

On somewhat higher ground, where floods reach only once in ten or fifteen years, Santa Rosinos often choose to grow plantains. Plantains are approximately equal to manioc in importance as a dietary staple and yield far better on *várzea* soils than on the *altura*. Plantain fields are virtually permanent fields; they can be maintained with little drop in production for fifteen years or so. After that span, if no flood has occurred, fields may be fallowed for two or three years. A significant flood will completely destroy plantain stands, although some varieties do withstand flooding for a limited time. Floods do, however, renew the soil and check weed growth. Farmers are also willing to face the risks of plantain farming because it demands little labor. Weeding is limited to one cursory slashing a year, and scheduling of maintenance tasks is highly flexible.

All the the *várzea* systems outlined above are monocultures, but farmers on the *várzea* do also plant mixed fields. Diverse swiddens, or *chacras,* where manioc is the dominant crop but where other species such as plantains, squashes, papayas, taro, various vegetables, and numerous other crops are grown tend to occupy the higher parts of levees. These fields are prepared using the methods of slashing, felling, burning, and dibbling that characterize swidden fields outside the floodplain. The advantages of farming the *restinga*—easier clearing, more fertile soils, and consequently higher yields—are bought at the risk of loss of the crops to flooding. Depending on the stage of growth, some crops such as manioc can be harvested early, processed, and stored when floods threaten. Other crops are lost. Few generalizations can be made about *restinga chacras* because variability is very high. However, in contrast to monocultural fields which tend to be market-oriented, most products of *chacras* are destined for household consumption. In the unlikely event that after many years a significant flood has not inundated a *restinga chacra,* such fields would be transformed to agroforestry stands, to plantain fields, or alternatively, they might be fallowed. The ratio of intensive production to fallowing times, however, is much higher on the levee than in *terra firme chacras*.

Where flood danger is minimal, agroforestry fields dominated by fruit trees and by timber species such as tropical cedar are occasionally planted. Immediately prior to and during our research period, a number of exceptionally high floods had largely destroyed some long-existing *restinga* agroforestry fields that had included many species not resistant to prolonged flooding. Faced with a number of such floods in quick succession, *restinga* farmers began to experiment with more flood-resistant fruits such as cacao, mango, breadfruit, *pomarosa* (*Syzygium malaccense*), and some citrus species. However, if a few years of lower inundations follow each other, we believe that some Santa Rosinos, especially those whose main residences are located in the floodplain, will doubtless again risk planting nonresistant varieties. Agroforestry fields commonly yield products that both contribute to the household diet and yield a surplus that is marketed.

Terra Firme Agriculture

The most common *terra firme* agricultural type in Santa Rosa is also the most well known in Amazonia: the upland multispecies swidden, or *chacra*. The *terra firme chacra* in Santa Rosa, like its *várzea* equivalent, tends to be dominated by manioc. Apart from this generalization, little else is true for all Santa Rosa swiddens. Crops, yields, management needs, and length of active management vary greatly from field to field. Most household staples tend to be produced in upland *chacras*. Production of *chacra* crops lasts for two to four years. Manioc is commonly harvested for two years; plantains and other crops can extend the period of *chacra* production to four. After this time, soil deterioration and weed invasion force a change in management to a concentration on agroforestry crops or an end to active management. Since *altura* lands suitable for *chacras* are abundant, every Santa Rosino can make a *chacra* if he or she chooses to. Unclaimed areas of high forest are now located at considerable distance from the village.

The recent availability of credit and a firm price for rice in the market has persuaded some agriculturists to plant rice wherever possible. Some Santa Rosa farmers who have no access to more productive rice lands such as *barreales, aguajales,* or *restingas* have chosen to plant the grain in *altura* fields. Local farmers distinguish rice planting from the making of *chacras,* since these fields are planted as monocultures. Rice produces for only one year on *altura* soils and yields a maximum of 700 to 800 kilograms per hectare. However, for those who have no place to produce a good cash crop, the *altura* rice field is an attractive option.

Following a year of rice cropping, *altura* sites are often switched to plantain production. Plantains yield less in *altura* sites than on the levees, but the risk of flooding is avoided. Permanent cropping of plantains is not possible on the poorer soils of the upland; four years usually marks the limit of *altura* production. Although the rice field–plantain field succession is rather common, it is not obligatory. Plaintains are also planted in the first year after clearing.

The last of the twelve agricultural types also tends to be a successional type. We have described swidden-fallow agroforestry systems in Santa Rosa in other articles (Padoch and de Jong 1987; Padoch and de Jong 1989), stressing their enormous variability as well as their long-term productivity. As we have discussed elsewhere, management of agroforestry plots varies greatly in intensity, technique, and duration. Agroforestry fields, however, no matter how they are managed, tend to require surprisingly low labor inputs for the return they give in subsistence and market products.

A Variety of Strategies

As we have mentioned, the twelve agricultural types that exist in Santa Rosa are not equally available to all households because each family does not have equal access to all landforms. *Barreales* are the most desired of sites but are unavailable to a large percentage of households. On the other hand, any farmer who wishes to make a *chacra* in the *altura* could do so. However, as the data presented in figure 8.2 demonstrate, in 1985 no agricultural type was included in the farming repertoire of every household. A significant proportion of Santa Rosa's farmers even chose not to make a *terra firme chacra,* the classic type of Amazonian farming, thought by many people to be the only kind used by indigenous Amazonians. The most commonly employed agricultural type in the village was the managed swidden fallow, and even that technique was used by only approximately 70 percent of all households. Differences in access to particular landforms were doubtless important in determining whether particular types of agriculture were employed, but other factors were also significant.

The data gathered show not only that there is no one type of agriculture used by every household in Santa Rosa, figure 8.3 also demonstrates that no universally employed strategy can be discerned from the data. All Santa Rosinos do not agree on the desirability of

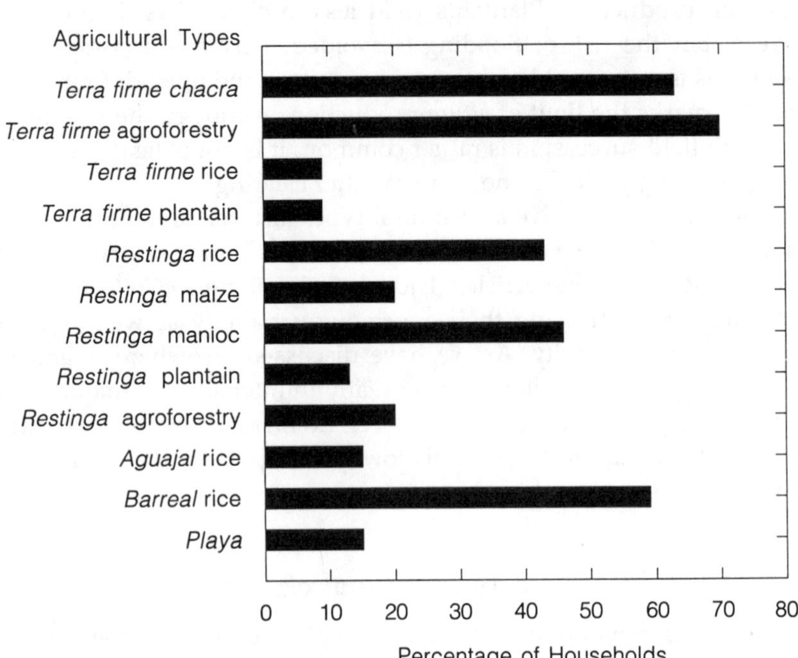

Figure 8.2

diversification of their endeavors, nor, alternatively, of specialization in one or two types of agriculture. Approximately half of Santa Rosa households used four different production types in 1985. A small but significant number—three—however, chose to employ only one production type. Of these, one household limited its farming to one of the riskiest types: a *restinga chacra* with manioc as its main product.

The data in figure 8.4 confirm this pattern of variation in farming strategies. Among the forty-six households of Santa Rosa, thirty-nine distinct ways of combining the twelve agricultural production types were found. The most popular option is the combination of an *altura chacra* with an *altura* agroforestry field and a *barreal*. However, only three of the community's households chose to farm in this fashion. Five other combinations were used by two households each. The remaining thirty-three combinations were each used by only one family.

Just as there is no agricultural technology that is typical of the village of Santa Rosa, there is also no typical strategy. Factors determin-

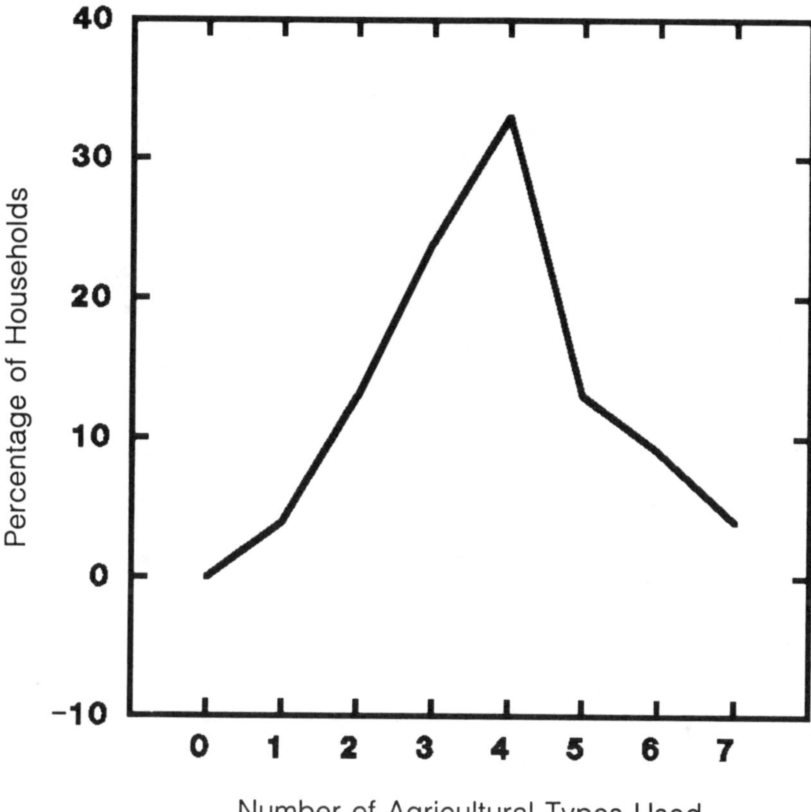

Figure 8.3

ing exactly which strategy is employed by any particular household are many and complex and will not be discussed here in detail. Generally, however, apart from differing access to land resources, several variables were important. These included differences in household labor, including the total amount of labor available, as well as the type of labor (e.g., a household with several young men is more apt to make new *altura* fields) and the seasonality of labor. Other important factors are differences in access to capital and credit and capital resources such as boats. Also significant in some cases is the family's recent farming history, including recent losses of crops to floods, as

Agricultural Types Managed by Santa Rosa Farmers

Household	*Terra firme chacra*	*Terra firme* agroforestry	*Terra firme* rice	*Terra firme* plantain	*Restinga* rice	*Restinga* maize	*Restinga* manioc	*Restinga* plantain	*Restinga* agroforestry	*Aguajal* rice	*Barreal* rice	*Playa*
1	■	■									■	
2	■	■									■	
3	■	■				■	■			■	■	
4	■	■			■		■					
5		■				■		■			■	
6	■	■							■		■	
7	■	■	■							■		
8	■					■		■			■	
9	■				■							
10	■			■							■	■
11	■	■				■	■				■	
12	■	■					■					
13		■				■	■	■			■	
14	■	■			■						■	
15		■	■	■	■		■	■			■	
16	■	■									■	
17			■		■							
18		■					■		■		■	■
19	■	■			■					■		■
20	■	■			■	■					■	■
21	■	■			■		■		■	■		
22		■			■				■			
23	■	■			■					■		
24	■	■			■		■			■		
25	■	■			■						■	
26	■	■			■							■
27		■										
28	■	■	■		■		■			■	■	
29	■	■										
30				■	■							
31					■		■		■		■	
32		■			■	■	■		■		■	
33	■	■							■		■	
34		■		■			■				■	
35	■							■			■	
36		■			■						■	■
37						■	■	■				
38	■						■				■	
39							■		■			
40	■	■					■				■	■
41	■	■					■				■	
42		■										
43	■						■				■	
44							■		■			
45	■				■						■	
46					■	■	■					

Figure 8.4

well as general preferences and attitudes. Some farmers are more actively interested in new plants, new techniques, and new products; others are less interested in these matters and might be considered more averse to taking risks.

Another important factor that must be considered in accounting for the great differences among Santa Rosinos is their different participation in nonagricultural activities. A few household heads concentrate much more on fishing for both subsistence and market than do others. Still others spend a good deal of their time in the town of Requena or the more distant city of Iquitos, where they frequently engage in wage labor as well as in petty trading. If all outside activities were included in our figures summarizing household strategies, the picture of great diversity would be even clearer.

Change in Agricultural Strategies

No environment remains static, and thus an ability to change is important for any farmer. The lower Ucayali floodplain, however, is characterized by such dramatic seasonal and annual changes that exploiting this environment demands exceptional flexibility. Several of the most important agricultural landforms of the region disappear under water each year. When they reappear, they may have changed radically in size, in elevation, in soil texture, in the presence of weeds, and in their spatial relationship with the river. Upland agricultural sites change as well: weeds grow, soils deteriorate, valuable fruit trees begin bearing. Changes in social factors are no less important. Credit may become more or less available for particular crops, market prices rise and fall precipitously, wage labor opportunities arise and disappear, and transportation facilities change. In addition, families change in composition, health of their members, and in aspirations, preferences, and obligations. Any of these factors may affect a household in its decisions on what to plant and where, what fields to continue to manage, and how to alter their management.

Approximately a year after we censused the households of Santa Rosa concerning their agricultural strategies, we reinterviewed the heads of twelve households to see how their strategies had changed. Figure 8.5 presents the rather surprising findings. Only three of the families sampled followed the same strategy in 1986 that they had followed during the previous year. Nine, or 75 percent, of the twelve

Changes in Agricultural Types Used by Santa Rosa Farmers

Household	*Terra firme chacra*	*Terra firme* agro-forestry	*Terra firme* rice	*Terra firme* plantain	*Restinga* rice	*Restinga* maize	*Restinga* manioc	*Restinga* plantain	*Restinga* agro-forestry	*Aguajal* rice	*Barreal* rice	*Playa*
1	■	■									■	
1a	■	■									■	
4	■	■			■		■					
4a	■	■			■		■					
5		■				■		■			■	
5a	■	■					■	■				■
8	■					■		■			■	
8a	■		■								■	
13		■				■	■	■			■	
13a		■				■	■	■			■	■
14	■	■			■						■	
14a	■	■			■	■	■				■	
15		■	■	■	■		■	■			■	
15a	■	■	■	■			■	■			■	
16	■	■									■	
16a	■	■				■					■	
28	■	■	■		■		■			■	■	
28a	■	■				■			■	■	■	
29	■	■										
29a	■	■										
34		■		■			■				■	
34a	■	■		■	■		■					■
40	■	■					■				■	■
40a	■	■					■				■	

Figure 8.5

households had adopted a new strategy. The magnitude of the changes and the reasons for them varied widely.

Complex Resource Use in a Diverse Environment

Our conclusion that traditional resource use as illustrated by the behavior of farmers of Santa Rosa is not simple nor uniform nor unchanging is hardly surprising. That this diversity, variation, and change can be seen not only over large distances and long time scales, but also in the smallest of samples and shortest of time spans is indeed surprising.

We do not believe that the variability we found is merely the result of applying overly fine distinctions, nor is the change we found an artifact of our artificially breaking apart cyclic systems into their constituent stages. Agriculture on the lower Ucayali is indeed far more complex than has ever been appreciated.

The Ucayali floodplain may offer an extreme example of landform and biotope diversity, agricultural variation, and rapidity of change. We believe, however, that researchers throughout the tropics have greatly underestimated these factors in traditional societies and have consistently ignored important production types such as agroforestry fields.

One of the reasons for this shortcoming is the common reliance on extremely small samples in research. We ourselves are guilty of arguing from pitifully small samples (Padoch and de Jong 1989). Warning readers and researchers of the inadequacy of sampling does not satisfactorily stop many from assuming not only that all Santa Rosinos farm in some particular fashion but that all *ribereños* or all floodplain dwellers manage resources in the same way. Although examining larger samples and restudying household strategies is extremely difficult and time-consuming, without such improvements in agricultural research, we can never hope to adequately understand how people make a living in Amazonia.

ACKNOWLEDGMENTS

This article is based on work done as part of a research agreement between the Institute of Economic Botany of the New York Botanical Garden and the Instituto de Investigaciones de la Amazonía Peruana. We would like to acknowledge the very valuable help of scientists and technicians from both institutions, particularly Dr. Jose Lopez Parodi, Marcio Torres, and Ruperto del Aguila. We would also like to thank Dr. Andrew Henderson and Dr. Elaine Elisabetsky for reading and commenting on the article and Carol Gracie for drawing Fig 8.1. Our greatest debt is to the women, men, and children of Santa Rosa for their kind aid and infinite patience. We hope they agree with our findings and are not displeased with this article.

LITERATURE CITED

Bergman, R. W. 1980. *Amazon Economics: The Simplicity of Shipibo* Indian Wealth. Ann Arbor, Mich.: University Microfilms.

Chibnik, M. and W. de Jong. 1989. Agricultural labor organization in *ribereño* communities of the Peruvian Amazon. *Ethnology* 28(1):75–95.

Denevan, W. 1984. Ecological heterogeneity and horizontal zonation of agriculture in the Amazon floodplain. In M. Schmink and C. H. Wood, eds., *Frontier Expansion in Amazonia,* Gainesville, Fla.: University of Florida Press.

Frechione, J., D. A. Posey, and L. F. da Silva. 1989. The perception of ecological zones and natural resources in the Brazilian Amazon: An ethnoecology of Lake Coari. In D. A. Posey and W. Balée, eds., *Resource Management in Amazonia. Advances in Economic Botany* 7:260–282. New York: New York Botanical Garden.

Hiraoka, M. 1985a. Mestizo subsistence in riparian Amazonia. *National Geographic Research* 1:236–246.

—— 1985b. Changing floodplain livelihood patterns in the Peruvian Amazon. *Tsukuba Studies in Human Geography* 3:243–275.

Meggers, B. 1971. *Amazonia: Man and Culture in a Counterfeit Paradise.* Chicago: Aldine Atherton.

Padoch, C. and W. de Jong. 1987. Traditional agroforestry practices of native and ribereño farmers in the lowland Peruvian Amazon. In H. L. Gholz, ed., Agroforestry: Realities, Possibilities, and Potentials, 179–195. Dordrecht, the Netherlands: Martinus Nijhof Publishers.

—— 1989. Production and profit in agroforestry practices of native and ribereño farmers in the lowland Peruvian Amazon. In J. O. Browder, ed., *Fragile Lands of Latin America,* 102–114. Boulder: Westview Press.

—— 1990. Santa Rosa: The impact of the forest products trade on an Amazonian village. In G. T. Prance and M. Baleck, eds. *New Directions in the Study of Plants and People. Advances in Economic Botany.* 8:151–158. New York: New York Botanical Garden.

Sioli, H. 1975. Amazon tributaries and drainage basins. In A. D. Hasler, ed., *Coupling of Land and Water Systems.* New York: Springer-Verlag.

Stocks, A. W. 1981. *Los nativos invisibles.* Lima, Peru: Centro Amazónico de Antropología y Aplicación Práctica.

NINE

The Logic of Extraction: Resource Management and Income Generation by Extractive Producers in the Amazon

ANTHONY B. ANDERSON AND EDVIGES MARTA IORIS

In frontier zones of Amazonia, the predominant forms of land use often bring tragic consequences for both people and the landscape. None of the major land uses in these zones—logging, shifting cultivation, and cattle ranching—utilize native forest ecosystems in a sustainable fashion. Even when carried out selectively, logging causes severe damage (Johns 1988) and leaves forests highly susceptible to subsequent destruction by fire (Uhl and Buschbacher 1985). Other major land uses require wholesale removal of the native forest cover. Although shifting cultivation incorporates forest regeneration during the fallow period, this period is often truncated as populations increase and/or available land decreases (Sioli 1973; Hecht, Anderson, and May 1988). With a useful life that typically ranges from four to ten years before abandonment, Amazonian pastures have been aptly characterized as a prolonged form of shifting cultivation (Buschbacher 1986). But when pastures are abandoned after intensive use, a scrubby, fire-resistant vegetation often takes over, and forest regeneration is brought to a standstill (Uhl, Buschbacher, and Serrão 1988).

Whether the land is initially subjected to logging, shifting cultivation, or cattle ranching, the end result is usually the same: widespread areas of one of the world's most biologically rich biomes are converted into landscapes that are ecologically degraded and economically impoverished (Velho 1972; Barbira-Scazzochio 1980; Fearnside 1983). As resources and economic opportunities dwindle, rural

Table 9.1
Production of Selected Extractive Products Derived from Native Plant Species in the Brazilian Amazon Region During 1974–1986

Product/Scientific Name	Vernacular Name	1974 Quantity (tons)	1974 Value (US $1,000)	1978 Quantity (tons)	1978 Value (US $1,000)	1982 Quantity (tons)	1982 Value (US $1,000)	1986 Quantity (tons)	1986 Value (US $1,000)
Natural Rubbers									
Hevea brasiliensis (Willd. ex A. Juss.) M. Arg.	*borracha*								
Coagulated latex		18,001	17,703	20,795	23,585	25,813	51,609	26,880	27,485
Liquid latex		896	377	1,062	708	1,005	1,008	1,520	954
Castilla ulei Warb.	*caucho*	154	133	1,074	837	914	1,313	200	136
Nonelastic Gums									
Manilkara bidentada (DC.) Chev. and other species	*balata*	279	152	407	426	216	249	22	12
Manilkara elata	*maçaranduba*	526	227	451	271	426	292	376	167
Couma utilis M. Arg.	*sorva*	3,787	1,338	5,555	2,373	5,461	3,202	3,002	1,160

Fibers									
Mauritia flexuosa L. f.	*buriti*	—	—	—	—	862	115	893	83
Leopoldinia piassabe Wall.	*piaçava*	2,360	732	2,321	779	38	15	303	127
Oils									
Orbignya phalerata Mart.	*babaçu*	1,354	186	254	57	48	16	43	7
Copaifera langsdorffii Desf.	*copaiba*	160	341	120	180	68	124	43	37
Dipteryx odorata (Aubl.) Willd.	*cumaru*	24	19	37	68	48	92	457	754
Scheelea huebneri Burret	*urucuri*	—	—	719	8	4,179	23	4,642	27
Foods									
Euterpe oleracea Mart.	*açai*								
Fruits		134	10	46,092	6,483	80,871	16,436	133,847	41,600
Palm hearts		21,246	524	20,573	1,871	95,084	6,991	124,315	6,423
Bertholletia excelsa HB	*castanha-do Pará* (Brazil nut)	35,276	7,791	40,244	15,596	36,419	14,595	35,563	6,990
TOTAL		—	29,513	—	53,242	—	96,000	—	85,962

Sources: FIBGE 1976, 1981, 1984, 1988. Brazilian Amazon region defined as including states of Amazonas, Pará, Rondônia, Roraima, Acre, and Amapá.

poor are compelled to migrate to new frontier zones, where the cycle of environmental degradation begins again.

Adoption of alternative forms of land use that are both ecologically and economically sustainable could break this relentless cycle. In humid tropical regions, long-term maintenance of forest cover is a prerequisite for attaining ecological sustainability (Goodland 1980). Yet land uses that provide forest cover and receive official support through research and extension programs—such as plantations of cacao, rubber, oil palm, or timber species—are only getting underway in Amazonia and require a level of capitalization beyond the reach of most rural producers.

There is one alternative form of land use that maintains forest cover and already has a large following in the region: traditional forest extraction systems based on products such as latex, fibers, and fruits. In the Brazilian Amazon alone, approximately 1.5 million people depend on such extraction for their living (IBGE 1982), generating revenues that approximate 100 million dollars per year (table 9.1). Market-oriented extraction of forest products has been carried out in the Amazon for centuries and in many instances appears to be ecologically sustainable.

Although traditional extractive systems are a promising form of land use for Amazonia, surprisingly little is known about how they function; such knowledge could provide a basis for enhancing their economic sustainability. Although many of the findings reported here about resource management and income generation by extractive producers in the Amazon estuary of Brazil are limited to a specific ecosystem and locale, we believe that they reflect a number of characteristics common to extractive populations throughout Amazonia. In addition, microeconomic studies in specific locations can provide clues for enhancing resource management and income generation among extractive populations.

Forest Extraction in Amazonia

Traditional systems of forest extraction have not only been neglected during recent initiatives to develop the Amazon; they are seriously disrupted by more intrusive forms of land use currently promoted throughout the region. In our view, such neglect and disruption are largely due to the invisibility of traditional extractive systems

(Hecht, Anderson, and May 1988), which in Amazonia are often carried out in extensive tracts of forest inhabited by a dispersed and economically marginal population. Planners and policy makers frequently view these lands as empty, unowned, useless, and abandoned—which facilitates their appropriation for other uses (Dove 1983).

The ongoing disruption of traditional extractive systems has been met with increasing resistance by rubber tappers in the Brazilian state of Acre, who formed a social movement currently supported by a broad coalition of indigenous and environmental groups, as well as by policy makers and government officials in and out of Brazil (Allegretti 1989). The rubber tappers' movement has called for the establishment of so-called "extractive reserves" in public lands designated for the specific purpose of exploiting forest products such as rubber and Brazil nuts on a sustainable basis. As defined by the Brazilian Ministry of Agrarian Reform, extractive reserves are forest areas inhabited by forest product collectors granted long-term usufruct rights to forest resources which they collectively manage (Schwartzman 1989). To date, three such reserves have been established in Acre, and others are being planned elsewhere in the Brazilian Amazon (figure 9.1), encompassing a total area of more than 2.5 million hectares (IEA 1988). The establishment of extractive reserves in Amazonia represents an important step toward promoting socially just forms of land use that potentially can reconcile economic development and environmental conservation.

Although these developments are promising, the long-term economic viability of extractive reserves—as well as traditional extractive systems in general—is questionable. In Amazonia, extractive economies are notoriously unstable and subject to disruption due to competitive displacement of the production system—as occurred in the case of rubber (Weinstein 1983)—or degradation of the resource base—as is currently taking place in natural stands of Brazil nut (Kitamura and Muller 1984). Of fifteen major extractive products for which data are available between 1974 and 1986 in the Brazilian Amazon, only four showed an increase in production, and a mere two showed increases in unit value (table 9.1).

While the macroeconomic outlook is far from promising, recent studies have shown that 1) a surprising diversity of extractive economies exist in Amazonia (IEA 1988), 2) extractive populations are ex-

Figure 9.1

tremely adaptable to both ecosystems and markets (Weinstein 1983; Almeida 1989; Schwartzman 1989), and 3) these populations not only harvest but actively manage forest resources (Fox 1977; Posey et al. 1984). These studies indicate that forest extraction is a far more resilient activity than was realized previously. The case study presented below corroborates these findings and also shows that under certain conditions forest extraction can be economically competitive with other forms of land use.

The Setting

Floodplain forests of the Amazon estuary cover an area estimated at 25,000 square kilometers (Calzavara 1972). Although this area has supported relatively dense human populations for millenia, the native forest structure is surprisingly intact, as revealed in a recently completed survey of forest cover in the Brazilian state of Pará (IBDF-SUDAM 1988). For example, in two municipalities (i.e., counties) lying mostly in the floodplain of the Amazon estuary, the percentage of total forest area altered was found to be only 1.7 percent (Afuá) and .8 percent (Gurupá) at the end of 1986. In contrast, the figures for altered forests were found to be far greater in more recently settled municipalities with relatively small areas of floodplain, such as Conceição de Araguaia (33.4 percent), Redenção (39.3 percent), Xinguara (47.6 percent) and São João de Araguaia (54.9 percent).

The most obvious cause for these differences in deforestation is that recent frontier expansion—in the form of road construction, land speculation, influx of settlers, etc.—has been minimal in the former municipalities and of major significance in the latter. We believe, however, that other factors account for the high degree of resource conservation that is characteristic of long-settled floodplain areas in the Amazon estuary. These factors, which include both ecological and cultural components, are briefly examined below.

The Estuarine Floodplain

The floodplain of the Amazon River and its whitewater tributaries have long been recognized as among the most suitable sites for agriculture in Amazonia, primarily due to their abundant and nutrient-rich sediments (Meggers 1971; Roosevelt 1980). This suitability, however, varies considerably along the course of the major whitewater rivers. For example, sediment concentrations as well as pH decline markedly between the Upper Amazon and the estuary (Lima 1956; Falesi 1974; Irion 1984): this decline can be attributed to the low suspension loads of most of the tributaries in the middle and lower sections of the Amazon. Pronounced variations in flood duration also occur between upriver areas and the estuary. As the river widens downstream and tidal influences increase, flooding becomes less pronounced but more frequent. The combination of these factors produces soils that are not only poor in nutrients compared with upriver areas but excessively wet during most of the year.

Conditions for agriculture thus appear to be less favorable in the estuarine floodplain, and agricultural production is, in fact, chronically low (Lima 1956). This precarious agricultural base supports a rural population that lives for the most part in isolated dwellings or in hamlets perched along the river banks.

Production Systems

Prior to European contact, human population densities along the major rivers of Amazonia were far greater than they are today (Carvajal, Rojas, and Azuna 1941; Denevan 1976). These populations appear to have been supported by a complex food production system characterized by intensive seed crop cultivation on bottomlands in the floodplain, manioc cultivation on the *terra firme*, and animal capture in both zones (Roosevelt 1980; Denevan 1984). These systems were in turn administered by relatively complex societies: the chiefdoms that once proliferated along the major rivers of the Amazon valley.

Nearly four centuries of colonization effectively destroyed these societies by turning a subsistence economy based on food production into a predominantly market economy geared toward extraction of forest and riverine products (Ross 1978). The contemporary residents of the floodplain—known as *caboclos*, or Portuguese-speaking rural inhabitants of Amerindian, Portuguese, and African descent (Parker 1985)—have adapted to the demands of an extractive economy. The scattered settlement pattern of contemporary inhabitants promotes efficient extraction of natural resources and at the same time effectively impedes agricultural development. Likewise, much of the indigenous knowledge of natural resources inherited by the *caboclo* contributes toward maintaining an essentially extractive economy.

The Floodplain Forest

In the floodplain of the Amazon estuary, not only the local inhabitants but their very environment seem to be eminently suited to the demands of an extractive economy. The intricate network of rivers and canals in this landscape has facilitated wood extraction for centuries (Huber 1943). The flooding cycle, which produces waterlogged soils over extensive portions of the year, also accounts for the peculiar composition and dynamics of floodplain forests in the Amazon estuary. In this ecosystem, lack of soil oxygen probably represents the key limiting factor to plant growth. Few species can thrive under waterlogged conditions, and as a result these forests exhibit relatively low biological diversity and pronounced dominance by a few tree species, many of which are of economic importance (Anderson 1989). Flooding also appears to have a dramatic effect on forest dynamics in this ecosystem. Shallow root systems due to impeded drainage, combined with constant soil movement associated with flooding, result in a high frequency of tree falls, and abundant light gaps provide ample opportunities for forest regeneration. With a high concentration of economic species and quick recovery following disturbance, these forests appear to be able to support short-cycle extraction on a sustainable basis.

Combu Island

The maintenance of forest cover in the floodplain of the Amazon estuary is thus to a large degree dependent on the ecology of this landscape and the nature of its inhabitants.

The case of Combu Island (1°27′S, 48°28′W), located 1.5 kilometers from the major port city of Belém on the Guamá River (figure 9.2), provides an example of how people and the forest interact at a specific site in the Amazon estuary. Both in the vicinity of Belém and near the port of Iquitos in the Peruvian Amazon (Padoch et al. 1985), local inhabitants have developed diversified agroforestry systems in response to dynamic markets. The climate at Combu, type Af in the Koppen system, is hot and humid throughout the year.[1] The tide-driven floods are the most striking environmental feature. Due to this factor, large portions of the island undergo daily inundations during the rainy season and, when peak tides occur (at full or new moons during the equinoxes), the island is completely flooded. The sediments transported by the Guamá River generally consist of well-weathered clays or silts. The predominant soil is a low-humic gley, characterized by high clay content, low base saturation, and pH of 4.5 to 5.

The vegetation on Combu Island is floodplain forest. As in other areas of the Amazon estuary (Anderson et al. 1985), this forest is characterized by high dominance of a few tree species, many of which are of economic importance and form the basis for extractive economies. Inspection on the ground and of satellite images indicates that forest

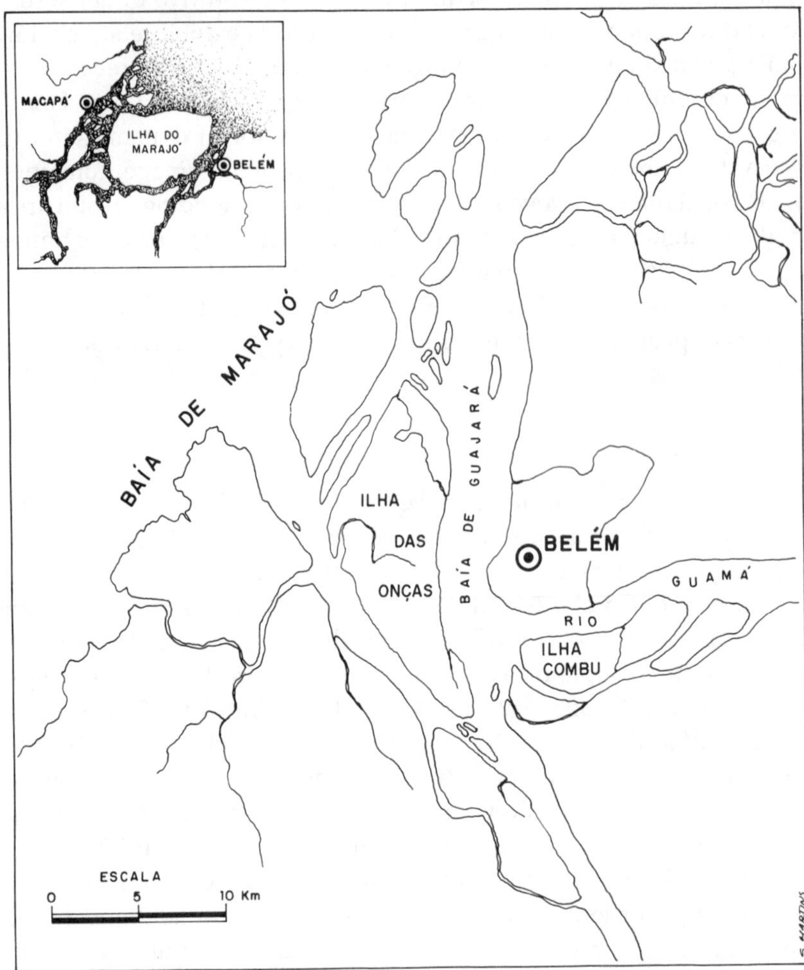

Figure 9.2 Location of Study Sites in the Vicinity of Belém. Brazil

clearings constitute less than 5 percent of the total land area of the island. Although all of Combu's contemporary forest cover has probably undergone some form of human-induced disturbance during the last centuries (if not millenia) of continuous occupation, today this cover is virtually complete and ranges in age and stature from young successional to mature forest.

A census[2] carried out in late 1988 revealed a total population of 645 inhabitants. As the total area of the island is 15 square kilometers, the population density is forty-three inhabitants per square kilometer. This density is exceptionally high compared with other rural areas of Amazonia and is especially noteworthy given the intact state of forest cover on the island. In frontier zones of Amazonia, the minimum size for family lots in government or privately sponsored settlement projects is typically 50 to 100 hectares; on Combu, the average landholding is 15.3 hectares per family. Approximately half (52 percent) of the resident families own land, 23 percent have free access to forest resources, and 24 percent have restricted access through contractual arrangements. This situation is in sharp contrast to consolidated frontier areas elsewhere in Amazonia, where land distribution is extremely regressive and access to natural resources far more restricted (Hecht, Anderson, and May 1988).

The population of Combu not only is dense but is remarkably sedentary: 54 percent of the resident families originated from the island, and 84 percent have lived there for at least ten years. Compared with frontier zones in Amazonia, the work force is also quite stable: only 16 percent are employed off the island, despite the nearby presence of the Amazon's largest city.

The vast majority (92 percent) of Combu's residents gain most of their income through extraction of forest products, and none is actively engaged in shifting agriculture. The predominance of forest extraction is a general phenomenon in floodplain forests of the Amazon estuary; the strategies that allow this dominance to be maintained are various.

Forest Management Strategies

Land-Use Units

Land-use patterns on Combu Island conform to those observed at a variety of locales in the Amazon estuary (Anderson et al. 1985; Anderson 1989). As elsewhere in the estuary, one can observe a recurring pattern of land-use units on Combu. A house garden (*terreiro*) typically surrounds the dwelling. This relatively open zone is used for raising domesticated animals and cultivating a wide variety of native and exotic plants for food, condiments, remedies, and ornamentation. The house garden, which is usually less than one hectare in area, contrasts sharply with the far denser and more extensive floodplain forest (*mata de várzea*), which is covered by a closed canopy of trees and used primarily for gathering forest products and for hunting. In contrast with the house garden, no discernible management takes place in this latter zone.

These two land-use units are easily distinguishable entities that are clearly designated by local terms. A third, less easily distinguishable zone occupies a variety of locations in relation to the household, originates by different means, and is designated by a variety of terms. The common element of this zone is that it is subject to long-term forhest management. For simplicity we refer to this zone as "managed forest."

Forest Management Objectives

On Combu Island, managed forests are frequently almost indistinguishable from the native floodplain forest and are composed almost entirely of native forest species. The management of these forests is subtle and often difficult to discern. The practice has two distinct but complementary objectives. The first objective is to eliminate or reduce plants that are undesirable, such as most species of vines or trees that are used exclusively for firewood. Large trees are frequently eliminated by ringing rather than felling, so as to minimize damage to the surrounding forest. Local inhabitants invariably eliminate undesirable species by selective thinning rather than by indiscriminate clearcutting, so as to maintain the structural integrity of the forest. Eliminating undesirable species indirectly favors desirable species by reducing competition.

The second objective of forest management is to favor desirable species directly by promoting their regeneration and/or production. The inhabitants of Combu carry out this objective in a number of ways. One frequently employed technique is to introduce the seeds of desirable species by conscious planting or seeding or, alternatively, by unconscious dispersal of propagules. Desirable species also may be introduced by planting or transplanting seedlings and cuttings. In addition to promoting regeneration, residents enhance the production of desirable species by actively protecting their seedlings, through weeding, or constructing improvised fences. Organic material is concentrated around the bases of favored plants; this practice is especially common in the case of introduced cultigens such as mango

(*Mangifera indica*) and coconut (*Cocos nucifera*), as well as native species such as cacao (*Theobroma cacau*) and *açaí* palm (*Euterpe oleracea*).

Finally, many species are neither eliminated nor subsequently favored by discernible management practices. These so-called "tolerated" species are often important sources of extractive products: the most noteworthy examples on Combu include rubber (*Hevea brasiliensis*), various sources of edible fruits (*Spondias mombin*, *Inga* spp., *Mauritia flexuosa*), and commercial timber species (*Carapa guianensis*, *Virola surinamensis*, *Cedrela odorata*).

Açaí

On Combu, *açaí* palm is the most important component of forest-based extraction systems, as well as of the entire island economy; in fact, this species has recently overtaken rubber as the highest revenue-generating extractive plant resource in the Brazilian Amazon (table 9.1). *Açaí* is also the most ecologically abundant tree species in the floodplain forests of Combú, as well as on topographically low sites over approximately 10,000 square kilometers of the Amazon estuary (Calzavara 1972).

In addition to a wide range of subsistence uses, *açaí* furnishes two products that have a significant role in regional markets: edible fruits and palm hearts. The fruits of the palm are used to make a thick beverage that serves as a staple food both on the island and throughout much of eastern Amazonia. *Açaí's* clumped habit makes palm heart extraction analogous to pruning: the process results in the demise of stems but not of individual palms, which include a cluster of stems that regenerate readily following selective cutting. Palm hearts are not part of the popular diet in the Amazon estuary, and a regional market for this product developed only in the 1970s, when other extractive sources of palm heart (i.e., nonclumping palm species) had been destroyed in southeastern Brazil.

In many areas of the Amazon estuary, palm heart extraction is undermining fruit harvesting of *açaí*, due to indiscriminate cutting of stems. Inhabitants on Combu Island and at numerous other locales in the estuary reconcile these two activities by applying some of the forest management practices described above: selective pruning (rather than clearcutting) of *açaí* stems. As described below, these practices result in substantial economic gains for extractive producers.

Cacao

On Combú Island, cacao production represents the second most important economic activity. Cacao production on Combú can be more appropriately characterized as a form of extraction rather than agriculture, due to the essentially nondomesticated nature of this resource. In the floodplain of the Amazon estuary, cacao does not appear to have been an original component of the flora; probably it was introduced during the colonial period from upland forest areas (where it is native) into extensive plantations in the floodplain. As these plantations were abandoned, the species has become semiwild and is capable of regenerating even in the absence of direct human intervention, although it is rarely (if ever) observed in mature floodplain forests. In the floodplain of the Amazon estuary, cacao thus appears to be an introduced species that requires minimal human intervention for its maintenance.

These characteristics make cacao ideally suited for the low-intensity forest management systems practiced on Combú Island and at numerous other sites in the estuary. In these systems, management typically consists of eliminating competitors, protecting seedlings, and tending mature trees; systematic establishment of plantations is the exception rather than the rule. Local inhabitants treat cacao like *açaí*, as an extractive resource that can be managed within the floodplain forest ecosystems.

Rubber

Rubber is the third most economically important forest resource on Combú Island. Declining prices for processed rubber in recent years (table 9.1), however, have reduced this species' relative contribution to revenue, and other products such as shrimp have assumed greater importance. To our knowledge, rubber is not consciously favored by management practices such as those described above. Yet on Combú and elsewhere, we have frequently observed a greater abundance of this species on sites subjected to management (Anderson 1989). This greater abundance is probably the result of long-term tolerance of seedlings and saplings, combined with selective elimination of competitors. Rubber thus appears to benefit indirectly from the subtle forms of forest management practiced in the Amazon estuary.

Economic Strategies

Patterns of Income Generation

On Combú Island, the harvest and sale of *açaí*, cacao, and rubber are carried out in a complementary fashion during the year. Data from five producers over a five-year period[3] show that production of *açaí* fruits is the primary economic activity from April through November and cacao from December through March (figure 9.3): these intervals correspond to the peak fruiting periods of the two species. Rubber production can be carried out at any time of the year; on Combú, it is concentrated at the end of the year (September through December), when *açaí* production is declining and cacao production is low.

Açaí fruits, cacao, and rubber represent only part of the range of natural resources effectively utilized by the residents of Combú in both market and subsistence economies. A more complete indication of this range is shown in figure 9.4, which illustrates monthly revenues from the sale of all natural resources by a single family on Onças Island (figure 9.2). Although close to Combú, Onças shows a distinc-

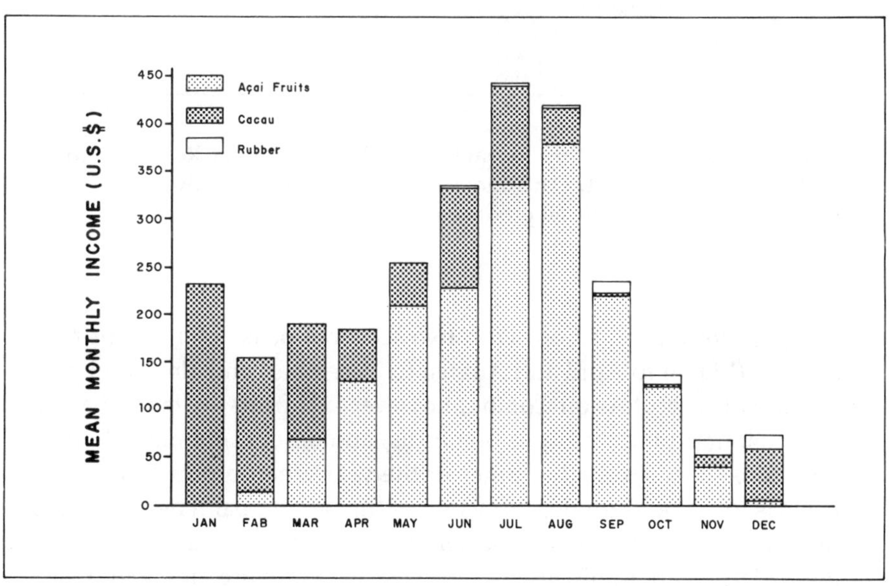

Figure 9.3 Mean Monthly Income from Sale of Forest Products per Producer on Combu Island. Data derived from five households over a five-year period (1984–88) and are based exclusively on sale of açaí fruits, cacao, and rubber.

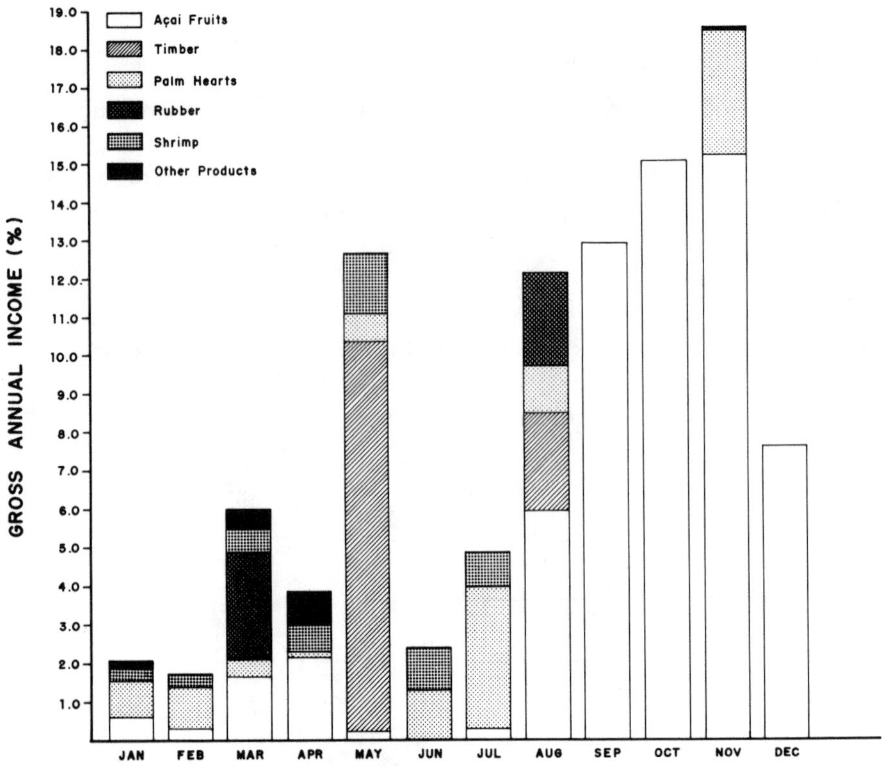

Figure 9.4 Relative Monthly Income Derived from Natural Resources for a Family on Oncas Island. Data derived from one household over a one-year period (1986) and include all produces sold.

tive pattern of production, with two seasonal peaks of *açaí* fruit production (March to April and August to December) and almost no production of cacao.[4] As on Combu, however, *açaí* fruits are far and away the most important resource on Onças, and other economic activities tend to be concentrated at times of the year when production of *açaí* fruits is low. For example, extraction of timber and palm hearts occurs primarily during the first semester, when labor demands for harvesting *açaí* fruits are minimal. Both of these activities are an integral part of the forest management practices described above. Timber extraction is especially favorable for the first semester, when peak floods occur and logs can be floated out of the forest (thus minimizing damage due to logging). Rubber is a nonperishable product that can be stored for sale when prices are higher or when alternative sources of income are minimal: in the household studied, rubber was sold in March and August. Shrimp production is a minor activity in absolute terms, but it assumes importance during the first semester, when income from other sources is low.

As these data illustrate, the floodplain forest provides a reservoir of economic opportunities for local inhabitants. The universe of potential economic products available in this forest includes edible fruits, palm heart, latex, timber, fibers, firewood, fertilizer, ornamental plants, honey, oilseeds, medicines, utensils, and game such as pacas, agoutis, porcupines, sloths, opossums, and feral pigs. Not all of these products are utilized: for example, rubber production has declined markedly on Combú Island. Furthermore, many of the resources utilized only find their way into subsistence economies and are not sold in local markets. Nevertheless, the wide range of products potentially available enables the inhabitants of Combú to adopt resilient responses to changing economic conditions.

Scale of Income Generation

Data on sale of *açaí* fruits, cacao, and rubber by five producers on Combú Island (table 9.2) indicate that reasonably high levels of income can be obtained from these products alone. Mean annual income from these products was found to be U.S. $3,171.56, which compares favorably with the mean annual income of U.S. $1,828.87 derived from the sale of all agricultural products by shifting cultivators near Tomé-Açu, Pará (table 9.3) . Not only is the gross income derived from extraction greater, but it requires less investment and (as indicated in Table 9.2) probably exhibits less variability from year to year, thus reducing risk—a crucial factor for small-scale producers. Furthermore, as illustrated above, maintenance of forest resources in extractive systems provides a reservoir of economic opportunities that can be drawn upon when market conditions change, thus enabling producers to avert future risk. Finally, as practiced in frontier zones of Amazonia today, shifting cultivation is ecologically unsustainable, whereas the extractive systems described in this paper apparently can be maintained indefinitely.

Table 9.2

Sale of Forest Products by Five Producers on Combu Island from 1984 through 1988

		Product						
		Açai		Cacao		Rubber		
Producer	Year	kg	US$	kg	US$	kg	US$	Total (US$)
Francisco Rosa	1984	705	115.62	507	596.28	216	200.87	912.77
	1985	3,885	514.33	1,056	1,290.78	291	233.60	2,038.71
	1986	9,660	1,935.45	1,161	1,465.06	0	.00	3,400.51
	1987	12,840	2,469.59	1,276	1,175.39	0	.00	3,645.32
	1988	9,780	1,580.60	1,302	1,086.58	0	.00	2,667.18
	mean	7,374	1,323.19	1,060	1,122.82	101	86.89	2,532.89
João Rosa	1984	2,940	501.32	458	546.42	373	333.60	1,381.34
	1985	4,305	608.91	971	1,093.00	51	41.74	1,744.16
	1986	9,645	2,315.81	929	1,173.27	0	.00	3,489.08
	1987	17,910	3,598.00	677	751.15	0	.00	4,349.22
	1988	27,210	4,304.38	703	589.64	0	.00	4,894.02
	mean	12,402	2,265.68	748	830.81	85	75.06	3,171.56

Nazaré Rosa	1984	4,695	759.27	593	697.00	270	241.31	1,697.58
	1985	4,890	704.76	946	1,064.14	6	5.56	1,774.46
	1986	10,215	2,261.72	861	1,094.58	0	.00	3,356.30
	1987	13,215	1,903.37	342	304.73	0	.00	2,208.10
	1988	13,215	1,903.37	342	304.73	0	.00	2,208.10
	mean	9,378	1,683.25	649	725.65	55	49.37	2,438.28
Oscar Quaresma	1984	705	116.78	588	694.76	172	157.55	969.09
	1985	3,750	505.66	1,278	1,496.76	112	88.53	2,090.95
	1986	9,300	1,852.27	1,262	1,560.37	0	.00	3,412.64
	1987	12,195	2,174.61	900	838.48	0	.00	3,013.09
	1988	5,745	867.18	840	665.08	0	.00	1,532.26
	mean	6,339	1,103.30	974	1,051.09	57	49.21	2,203.60
Raimundo Rosa	1984	4,125	726.16	696	824.45	157	142.43	1,693.04
	1985	8,205	1,073.37	978	1,218.19	63	50.35	2,341.91
	1986	15,675	3,518.79	1,104	1,400.12	0	.00	4,918.91
	1987	21,735	3,851.85	339	320.16	0	.00	4,171.76
	1988	24,135	3,372.16	120	106.95	0	.00	3,479.11
	mean	14,775	2,508.46	647	773.97	44	38.55	3,320.99
Overall Mean	—	12,402	2,265.68	748	830.81	85	75.06	3,171.56

Table 9.3

Structure of Agricultural Production in Four Families of Shifting Cultivators near Tomé-Açu, Pará, in 1981

Property No.	Area (in hectares)									Value of Production (in US$)				
	Occupied	*Cultivated*	*Manioc*	*Rice*	*Corn*	*Beans*	*Malva*	*Black Pepper*	*Other Crops*	*Flour*	*Pepper*	*Other Crops*	*Livestock*	*Total*
1	106	12.0	10.1	2.0	2.0	1.0	—	1.8	0.1	2,783.29	234.61	586.52	319.92	3,924.34
2	3	3.0	0.5	2.5	2.5	0	—	—	—	31.99	—	330..58	95.98	458.55
3	25	4.5	2.3	—	2.1	—	0.6	0.6	0.1	1,450.30	149.30	490.54	234.61	2,324.75
4	272	5.4	3.0	3.0	3.0	0.5	—	2.3	—	21.33	202.61	227.26	106.64	607.84
Mean	101	6.2	4.0	1.9	2.4	0.4	0.1	1.2	0.1	1,071.73	146.63	421.23	189.29	1,828.87

Adapted from Flohrshutz 1983

As carried out on Combú and at numerous other locations in the Amazon estuary, forest extraction involves a considerable degree of resource management. What does this management mean in economic terms? In an experiment designed to test the effects of management practices described earlier in this paper on the *açaí* palm (Anderson and Jardim 1989), it was found that the value of annual fruit production increased 58 percent from U.S. $235.25 per hectare of unmanaged floodplain forest to U.S. $371.58 per hectare under management. Discounting estimated costs of implementing management, net annual gains for managing (as opposed to not managing) each hectare were found to be U.S. $109.83. This analysis does not take into account the value of other forest products extracted during thinning and pruning operations, such as palm heart and timber. Nor does it consider the greater facility of obtaining these and other forest products on managed sites in comparison to sites not subjected to management. Finally, the analysis only covers the first year; in subsequent years, site preparation costs should decline while returns from fruit yields are likely to remain constant.

We believe that the management currently practiced on Combú Island offers opportunities for enhancing economic returns to the resident population. With little extra effort, a wide range of economic species currently tolerated in the floodplain forest could be managed more intensively to increase their abundance and yields. In addition to rubber, local inhabitants could subject various species that provide edible fruits, oilseeds, and high-quality timber to management without changing local patterns of land use or undermining the resource base.

The case of Combú Island illustrates the ecological sustainability and economic resilience that we believe is common to traditional forest extraction systems throughout Amazonia. Furthermore, in the specific case of Combú, forest extraction can be a lucrative activity when compared to land-use alternatives such as shifting cultivation. Because this profitability is largely due to a high-value, perishable product (*açaí* fruit) that can be readily absorbed by a large, easily accessible market (Belém), it is unlikely that the economic returns on Combu are representative for extractive producers elsewhere in Amazonia. As in other extractive systems (Hecht, Anderson, and May 1988), however, a large share of the economic benefits is not readily measurable and derives from the forest itself. We suspect that a holistic cost-benefit analysis of these benefits within rural households—whether on Combu or elsewhere—will show that traditional extractive systems are not only lower in risk and higher in sustainability but also generate greater economic returns than other forms of land use currently promoted in Amazonia.

ACKNOWLEDGMENTS

This study is part of an ongoing research and training project in the Amazon estuary funded by a grant from the Ford Foundation to the Museu Goeldi/CNPq; we gratefully acknowledge the support of both institutions. We also thank Sandoval Martins for preparing the drawings, Rebecca Abers for assisting with the data analysis, and Luz Maria Araujo e Souza for typing the tables.

This paper is dedicated to the residents of Combu Island, who consistently provided insightful answers to our incessant questions.

NOTES

1. Rainfall in the adjacent city of Belém averages 2,732 millimeters per year, with highs (more than 200 millimeters per month) from January through May and lows (less than 100 millimeters per month) in October and November; mean monthly temperatures vary from 25°C in February to 26.3°C in November (SUDAM 1984).

2. General data on the island's population and economic activities were obtained through application of a questionnaire in eighty-seven of ninety-eight households present on the island in November-December 1988.

3. These producers represent five households established on the property of a landowner, who has kept production records since 1984. The holding of the landowner is approximately 280 hectares, which is large by Combu standards. A total of nine families live on the property, but complete records exist for only five. These producers are obliged to give half their production of *açaí* fruits, cacao, and rubber to the landowner in exchange for usufruct rights. As discussed above, such restricted access applied to approximately 23 percent of the households on Combu. The five producers have access to areas ranging from 20 to 60 hectares, which is also large by Combu standards. Yet there is remarkably little variation between producers (Table 9.2), which suggests that each producer effectively utilizes a smaller area.

Despite these qualifications, the material welfare of the five producers was similar to that of other families on Combu, and we believe that the data presented here are representative of the productive capacity of households in general on the island.

4. For the family on Onças Island, cacao production was negligible and was included under "other products" in Figure 9.4.

LITERATURE CITED

Allegretti, M. H. 1990. Extractive reserves: An alternative for reconciling development and environmental conservation in Amazonia. In A. B. Anderson, ed., *Alternatives to Deforestation: Steps Toward Sustainable Use of the Amazon Rainforest*, 252–264. New York: Columbia University Press.

Almeida, M. B. de. *Seringais e trabalho na Amazônia: O caso do Alto Juruá*. Anuario Antropológico. Brasília, Brazil: University of Brasília.

Anderson, A. B. 1990. Deforestation in Amazonia: Dynamics, causes, and alternatives. In A. B. Anderson, ed., *Alternatives to Deforestation: Steps Toward Sustainable Use of the Amazon Rainforest*, 3–23. New York: Columbia University Press.

Anderson, A. B., A. Gély, J. Strudwick, G. L. Sobel, and M. G. C. Pinto. 1985. Um sistema agroflorestal na várzea do estuário amazônico (Ilha das Onças, Município de Barcarena, Estado do Pará). *Acta Amazonica* 15(1–2): 195–224.

Anderson, A. B. and M. A. G. Jardim. 1989. Costs and benefits of floodplain forest management by rural inhabitants in the Amazon estuary: A case study of açaí palm production. In J. Browder, ed., *Fragile Lands in Latin America: The Search for Sustainable Uses*. Boulder, Colo.: Westview Press.

Barbira-Scazzochio, F. 1980. From native forest to private property: The development of Amazonia for whom? In Barbira-Scazzochio, ed., *Land, People, and Planning in Contemporary Amazonia*, iii–xvi. Occasional Publication No. 3. Cambridge, England: Center of American Studies, Cambridge University.

Buschbacher, R. J. 1986. Tropical deforestation and pasture development. *Bioscience* 36(1):22–28.

Calzavara, B. B. G. 1972. As possibilidades do açaizeiro no estuário amazônico. *Boletim da Fundação de Ciências Agrárias do Pará* 5:1–103.

Carvajal, G. de., A. de Rojas, and C. de Azuna. 1941. *Descobrimentos do Rio Amazonas*. São Paulo, Brazil: Editora Nacional.

Denevan, W. M. 1976. The aboriginal population of Amazonia. In W. M. Denevan, ed., *The Native Population of the Americas in 1492*, 205–234. Madison, Wisconsin: University of Wisconsin.

—— 1984. Ecological heterogeneity and horizontal zonation of agriculture in the Amazon floodplain. In M. Schmink and C. H. Wood, eds., *Frontier Expansion in Amazonia*, 311–336. Gainesville, Florida: University of Florida Press.

Dove, M. 1983. Theories of swidden agriculture and the political economy of ignorance. *Agroforestry Systems* 1:85–99.

Falesi, I. C. 1974. Soils of the Brazilian Amazon. In C. Wagley, ed., *Man in the Amazon*, 201–229. Gainesville, Florida: University of Florida Press.

Fearnside, P. M. 1983. Development alternatives in the Brazilian Amazon: An ecological evaluation. *Interciencia* 8(2):65–78.

Flohrschutz, G. H. H. 1983. Análise econômica de estabelecimentos rurais do Município de Tomé-Açu, Pará: Um estudo de caso. Empressa Brasileira de Pesquisa Agropequaria, Documentos 19. Belém, Brazil.

Fox J. J. 1977. *Harvest of the Palm.* Cambridge, Mass.: Harvard University Press.

Goodland, R. 1980. Environmental ranking of Amazonian development projects in Brazil. *Environmental Conservation* 7(1):9–26.

Hecht, S. B., A. B. Anderson, and P. May. 1988. The subsidy from nature: Shifting cultivation, successional palm forests, and rural development. *Human Organization* 47(1):25–35.

Huber, J. 1943. Contribuição à geografia física da região dos furos de Breves e da parte ocidental da Ilha do Marajó. *Revista Brasileira de Geografia* 5(3): 449–474.

IBDF/SUDAM. 1988. *Levantamento da Alteração de Cobertura Vegetal Natural.* Belém, Brazil: Instituto Brasileiro de Desenvolvimento Florestal e Superintendência de Desenvolvimento da Amazônia.

IBGE. 1978. *Produção extrativa vegetal 1976.* Rio de Janeiro, Brazil: Fundação Instituto Brasileiro de Geografia e Estatística.

—— 1981. *Produção extrativa vegetal 1978.* Rio de Janeiro, Brazil: Fundação Instituto Brasileiro de Geografia e Estatística.

—— 1982. *Censo demográfico 1980.* Rio de Janeiro, Brazil: Fundação Instituto Brasileiro de Geografia e Estatística.

—— 1984. *Produção extrativa vegetal 1982.* Rio de Janeiro, Brazil: Fundação Instituto Brasileiro de Geografia e Estatística.

—— 1988. *Produçaõ de extração vegetal e da silvicultura 1986.* Rio de Janeiro, Brazil: Fundação Instituto Brasileiro de Geografia e Estatística.

Instituto de Estudos Amazonicas. 1988. *Seminário: Planejamento e a gestão do processo de criação de reservas extrativistas na Amazônia.* Curitiba, Brazil: Instituto de Estudos Amazônicos.

Irion, G. 1984. Sedimentation and sediments of Amazonian rivers and evolution of the Amazonian landscape since Pliocene times. In H. Sioli, ed., *The Amazon: Limnology and Landscape Ecology of a Mighty Tropical River and its Basin,* 201–214. Dordrecht, The Netherlands: Dr. W. Junk Publishers.

Johns, A. 1988. Effects of "selective" timber extraction on rain forest structure and composition and some consequences for frugivores and folivores. *Biotropica* 20(1):31–37.

Kitamura, P. C., A. K. O. Homma, G. H. H. Flohrschutz, and A. I. M. dos Santos. 1983. A pequena agricultura no nordeste paraense. Empressa Brasileira de Pesquisa Agropequaria. Documentos 22. Belém, Brazil.

Kitamura, P. C. and C. H. Muller. 1984. *Castanhais nativos de Marabá (PA): Fatores de depredação e bases para a sua preservação.* EMBRAPA/CPATU. Documentos 30. Belém, Brazil.

Lima, R. R. 1956. *A agricultura nas várzeas do estuário Amazônico.* Boletim Técnico do Instituto Agronômico do Norte 33. Belém, Brazil.

Meggers, B. J. 1971. *Amazonia: Man and Culture in a Counterfeit Paradise.* Chicago: Aldine-Atherton.

Padoch, C., J. Chota Inuma, W. de Jong, and J. Unruh. 1985. Amazonian agroforestry: A market-oriented system in Peru. *Agroforestry Systems* 3: 47–58.

Parker, E. P. 1985. Cabocloization: The transformation of the Amerindian in Amazonia 1615–1800. In E. P. Parker, ed., *The Amazon Caboclo: Historical and Contemporary Perspectives*, xvii–li. Studies in Third World Societies Publication Series, vol. 29. Williamsburg, Va.: William and Mary Press.

Posey, D. A., J. Frechione, J. Eddins, and F. S. Silva with D. Myers, D. Case and P. Macbeath. 1984. Ethnoecology as applied anthropology in Amazonian development. *Human Organization* 43(2):95–107.

Roosevelt, A. C. 1980. *Parmana: Prehistoric Maize and Manioc Subsistence along the Amazon and Orinoco*. New York: Academic Press.

Ross, E. B. 1978. The evolution of the Amazonian peasantry. *Journal of Latin American Studies* 10(2):193–218.

Schwartzman, S. 1989. Extractive reserves: The rubber tappers' strategy for sustainable use of the Amazon rain forest. In J. Browder, ed., *Fragile Lands in Latin America: The Search for Sustainable Uses*. Boulder, Colo.: Westview Press.

Sioli, H. 1973. Recent human activities in the Brazilian Amazon region and their ecological effects. In B. J. Meggers, E. S. Ayensu, and W. D. Duckwork, eds., *Tropical Forest Ecosystems in Africa and South America: A Comparative Review*, 321–334. Washington, D.C.: Smithsonian Institution Press.

SUDAM. 1984. *Atlas Climatológico da Amazônia Brasileira*. Belém, Brazil: Superintendência de Desenvolvimento da Amazônia.

Uhl, C. and R. Buschbacher. 1985. A disturbing synergism between cattle ranching burning practices and selective tree harvesting in the eastern Amazon. *Biotropica* 17(4):265–268.

Uhl, C., R. Buschbacher, and A. Serrão. 1988. Abandoned pastures in eastern Amazonia. Part I: Patterns of plant succession. *Journal of Ecology* 76:663–681.

Velho, O. C. 1972. *Frentes de Expansão e Estruture Agrícola*. Rio de Janeiro, Brazil: Zahar Editores.

Weinstein, B. 1983. *The Amazon Rubber Boom 1850–1920*. Stanford, California: Stanford University Press.

PART III

Case Studies of Resource Management Projects in Protected and Unprotected Areas: Institutional Perspectives

Introduction

KENT H. REDFORD AND
CHRISTINE PADOCH

Part 3, "Case Studies of Resource Management Projects in Protected and Unprotected Areas: Institutional Perspectives," examines several attempts that are underway to limit forest destruction, conserve biodiversity, and improve the livelihoods of traditional peoples. In many cases these projects are based on traditional resource use patterns such as those described in earlier chapters.

One of the strengths of this section is its unique emphasis on institutional experiences—represented are contributors from the Programme for Belize, Conservation International, World Wildlife Fund, the Ecuadorean National Park Service, Cultural Survival, and the University of Florida. Each author has written of the experiences his or her institution has had in trying to implement novel ideas. These experiences are the raw material from which future progress will be made.

The essays in part 3 differ from those in the rest of the book in style and content. Rather than being written by research scientists, they are written by representatives of the organizations sponsoring or supporting the projects. Given this, the reader will find not a critical review of the successes and failures of these projects, but a representation of the institutional constraints and points of view that mold the final project plans. This institutional perspective is a vital one, since virtually all of the major projects in the Neotropics are initiated and funded by organizations. An outside evaluation of the successes and failures of each project would be valuable; however, such evaluations are rarely done, and even more rarely published. Our conviction is that academics, students, and practitioners will all gain from an appreciation of the institutional perspectives that will continue to frame the evolution of much of the forested landscape of the Neotropics.

The first article in this section serves as a bridge to the previous section on resource use by folk societies. In it, James D. Nations presents a study of forest product use by peoples inhabiting the Petén region of Guatemala—an area that is to be called the Reserva de la Biósfera Maya. This reserve consists of a ring of national parks along Guatemala's border with Mexico and Belize and, in the center, a 5,500-square-kilometer multiple-use reserve—what conservationists are calling an "extractive reserve." From this extractive reserve the people of the region harvest forest products—chicle, *xate* palm, and allspice—for the international market. Nations discusses the biology, anthropology, and economics of these products, making the point that, combined, these products produce between US $4 million and US $7 million per year in export revenues for Guatemala without destroying the natural tropical forest cover.

Nations states that the harvest of *xate,* chicle, and allspice "promotes the conservation and sustained use of the Petén tropical forest" because they produce an income of many millions of dollars a year. This is an assumption made by many other writers who feel that by placing a monetary value on the forest, people will conserve it. Yet this is an assumption that in many cases has gone untested. Historically, when a particular natural resource becomes valuable on international markets, it is often exploited to commercial extinction. Hopefully, with the creation of Guatemala's Reserva de la Biósfera Maya, such destructive exploitation will be prevented. Some suggestions toward this end are advanced by Nations at the end of his paper.

The next three essays examine institutional efforts to develop new approaches to resource use in tropical forest settings. All three case studies are creative attempts to reconcile the needs of human populations with biological conservation. The first of these chapters, by William Burley, details the establishment of the Rio Bravo Conservation and Management Area in the portion of the Petén Region located in Belize. This innovative project integrates the efforts of the Belizean government, a multinational business, and an international nongovernmental organization, and is intended to be a self-sustaining conservation and economic development program with extraction of wood and nonwood forest products as important activities. The Rio Bravo Conservation and Management Area is intended to pay its way—with a subsidy from commercial sales being used for research and training as well as management.

Burley's essay provides an interesting contrast with the Nations article. Even though the forest is similar in both areas, the political and demographic conditions are very different in Belize and in Guatemala, resulting in different approaches to the shared end of sustainable resource harvesting and maintenance of tropical forest cover. In the Rio Bravo area, management plans are based on commercial timber extraction from largely unpopulated lands, while in the Reserva de la Biósfera Maya, local population levels are much higher and the resources to be exploited require high labor costs and extensive local participation.

The next contribution, by Liliana C. Campos Dudley, details another innovative program, this time in the Beni region of Bolivia. Here, Conservation International has begun an effort to involve all sectors of society, including indigenous groups, producer cooperatives, national governmental agencies, and international groups, in the implementation of a sustainable development project centered around the Beni Biosphere Reserve. From its original conception, the project was designed with the premise that long-term conservation in the tropics will only succeed when three conditions are met: local institutions take the lead; local knowledge is used and integrated; and social needs are fulfilled.

The Chimane Conservation Program that Campos Dudley describes has become very well known, due to the extremely complicated local, national, and international politics surrounding its creation and implementation. In her article Campos Dudley outlines the philosophy underlying Conservation International's creation of the program; the various steps taken in its creation; and some of the events that have shaped the current program.

In his paper, Flavio Coello Hinojosa outlines the mandate of the Cuyabeno Wildlife Production Reserve in the Ecuadorean Amazon. The management plan intends to improve the quality of life of the neighboring Indian and colonist populations and reduce the pressure on natural resources through development of tourism, wildlife management, harvesting of forest products, and sponsorship of scientific research. The Cuyabeno is one of the few conservation areas in Latin America that explicitly addresses the sustainable production of animal products for consumption by traditional peoples inhabiting the area, as well as for sale in outside markets.

Most of the ideas outlined in Coello's essay have not progressed past the early planning stages, yet the ambitious effort outlined here gives an excellent idea of the myriad ways in which natural resource harvesting can be incorporated with the needs of local people.

Burley, Campos Dudley, and Coello's case studies are all examples of efforts to successfully incorporate local peoples—indigenous and nonindigenous—into conservation areas, a goal shared with MAB-sponsored biosphere reserves. In the next paper in this section Dennis Glick and Michael Wright discuss the Wildlands and Human Needs Program run by the World Wildlife Fund and the U.S. Agency for International Development. This program supports more than twenty projects worldwide that link wildland protection with participatory rural development projects in areas adjacent to parks and reserves. Begun in 1985, this cooperative venture between a development-oriented agency and a nature conservation group expanded cooperation between these two groups. In their paper the authors outline the Latin American projects funded under the initiative and discuss the lessons learned so far.

As Glick and Wright point out, wildland management, or what has traditionally been called "conservation," has had to abandon its reliance on traditional methods derived from the natural sciences and begin to address such things as land titling, agricultural credit, artisan cooperatives, ecotourism, appropriate technologies, sustainable agriculture, and rural community leadership training—methods derived from rural development. The authors end by elucidating a concern too rarely expressed by representatives of conservation organizations: "Will our underlying goal of preservation of biological diversity be lost in the rush to embrace this currently popular approach to conservation?"

Glick and Wright suggest that the greatest successes of the Wildlands and Human Needs Program have been where an existing counterpart institution has been the implementing organization. They suggest that institution building is an important part of any conservation and development project. The final essay in part 3, "Building Institutions for Sustainable Development," by Marianne Schmink, picks up this theme and serves as a bridge to the next section by examining a new form of institutional cooperation. In 1986 the University of Florida and the University of Acre in Brazil began a technical cooperation program focused on forest extraction designed to strengthen the technical capacity of participating institutions through a research-and-training program based on the methodology of "adaptive research." Describing the process by which this program was initiated, Schmink provides a template for other innovative cooperative efforts that seek to develop institutional and scientific methods to address the needs of local peoples who use and manage forest products.

TEN

Xateros, Chicleros, and Pimenteros: Harvesting Renewable Tropical Forest Resources in the Guatemalan Petén

JAMES D. NATIONS

As part of the land bridge between North and South America, Guatemala is home to some of Latin America's most impressive tropical forests and enjoys great biological diversity (Nations et al. 1988). Guatemala's conservationists are aware of these facts and have dedicated themselves to conserving this rich natural heritage. During early January 1989 they achieved a major success when the Guatemalan National Congress created the Sistema Guatemalteco de Areas Protegidas (Congreso 1989). The law Congress passed gives legal protection to fifty new national parks, wildlife reserves, and multiple-use reserves, providing protected status to 14.5 percent of Guatemala's national territory.

Nonetheless, providing legal protection for wildlands is only the first of several crucial steps in their long-term conservation. The more difficult task is managing these areas, and conservationists in Guatemala are now turning their attention to that task.

The newly created parks and reserves cover the range of Guatemala's ecosystems—cloud forests, mangroves, wetlands, rivers, and lakes. But in extent of territory, the focus of the new system is Guatemala's tropical forest. The largest of the new protected areas lies in the northern third of Guatemala called the department of the Petén.

The Petén tropical forest was home to a civilization that flourished in the region for a thousand years. The Classic Maya (A.D. 250–A.D. 900) used the diversity of the Petén tropical forest to develop a trade network that reached throughout a major part of Mexico and Central America. They built huge cities that we know today by the names Mirador, Tikal, and Uaxactún. Ten centuries after the disintegration of Maya civilization, these cities have long since been reclaimed by the tropical forest that served as the subsistence base for several million people.

In recognition of the Petén's history and its current biological importance, Guatemalan conservationists have proposed almost the entire northern half of the department of the Petén as a biosphere reserve. The law that establishes this reserve was passed by the Guatemalan National Congress in 1989 (Ley 4–89). Once legalized, the reserve will be formally proposed to the UNESCO Man and the Biosphere headquarters in Paris.

The new reserve will be called the Reserva de la Biósfera Maya, the Maya Biosphere Reserve, and it will incorporate five of the new national parks and one multiple-use reserve created by the Guatemalan National Congress.

One interesting aspect of this reserve is that it reverses the usual biosphere reserve design, in which a core protection area is surrounded by concentric rings of increasingly intensive human use (Gregg and McGean 1985:42). In contrast to this, the Maya Biosphere Reserve will have a ring of national parks along Guatemala's border with Mexico and Belize, and in the center of these parks, the 5,500-square-kilometer Uaxactún-Carmelita Multiple-use Reserve. In addition to including half a million hectares of tropical forest and two communities of several hundred people, the Uaxactún-Carmelita Reserve also has a thriving export economy based on the extraction of renewable tropical forest resources.

In Guatemala, this multiple-use reserve is already being called an extractive reserve, following the model of the Brazilian rubber tappers and Brazil nut gatherers. This extractive reserve is the second important point about the Maya Biosphere Reserve. The Petén multiple-use reserve, or extractive reserve, focuses on three natural forest products: chicle gum from the tree *Manilkara zapota,* allspice from the *pimienta gorda* tree (*Pimenta dioica*), and *xate* palm (pronounced "sha te"), from several species of *Chamadorea.* All three of these products are renewable forest resources. Combined, the three produce between US $4 million and US $7 million per year in export revenues for Guatemala—all without destroying the natural tropical forest cover.

Xate Palms

More than 100 million stems of *xate* palm are exported from the Petén each year, destined for the United States, Switzerland, Germany, and the Netherlands for use in the floral industry (Heinzman and Reining 1988). Florists in the United States and Europe use the fronds of *xate* palm as a green backdrop for cut flowers. Usually, they spread the fronds out as a flat screen in flower arrangements, then lay cut flowers in front of them. The resulting flower arrangements—wreaths and sprays—are used most frequently in funerals and weddings.

Xate is the leaf of several species of *Chamadorea* palm that occur naturally in the lowland tropical forest of the Guatemalan Petén, as well as in neighboring Tabasco and Chiapas, Mexico, and in Belize. Although more than one hundred species of *Chamadorea* grow in Mexico and Central America (Standley and Steyermark 1958), only four are exported. Of these four species, two make up more than 95 percent of all exports. Known by the names *jade* palm and *hembra,* the two are respectively, a dark-green species with broad leaflets (*C. oblongata*) and a light-green species with thin, delicate leaflets (*C. elegans*).

Guatemala is the second major producer of *xate* in the world, just after Mexico. But Guatemala may soon be the primary exporter because lowland tropical forests will not last much longer in Mexico. Mexico's small remaining areas of *xate*-producing tropical forest are being replaced by farmland and cattle pasture at a rate far surpassing that in Guatemala.

The appeal of *xate* as a forest resource and as the basis for the Petén extractive reserve is the fact that the palms can be harvested every three months without damage. *Xate* harvesters, called *xateros,* search for the palms in the understory of the tropical forest and cut only one or two leaves from each plant. Using a pocket knife, they ideally take only mature green leaves. Experienced *xateros* are careful not to cut leaves that are spotted or torn and always leave the tender, new leaves so that the palm remains alive.

Three months after harvesting the usable leaves from a *xate* palm, the *xatero* can return to the same palm and harvest another one or two leaves. In such a fashion, *xateros* can harvest *xate* leaves from the same palm and from the same area of forest throughout the year. The fact that the tropical forest is not degraded by *xate* harvesting means that the process can continue indefinitely.

Guatemala began to export *xate* during the 1950s. Even today, though, many Guatemalans—including many *xate* harvesters—do not know what the plant is used for in the United States and Europe. Incorrectly confusing the use of *xate* with that of the *barbasco* vine (*Dioscorea* sp.), which comes from the same tropical forest, some xateros guess that the leaves are used to make contraceptive pills. And several *xateros* reported that the United States imports *xate* "so that hippies can smoke it."

On the other end of the industry, few citizens of the United States or Europe know what *xate* really is or where it comes from. In the United States, florists call the leaves "evergreen fern" and say simply that they buy it from brokers in Miami or New Orleans.

Xateros are organized into harvesting teams by contractors based in the town of Santa Elena, Petén. The contractors provide the *xateros* with rice, corn flour, sugar, coffee, powdered milk, and soups, then take them by truck to a tropical forest region where *xate* is known to grow. The *xateros* set up a camp and begin to work the forest. Every other day the contractor returns to the camp, bringing in more food and taking out the *xate* the workers have cut. The contractor then rushes the *xate* to processing warehouses in Santa Elena, Petén, or takes it to the Petén airport for shipment to processors in Guatemala City.

Xate harvesters are paid by the number of leaves they cut during their stays within the forest. During 1988, *xateros* were receiving $.20 per fifty leaves of *jade* or one hundred leaves of *hembra*. The *xateros'* goal in cutting *xate* is to get through the forest as quickly as possible, cutting as many leaves as they can, but leaving the palms alive for a subsequent work session in the area.

On an excellent day, a good *xatero* makes as much as $32. But, depending on the amount of *xate* growing in the area, he may also make only $16 or $12. Over the year, the average take is about $8 per day, an amount that is still much better than the $2 to $3 a day that laborers in the Petén make working on cattle ranches or harvesting corn.

The unsung heroes of the *xate* industry—and of the chicle and allspice industries as well—are mules. Mules are used to haul the cut *xate* out of the forest to the nearest road so it can be transported by truck to the warehouse. The obvious question is: What do mules live on inside a tropical forest? The answer is the leaves, fruit, and bark of the breadnut *ramón*, the *Brosimum alicastrum* of Maya subsistence fame (Peters and Pardo-Tejeda 1982). The mule drivers climb and coppice five *ramón* trees for each team of six mules every single day. The *ramón* tree later resprouts, and the mule driver can harvest the same tree four to five years later. Because up to 20 percent of the Petén forest is made up of *ramón* trees, there is little danger of exhausting this renewable resource by feeding mules (Wagner 1964).

After the *xate* is delivered to processing warehouses, workers grade it, sort it, and discard unusable leaves—those that are discolored, poorly formed, or too small for export. *Xate* buyers in Guatemala City report that up to 40 percent of the *xateros'* harvest is discarded at this stage, a clear indication that the harvesting process must be improved. The 60 percent of leaves that survive the selection process are wrapped in packages of thirty thousand leaves each and transported by refrigerated truck either to the Guatemala City airport or to the Guatemalan port city of Santo Tomás del Castillo, on the Caribbean.

Xate headed for European countries—all of it of the *hembra* type—flies out from the Guatemala City airport in cargo containers loaded on commercial airliners. The rest, almost all of which is the broader-leaved *jade* type, is packed into refrigerated containers and loaded onto cargo ships for export to Miami and New Orleans. There, U.S. brokers distribute it to florists throughout the country.

About 78 percent of Guatemala's *xate* is exported by air to European countries. The proceeds to Guatemalan exporters surpass $2.9 million per year. United States buyers purchase another $820,000 worth of leaves per year. Combined, sales of these two species of *xate* reach almost $4 million per year (Heinzman and Reining 1988).

Xate production also creates jobs in a country plagued by underemployment. At least six thousand professional *xateros* live in the Guatemalan Petén today, and another two hundred professional sorters work year round in processing warehouses in the Petén and Guatemala City (Sagastume 1987).

In addition, several thousand Petén farmers gather *xate* during a few months each year to earn additional income. University of Delaware anthropologist Norman Schwartz gathered information on farmers in communities of the central Petén during 1985 and found that at least half earned additional income harvesting *xate*. Schwartz reported that more than a quarter of household heads in some Petén communities were supporting themselves exclusively as *xate* gatherers (Letters, Fall 1988).

The key point in this discussion is that *xate* harvesting allows rural families to produce an export product and solid income from a renewable natural resource within the tropical forest. *Xate* production also aids the protection of biological diversity because *xateros* are economic allies in tropical forest conservation. They understand that their families' economic future grows in the understory of the tropical forest and have, therefore, become some of the forest's most vocal defenders. Finally, *xate* is an important export product for Guatemala, bringing in more foreign exchange in the long term than what the Petén would produce if cleared for farmland and cattle pasture (Rose 1988).

Allspice

Xate is only one of the many natural forest products the Petén produces. A second important product is allspice, the dried, unripe berries of the tropical forest tree known in Guatemala as *pimienta gorda* (*Pimenta dioica* or *Pimenta officinalis*). Allspice is one of the few important tropical spices indigenous to the Western Hemisphere (Rosengarten 1975:100). Exported to the United States and Europe, it is used in fruitcakes, pies, sausages, meat broths, gravies, and pickling liquids. It is one of the chief flavoring agents in pumpkin pie.

In Guatemala's Petén forest, allspice is gathered each year during June, July, and August by coppicing seed-bearing trees and boiling and drying the harvested seeds. Although the practice sounds destructive, allspice trees soon sprout new branches, and the tree can be harvested again after six years of regrowth. Enough trees exist within the Petén forest to support the export of almost one million pounds of allspice per year.

Guatemala's allspice exports are worth $16 million retail in the United States and Europe, although only part of this total accrues to Guatemalans. Guatemala currently has about 30 percent of the international allspice market, trailing Jamaica, which benefits from plantation production (Heinzman and Reining 1988). In the Petén, allspice harvesters receive $.52 per pound of dried berries. Almost as if by plan, the allspice season comes during the summer months, when *xate* is at its lowest seasonal demand.

Chicle Gum

Many Petén *xate* harvesters also work February to June collecting chicle gum, a latex harvested by cutting cross-sections in the outer bark of the chicle tree. This tree also grows wild in the Petén tropical forest. Properly cut, chicle trees are not harmed by harvesting. However, inexperienced harvesters may cut too deeply into the tree, penetrating its core and creating wounds that are exploited by insects, leading to the death of the tree.

Chicle is exported mostly to Europe and Japan for use in the chewing gum industry. Italy especially is currently undergoing a surge in the use of "natural chewing gum." The chicle industry in Guatemala had previously bottomed out due to industry substitution of *sorva* gum from Brazil. During 1985, Brazil exported almost $29 million of *sorva* gum from Manaus alone (Ghillean Prance, pers. comm., 1988).

Official statistics indicate that the value of exploiting chicle has reached as high as $1.4 million per year (INGUAT 1981), but official statistics on forest products from the Petén are notoriously underreported, due to attempts to prevent paying taxes on the extracted products. Chicle production for 1988 was estimated to reach only 300,000 pounds, indicating that the year's harvest was worth one-third of a million dollars.

Benefits of Extractive Reserves

The harvest of these three renewable resources—*xate,* allspice, and chicle—promotes the conservation and sustained use of the Petén tropical forest by producing a combined income of four to seven million dollars per year. Knowing that their economic future lies in the sustained use of these products, the families who harvest these resources are strong promoters of forest protection.

Natural resource economist Dietmar Rose, of the University of Minnesota, carried out cashflow analyses for three alternative land uses in the Guatemalan Petén using data from Heinzman and Reining

(1988). Comparing cattle ranching, colonist agriculture, and resource extraction, and using a planning horizon of twelve years with a real discount rate of 5 percent, Rose noted that use of the Petén for extraction "is strongly supported by the economic analyses" (Rose 1988:49). Cattle ranching and agriculture, Rose concluded, are nonsustainable land uses, with negative net present value and benefit-cost ratios.

Another advantage of the Petén extractive industry is that by protecting the forest for renewable resources, the families also are protecting the region's income from tourism. The Classic Maya civilization left behind ruins that are today one of the most important tourism destinations in Latin America. The Petén receives seventy five thousand foreign visitors each year, who leave behind $22 million (Nations et al. 1988).

They come to the region to see the rare combination of Mayan ruins and tropical forest. The Petén, along with Belize and a few small spots in Mexico, is one of the few places where tourists can look up from a thousand-year-old ruin and see spider monkeys scampering through the forest canopy. Protecting the Petén tropical forest for renewable forest products also protects the region for its most important income producer, the growing industry of tourism.

Extractive reserves also have benefits beyond simple economics. In addition to producing income for local communities and for their host-nations at large, they also help organize groups of families for the protection of tropical forest, and they effect the conservation of tropical forest areas for the future development of other, perhaps even more valuable, resources. As Hecht and Schwartzman have written, extractive reserves represent "an innovative alternative to livestock and colonist settlement" (1988:5).

Problems

Extractive reserves are not without problems. One problem is economic—namely, that markets change. Substitutes for products are discovered, much as tapping *sorva* trees in Brazil led to the temporary collapse of the chicle industry in Guatemala during the early 1980s. However, one advantage of extractive reserves is that they do not have to depend on only one or two products. Economic diversity can emerge from the biological diversity of the forest. Scores of commercial products exist in the tropical forest, and new products can be developed to add to the income generated by products already being exported.

Another problem is preventing overexploitation of the resources. *Xate* and allspice production in the Guatemalan Petén is already threatened by inexperienced harvesters entering the industry. Not knowing how to harvest *xate,* they cut all the leaves on a palm and end up killing the plant. Inexperienced or lazy allspice gatherers cut down the trees and harvest the seeds in a onetime rip-and-run operation. Halting this onetime harvest of what should be a renewable resource is one of the most serious challenges in the Petén.

One proposed answer to this problem has been government licensing, by which experienced *xateros* and allspice gatherers would train new harvesters in how to properly harvest the products. After completing a short course, individuals would receive a license that would allow them to harvest the products in designated areas.

Another problem with extraction in the Petén is that the harvesters are not yet organized into cooperatives as are the Brazilian rubber tappers. As a result, the individuals who are making the real money in the *xate,* allspice, and chicle industries are the contractors and exporters.

Yet another problem is how to guarantee families continued access to the resource during future years. How do we prevent the tropical forests that produce *xate,* chicle, and allspice from being destroyed for other uses? In Latin America, a forest that provides income to several thousand local families can quickly become the weed-choked pasture of a cattleman with political connections and a government subsidy. Loggers may gain rights to move onto supposedly "unused" forest land for their own benefit. And, finally, landless colonists from other regions of the country may move in and colonize the forest for farming. Having no experience in sustained-yield forest extraction, they end up destroying the economic future for three or four years of corn production.

One way to prevent this problem of future access to the resource is to create private land holdings dedicated to the extraction of natural resource products, with guarantees of protection from the state or federal government. This has partly been the case with the Brazilian rubber tappers (Hyman 1988). A second alternative is for central governments to declare forest areas extractive reserves but without providing titles to individual families. This is the route being taken by Guatemala. At the foundation of either choice, however, lies the need to convince central governments that forest land is worth more in its natural state than it would be cleared and transformed into cornfields and cattle pastures.

Not surprisingly, the move toward sustainable forest use is opposed by some loggers, cattlemen, and land speculators. During summer 1988, demonstrations by rubber tappers protesting illegal deforestation in Acre, Brazil, ended in the death of one rubber tapper and the wounding of two others. During December 1988 Chico Mendez, the leader of Brazil's rubber tapper movement, was murdered, probably by local cattlemen (RAN 1988; House 1988). The Rainforest Action Network called the violence "a clear attempt to stop the rubber tappers' resistance to deforestation and crush their fight for land rights" (RAN 1988).

What this violence points to is that simple logic about the economic superiority of extractive reserves is not enough. Cattlemen in the tropical forest are not thinking about the national economic good, or they would not be there in the first place. They are thinking about their own individual profits, and many of them do not seem to care if those profits come at the expense of tropical forest. Cattlemen are in clear competition with sustained-yield harvesting, and, together with colonist farmers, they constitute the most serious threat to extractive reserves anywhere in Latin America.

The problem is both economic and political. But by demonstrating the economic superiority of extractive reserves over short-term uses like farming and cattle ranching, we can wipe out the myth that cattle ranching and corn production in the tropical forest is good for the country.

As allies in this effort Guatemala and other Latin American countries have the support of thousands of rubber tappers, *xateros,* and harvesters of whatever forest product is involved. As these families demonstrate that the tropical forest is worth more to the people of the forest and to the nation at large than short-lived, onetime profits, they can gain the support of international agencies and their own national governments. One of conservationists' primary objectives should be to support these families scientifically, politically, and, where necessary, financially.

New problems and new products are sure to emerge as the idea of extractive reserves spreads to include tropical forests in more and more countries. But as this pattern of harvesting renewable natural resources expands, so also will its contribution to the wise use and protection of the world's tropical forests.

ACKNOWLEDGMENTS

The information reported here is the product of an interdisciplinary research team that includes Conrad Reining and Robert Heinzman of the Yale School of Forestry and Environmental Studies; Juan José Castillo of the Herbarium; Universidad de San Carlos de Guatemala; and Barbara Dugleby of Duke University, as well as the author. Special thanks go to Nancy A. Smith and Karen Buehler for logistical and photographic support in the field. Financial backing for the study came from a Fulbright Research Grant, the United States Agency for International Development/Guatemala, and the Plant Conservation Program of the World Wildlife Fund (U.S.).

LITERATURE CITED

Ante-Proyecto de Acuerdo Gubernativo de Declaratoria de Reserva de la Biósfera Maya. 1989. Guatemala: Consejo Nacional de Areas Protegidas.

Congreso de la República de Guatemala. 1989. Ley de Areas Protegidas. Ley Número 4–89. Guatemala City, Guatemala.

Gregg, W. P., Jr., and B. A. McGean. 1985. Biosphere reserves: Their history and their promise. *Orion Nature Quarterly* 4(3):40–51.

Hecht, S. B. and S. Schwartzman. 1988. The good, the bad, and the ugly: Extraction, colonist agriculture, and livestock in comparative economic perspective. Manuscript, Graduate School of Architecture and Urban Planning, University of California, Los Angeles.

Heinzman, R. and C. Reining. 1988. Sustained rural development: Extractive forest reserves in the northern Petén of Guatemala. USAID/Guatemala Contract 520–0000–0–00–8532–00. Guatemala City, Guatemala. Final report.

House, R. 1988. Leading Brazilian ecologists murdered at home in Amazon. *Washington Post* December 1988: 24: A10.

Hyman, R. 1988. Rise of the rubber tappers. *International Wildlife* 18(5):24–28.

Instituto Guatemalteco de Turismo (INGUAT). 1981. *Plan Maestro para la Ordenación Turística de Flores, Petén.* Guatemala City, Guatemala: Instituto Guatemalteco de Turismo.

Nations, J. D., B. Houseal, I. Ponciano, S. Billy, J. C. Godoy, F. Castro, G. Miller, D. Rose, M. R. Rosa, C. Azurdia. 1988. *Biodiversity in Guatemala: Biological Diversity and Tropical Forests Assessment.* Washington, D.C.: Center for International Development and Environment, World Resources Institute.

Peters, C. M. and E. Pardo-Tejeda. 1982. *Brosimum alicastrum* (Moraceae): Uses and potential in Mexico. *Economic Botany* 36(2):166–175.

Rainforest Action Network (RAN). 1988. *Rubber Tappers Shot, Killed, After Bid to Protect Threatened Rainforest.* Rainforest Action Network, Action Alert #28, August 1988. San Francisco, California: Rainforest Action Network.

Rose, D. 1988. Economic assessment of biodiversity and tropical forests. Background Paper for Guatemala Biodiversity and Tropical Forest Assessment Project. Washington, D.C.: Center for International Development and Environment, World Resources Institute.

Rosengarten, F., Jr. 1975. *The Book of Spices.* New York: Jove Books.

Sagastume Aldecoa, F. 1987. Diagnóstico de la comercialización de la palma silvestre "shate" (*Chamadorea* sp.) en los Municipios de Flores, Santa Elena, y San Benito del Departamento del Petén. Universidad de San Carlos de Guatemala, Santa Elena de la Cruz, Petén.

Standley, P. C. and J. A. Steyermark. 1958. *Flora of Guatemala.* Fieldiana: Botany 24, Part I. Chicago: Natural History Museum.

Wagner, P. L. 1964. Natural vegetation of Middle America. In R. C. West, ed., *Handbook of Middle American Indians,* vol. 1, 216–263. Norman, Okla.: University of Oklahoma Press.

ELEVEN

The Rio Bravo Conservation and Management Area of Belize

F. WILLIAM BURLEY

Aldous Huxley once ventured the opinion that if the world had any ends, British Honduras surely would be one of them (1934). Huxley was probably dealing unsuccessfully with the heat and mosquitoes of the Belize lowlands or the squalor of the back streets of Belize City. He can be forgiven for not seeing the potential of this little land earlier in this century. Were he to visit Belize today, however, he would probably share with many people the feeling that Belize is changing rapidly and that its enormous potential may yet be fulfilled.

The Rio Bravo area of northwestern Belize is part of the Petén region, a large biogeographically distinct area found in southeastern Mexico, northeastern Guatemala, and Belize.

Today, most of the Rio Bravo area is comparatively undisturbed by recent development. However, all parts of the entire area once were occupied by the Maya, and the vestiges of their cities and homesites are found throughout Rio Bravo.

In the last two hundred years, much of the forest has been selectively cut for mahogany, cedar, and other hardwoods. More than 700,000 acres of the greater Rio Bravo region were owned and managed for over a century by the Belize Estate and Produce Company, a London-based organization engaged in logging and trade in dry goods. The company greatly influenced the history and development of Belize. Until recently, it was the largest single landowner in Belize.

In 1988 the Programme for Belize, a nonprofit Belizean corporation, purchased 110,000 acres of this forest land in what has become known as the Rio Bravo project area. Under a formal agreement, this land will be turned over to the government of Belize no later than 1998. The Programme for Belize is managing the Rio Bravo site for the government as a major natural resource management and conservation area. In addition, 42,000 acres of contiguous land have been donated to the Programme by Coca-Cola Foods Corporation and are being managed along with the larger parcel. With the cooperation of other landowners nearby, a total forest area exceeding 500,000 acres is being developed as an innovative regional land-use planning and conservation effort. Resource surveys needed for land-use planning are under way in Rio Bravo, with international assistance from Great Britain and other sources.

The Rio Bravo Area

Geology

Rio Bravo lies in the southeast corner of the Yucatán Peninsula. The Yucatán Platform, which during the early Eocene (55 million years ago) was submerged beneath the sea, underlies the peninsula. Marine sediments deposited here consolidated into limestone. Topography, soils, and hydrology at Rio Bravo are similar to adjacent areas in northern Belize, Mexico and Guatemala; all are located on the same parent Eocene rock formations.

Northern Belize emerged from the sea much later, in the Pliocene (2 to 13 million years ago), and the coastline of Belize was west of its current position, near present-day Rio Bravo. A series of terraces developed over geological time, resulting in several distinct escarpments which run northeast-southwest through Rio Bravo. These escarpments break up Rio Bravo's generally flat or rolling terrain and are easily recognized on the ground and in the air. Often a sharp break in vegetation type occurs, resulting in areas (ecotones) of great interest to ecologists and other scientists.

Vegetation

The forests at Rio Bravo are interesting for many reasons, including their great species diversity, variation in canopy and structure of the vegetation, and very noticeable transitions from one vegetation type to another. The dominant vegetation formations are subtropical broadleaf forests on limestone substrates. Within the Rio Bravo area, more than a dozen floristically distinct vegetation types can be seen, although at present five general types are defined and being studied for management purposes: upland broadleaf forest, swamp forest, palm forest, savanna, and marsh (Brokaw and Mallory 1988).

Two-thirds of the Rio Bravo area is in upland broadleaf forest. This vegetation "type" in itself varies greatly in canopy height, structure, and species composition. For years it has been an important source of mahogany, cedar, and a dozen other commercially exploited hardwoods. It is safe to assume that virtually all of Rio Bravo has been cleared (although not all areas at once) either during the period of Maya occupation or, more recently, under the tenure of the Belize Estate and Produce Company.

Today, however, the Rio Bravo forests are in various stages of development, with large areas apparently ecologically mature despite natural disasters, such as hurricanes and human-caused disturbance such as selective removal of several timber species. There is no clear-cutting for timber in Belize; logging is highly selective, and most timber species are widely scattered in the forest.

Wildlife

Wildlife seems abundant in Rio Bravo, due in part to the remoteness of the region and to the fact that hunting has been discouraged actively or eliminated in recent years. Five cat species occur here: jaguar, puma, jaguarundi, ocelot, and margay, and all seem to have healthy populations. Other mammals seen frequently include tapir, tayra, anteaters, peccaries, the Mexican porcupine, and the gray fox. More than 80 species of bats make up 60 percent of the mammal fauna.

Ornithologists calculate that up to 367 species of birds could occur in the Rio Bravo lands, and more than 200 species have been recorded in preliminary surveys. Some game birds seem very prevalent, including ocellated turkey, great curassow, and crested guan. Raptors appear to be numerous also.

Distinct wet and dry seasons in Belize affect wildlife, and in Rio Bravo there is a noticeable increase in activity at the beginning of the rainy season, usually in June. Reptiles and amphibians are much more abundant and easily found, and the activity of birds and mammals also increases appreciably. The dry season is a time of food scarcity and noticeable stress and mortality, and it is well known that animals resort to a few primary plant resources that are dependable food sources when others are not producing. It is imperative, therefore, that logging and other exploitative activities at Rio Bravo take these factors into account as much as possible, to avoid ecological disruptions that are not easily detected in the early stages of exploitation.

Present Land Use

Most of the Rio Bravo area today is dense forest that is relatively little used and not being heavily disturbed by logging or other activities. A multiproduct agricultural enterprise is getting under way at Gallon Jug, a few miles south of the Programme for Belize lands, but this development occupies fewer than two thousand acres, and it is not expected that much additional land will be cleared. Timber extraction (not sustainable forest management) is occurring on lands adjacent to Rio Bravo in all directions, and there are good opportunities to improve the logging methods in these areas. To the north and east, communities of Mennonites continue to clear land for agriculture and to log forests within their land grants, using logging practices that are not particularly sensitive to long-term forest management needs. A small amount of *milpa* farming (shifting or slash-and-burn agriculture) occurs on the northern edge of Rio Bravo, as described below. In nearby areas, numerous small-scale activities could, with care and planning, be expanded into Rio Bravo. These activities include beekeeping, pine and other softwood tree plantations, various agricultural projects, and freshwater shrimp farming.

Management of Rio Bravo

The Rio Bravo project is intended to be a self-sustaining conservation and economic-development project. It is a large area of tropical forest that has significant and widely recognized economic potential—if the resource base is managed carefully. From the beginning of the project, the Programme for Belize has intended that Rio Bravo "pay its way" by exploiting forest resources sustainably. Further, a portion of the profits from timber exploitation and other enterprises is intended to cover most of the costs of nonrevenue-generating activities, including management planning, some research, training, and environmental education projects supported by the Programme.

Several management objectives for the Rio Bravo area have been defined. They include, but are not necessarily limited to, the following:

1. Maintenance of small-scale subsistence agriculture and experimental programs to stabilize several existing farms and to yield methods and results that could be useful elsewhere in Belize and Central America.
2. Natural forest management, primarily for mahogany, cedar, and several other hardwoods, with timber extraction tailored to specific sites, not exceeding mean annual increment of wood volume, and done by methods that minimize ecological disturbance and disruption of wildlife.
3. Extraction of nonwood products, including edible fruits and chicle (for chewing gum) which is a long-standing industry in the region.
4. Conservation of biological diversity and management of wildlife, through zoning, designation of strictly protected areas, and careful monitoring of animal populations.
5. Ecological research, biological inventories, and long-term ecological monitoring, centered around a major biological field station and attracting scientists from many disciplines and countries.
6. Archaeological investigations of the region, especially the Las Milpas site, a major classical Maya site, but also including extensive archaeological surveys of the greater Rio Bravo region.
7. Natural history tourism development, including small-scale jungle lodges, camping areas, and other facilities for visitors interested in wildlife and natural history.

Management planning is in the early stages at Rio Bravo. The first general management plan necessarily will have to be incomplete and flexible enough to accommodate major modifications as inventory information becomes available and the area becomes better known. For example, small-scale, low-impact logging is under way already to provide lumber for construction of the research station and modest visitor facilities, but also to demonstrate that the area truly is planned for production and economic benefits in addition to conservation to wildlife. Larger-scale logging and timber extraction in some areas of Rio Bravo, however, will have to await sufficient forest inventory data and development of a complete forest harvesting plan.

At present, planning for Rio Bravo is being done by the staff of the Programme for Belize, in collaboration with the Belize government and with the assistance of various professionals under contract. The management plan is expected to be revised periodically, and the Programme's board of directors has final approval over this process. An important component of Rio Bravo management is the plan for harvesting hardwoods on a sustained-yield basis. Timber harvesting and most aspects of forest management likely will become the responsibility of the managers of a for-profit subsidiary of the Programme for Belize.

Under the present agreement with the government, the Programme for Belize is acquiring and managing Rio Bravo, and various facilities will be constructed. These include a research station adequate for a full range of biological inventory work and ecological research. Permanent plots and transects already are being established throughout Rio Bravo, and a long-term research program is being designed.

The archaeological site of Las Milpas is located in the northern part of Rio Bravo. It is a large, unexcavated Mayan site which has enormous potential as a tourist attraction. Two archaeological teams have prepared a preliminary map of the site, but a decision whether to proceed with major excavation has not yet been made.

Tourism facilities are being planned that will accommodate as many as fifty visitors interested in natural history or in assisting the research staff at the field station. These facilities will be part of a larger ecotourism network being developed also by a for-profit subsidiary of the Programme. The concept of La Ruta Maya has been helpful in envisioning a tourism network involving hotels or lodges in several countries of the region, all linked by an appropriate transportation network.

Development and Management Constraints

Beyond the usual constraints of adequate staff and available funding for Rio Bravo, several conditions or situations exist or can be expected to occur that must be addressed. Illegal occupation of the site along the northern boundary currently involves six families and a dozen single men. Collectively, they have cleared about 150 acres in the last four years. Attempts have been made to stabilize these *milpas* and have met with only limited success. The government has agreed to offer titled land parcels of five to ten acres each to the *milperos* so that they may move out of Rio Bravo onto Government land nearby.

It is possible that other landless farmers will attempt to move into Rio Bravo, either for subsistence farming or for growing the most profitable cash crop, marijuana. Traffic in drugs and marijuana production are a constant and possibly ineradicable problem in Belize. Given the large size of Rio Bravo and its lack of roads, it is nearly impossible to detect small patches of marijuana located deep in the forest. Ultimately, the only way to minimize this problem, aside from eliminating demand, is to establish a greater "presence" for the Programme in the area with tourism, research, logging, and monitoring.

Timber theft, always a serious issue in accessible tropical forests, has occurred to a minor extent at Rio Bravo recently. Theft will continue to be a management problem until the greater presence mentioned above is established and until all boundaries are properly surveyed, marked, and monitored. To minimize this and similar problems, Programme staff are working with the local communities to explain the objectives of the Rio Bravo project and to elicit support for making Rio Bravo work as a regional development project. For example, logging in Rio Bravo is being done by local men who have spent most of their lives in the bush, either logging or tapping sapodilla trees for chicle (the chewing gum precursor).

A significant constraint to management and planning for Rio Bravo is the persistent image that these resources are being "locked up" and therefore are not available for exploitation or use by Belizeans. This misconception can be found at all societal levels in Belize, from local farmers to top government officials. A solution is better and more active presentation of the project's goals to the public. Even more effective will be actual demonstrations of use and exploitation of the area's resources, producing measurable employment, revenues, and profits. The fact that significant employment of local people now is possible at Rio Bravo also is helping to change the unfortunate perception by some that Rio Bravo is ultimately a playground for wealthy foreigners.

Finally, the most serious management constraints may be the sociopolitical realities of the region. These include the major, unpredictable factors of population growth, influx of refugees, political unrest, and the massive infusion of foreign investment, along with payoff and graft and traffic in drugs. The potential effects of these factors on Rio Bravo are impossible to determine at this early stage of development. The most effective way to guard against them is to develop a large enough Belizean constituency for Rio Bravo: Belizeans and their government officials must have a great enough vested interest in Rio Bravo's success to support and ensure its development and survival. The more integral a part of regional development planning in Belize Rio Bravo can become, and the sooner this can happen, the more likely its management goal and objectives will be met.

LITERATURE CITED

Brokaw, N. V. L. and E. P. Mallory. 1988. Natural history of the Rio Bravo Resource Management and Conservation Area. Report to the Programme for Belize, November 7. First draft.

Huxley, A. 1934. *Beyond the Mexique Bay.* New York: Harper and Bros.

R. Nicolait and Associates, Ltd. 1984. *Belize: Country Environmental Profile.* Belize City, Belize: Robert Nicolait and Associates.

Wright, A. C. S., D. H. Romney, R. H. Arbuckle and V. E. Vial. 1959. *Land in British Honduras.* Colonial Research Publication Number 24. London: Her Majesty's Stationery Office.

TWELVE

The Chimane Conservation Program in Beni, Bolivia: An Effort in Local Participation

LILIANA C. CAMPOS DUDLEY

The strategies that international conservation organizations have used to encourage conservation in the Neotropics and around the world have evolved through time. These strategies have been shaped by a history of change in the perception of complex biological, economic, and political facts.

Conservation programs initially designed to benefit a single species evolved into programs aimed at protecting the ecosystems in which these species lived. Gradually, humans and their needs were incorporated into such programs. Eventually, international conservation organizations came to question their own participation in the internal conservation scene of individual countries: Weren't they foreigners in a foreign land? Shouldn't their role be to encourage local residents to conserve their own resources? In response to these concerns, for many years some conservation organizations have been actively engaged in building institutions within tropical countries. As a result, many Latin American nongovernmental organizations (NGOs) have been created and nurtured. In addition to the standard conservation tools, increasingly conservation organizations have incorporated strategies used by rural development groups, in some cases actively supporting grassroots organizations themselves.

The most successful of these national organizations are now playing an important role as "national brokers," efficiently lobbying for policy adjustments within their own countries and directly tapping international sources of funding and technical support.

Despite the enormous (some say modest) amounts of time and money spent in conservation projects in tropical America during the last few decades, destruction of Neotropical landscapes is proceeding at an accelerating rate. New conservation modes need to be developed. For instance, many of the conservation projects supported though the new Latin American NGOs responded to priorities set by funding agencies. In some cases, the projects were completely divorced from the local peoples and their governments. Such projects were usually better known and appreciated outside their countries than within. There was a basic contradiction between the philosophy used to create them and the methodology used to implement them; the fate of such projects was to wither on the vine as soon as external funding was halted.

What is now needed is a new way of looking at conservation programs, not as isolated investments, but as interdependent programs that involve all sectors of society and numerous aspects of the biological and social sciences. The Regional Ecosystem Conservation Program, a key strategy for cooperative international projects developed by Conservation International (CI), a United States–based organization, constitutes just such an approach—the focal concepts of which are multisectorial participation and effective flow of information across disciplines.

Philosophical Framework

The Chimane Conservation Program was founded on the basic premise that conservation should go hand in hand with development. Important consideration was given to the belief that long-term conservation in the tropics will only succeed when three conditions are met: local institutions take the lead, local knowledge is effectively used and integrated, and social needs are fulfilled. In this scenario, foreign conservation organizations play important roles as purveyors of technical assistance and catalytic financial support.

Cornerstones of the Chimane Ecosystem Conservation Program are:

1. Program sustainability: Key actions are being taken to effectively

tie the project to regional, national, and international support systems in order to secure the program's continuity. Important components of these support systems are pertinent governmental agencies, local federations, production sectors, national NGOs, and academic institutions.

2. Program self-determination: In order to translate rhetoric into action, decisions on the program's priorities for implementation are being made by a consortium of organizations representing legitimate decision makers in the program area. This requires a major shift in attitude on the part of both governmental and international agencies. Decision-making capacity brings along local commitment, and the program becomes less dependent upon outsiders.

3. Information systems integration: Data concerning the resource base (biodiversity and nonrenewable resources), resource use, and human populations and their impacts on the resources (and vice versa) need to be efficiently integrated. The type of knowledge and research required to make development successful while maintaining its ecological soundness—in other words, to attain sustainable development—should respond to local ecological and cultural conditions. This implies considerable investment in local research and specialized information management to allow for interdisciplinary exchange of information.

4. Process replication: The Chimane program is providing on-the-job training for an array of Bolivian professionals and institutions, who will then replicate what they learned in other parts of Bolivia.

Such an approach requires not only creativity and expertise, but a great deal of specific information concerning the local environment, including local customs, local patterns of political influence, and much more of what Dahlberg (1977) called "contextual" thinking: thinking that relates to specific groups in their natural and historical environments, thinking that includes complexities feared by international organizations.

The Region

Beni Province, set in the eastern lowlands of Bolivia, is part of the Amazon basin and contrasts sharply with the highland plateaus and subtropical valleys to the west. Second largest of Bolivia's nine provinces, the Beni has an area of 132,708 square miles, about 20 percent of the national territory. The region is sparsely populated, with an estimated population density of one inhabitant per square kilometer. The Beni is a complex blend of eastern Andean slopes, seasonally flooded savanna, forest, lakes, and rivers. Isolated from Bolivia's population centers by a lack of roads, the area is biologically rich and is devoid of parks or protected areas.

William Denevan, a geographer who studied the remnants of preconquest aboriginal civilizations, speculates that as many as five hundred thousand people may have inhabited this area before its discovery by Spanish explorers (Denevan 1963; Denevan 1966). Perhaps 100,000 linear drained fields, numerous canals, and large and small mounded fields are still seen in the Beni savannas, where there is seemingly little opportunity for agriculture because of seasonal flooding.

The Program

The Chimane Conservation Program in the Beni is an example of regional planning for conservation and development. It is a product of the long-term commitment to conservation by a group of Bolivian institutions, as well as by professionals now with Conservation International.

During the initial four years of intense conservation work, a pilot project was designed and four lines of action were emphasized, with the main goal being to increase conservation capacity within Bolivia:

1. Institution building and development;
2. Knowledge-base expansion and information management;
3. Resource protection and management; and
4. Environmental education and training.

The program was designed to grow from the ground up within a flexible technical framework allowing for incorporation of local needs and demands.

Activities during the years that followed have integrated the Beni Biosphere Reserve into its regional context; they have effectively influenced land use planning in the areas surrounding the reserve and the larger conservation movement in Bolivia as well.

The Beni Biosphere Reserve

The Beni Biosphere Reserve (BBR) was chosen as the initial geographic focus for conservation efforts. The BBR, home to many Chimane Indians, was initially named the Beni Biological Station (EBB) when it was created in 1982 under the supervision of the Bolivian Academy of Sciences. The reserve (135,000 hectares) is an area of diverse forest formations bounded on the northeast and west by vast expanses of open, seasonally flooded savanna grasslands. The region supports thirteen of Bolivia's eighteen endangered tropical animal species, and a preliminary survey identified more than five hundred species of birds, eight species of primates, five species of cats, and forty species of bats.

The EBB was recognized by UNESCO's Man and the Biosphere Program as a biosphere reserve in 1986. Biosphere reserves are protected areas representative of a particular ecosystem that have multiple functions: research, monitoring, conservation, education, and training. Most importantly, biosphere reserves consider human populations an integral part of the ecosystem to be protected. Each biosphere reserve functions on the basis of a detailed management plan, which establishes guidelines for achieving the goals of the reserve.

For five years, an interinstitutional team consisting of members of the Bolivian Academy of Sciences, the Institute of Ecology (IE) in the San Andrés National University, the Bolivian Conservation Data Center (CDC), the Beni Interdisciplinary Center for Development (CIDE-BENI), the Bolivian Ministry of Agriculture and Peasant Affairs, and professionals in Conservation International (CI) worked to achieve three main objectives: to begin building the Beni Biosphere Reserve's conservation capacity at all levels, to formulate the reserve's management plan, and to efficiently integrate the BBR into its regional context.

These objectives are being accomplished. Following BBR's mandate, a sequence of technical tasks was implemented. An exhaustive biological inventory of the area was initiated by young Bolivian biologists through a small grants program. Base maps and preliminary vegetation maps of the reserve were developed, and initial human population surveys were carried out. Additional assistance was provided by the Bolivian Museum of Ethnography and Folklore, the Smithsonian Institution's BIOLAT Program, and the University of

Florida. The data gathered in these studies were analyzed and integrated in the preliminary zoning of the reserve.

During a period of approximately eighteen months, a multidisciplinary team worked on the reserve's preliminary management plan, which was reviewed by an international advisory committee and completed in 1989. This plan established guidelines for short-, medium-, and long-term development of the reserve in the areas of resource management, resource protection, research, human needs, and public use (education and conservation awareness), as well as the supervision and implementation of an efficient administrative infrastructure.

The BBR's Environmental Education Program began in 1986. Its first phase determined educational channels that could be used to influence resource use among local people. In its second phase, it is focusing on the two most important population centers near the reserve, El Totaizal and San Borja, and working to raise awareness in these areas through formal and informal education systems. Later phases will focus on public campaigns for environmental awareness.

A biosphere reserve differs from other categories of protected areas in that it includes human components within the reserve and allows for their full participation in the planning and management of the reserve. The BBR's present challenge is the integration of the needs of its own inhabitants, Chimane Indians and mestizos, into the reserve management guidelines and services. Most important, the BBR needs to create appropriate forums and decision-making levels where scientists and local inhabitants can exchange information and make joint decisions.

Institutionally, the BBR has grown and currently is run by a well-trained staff. A three-person office in La Paz allows for international and national networking. At El Porvenir ranch, the Biological Station's headquarters, reside a scientific coordinator, an administrator, and park guards. The accommodations in El Porvenir are also used by full-time researchers, project personnel, and visitors.

The Bolivian government has great interest in the reserve as a baseline for development and colonization and the Regional Forest Service (CDF-RN) has appropriated 7 percent of timber-production taxes directly for conservation work in the BBR. The progress of the activities in the BBR have been presented and cited in many Latin American and international forums and exchange of information with other programs and biosphere reserves around the world is taking place.

The Chimane Forest Area

Regional Ecosystem Conservation Programs have as a cornerstone the integration of protected areas such as the BBR into their regional contexts, serving local concerns through education and research. During the latter part of 1986 an event further challenged the BBR's capacity to act and influence regional policies.

In November 1986 the Forest Reserve surrounding the BBR changed status: it was opened to immediate extraction of timber. This change of status was said to be part of the regional government's effort to monitor and tax timber extraction within the area, which was said to have been going on informally for a number of years, and it was intimately tied to local pressures and governmental operational budgets.

The area in question was broadly known as the Chimane Forest, traditionally inhabited by four ethnic groups known as the Chimane, the Moxeño, the Yuracaré, and the Movima, all of them at different levels of cultural assimilation. The area is also inhabited by a comparatively large population of mestizos, comprising long-term cattle ranchers, newly arrived colonists from the Andes, and merchants.

By February 1987, the Chimane Forest had become a mosaic of restricted-use sectors, determined during a meeting of the Comité Cívico del Beni. These sectors were initially determined based on policies already in place (i.e., regional government's decision to grant timber concessions in a sector of the forest) and land-use potential. Population distribution within these areas and provincial boundaries were not taken into account in this initial land-use zoning.

The new zoning (see figure 12.1) consisted of 1) the Chimane Permanent Production Forest (590,000 hectares), where mahogany (*Swietenia macrophylla*) and cedar (*Cedrela lilloi*) were being extracted selectively; 2) the Yacuma Regional park (130,000 hectares), an area of savanna where cattle ranching is prevalent; 3) an area set aside for watershed protection (225,000 hectares) with virtually untouched and unknown forests, and 4) the Beni Biosphere Reserve (135,000 hectares).

During 1987 and 1988 the people of the Beni showed strong politi-

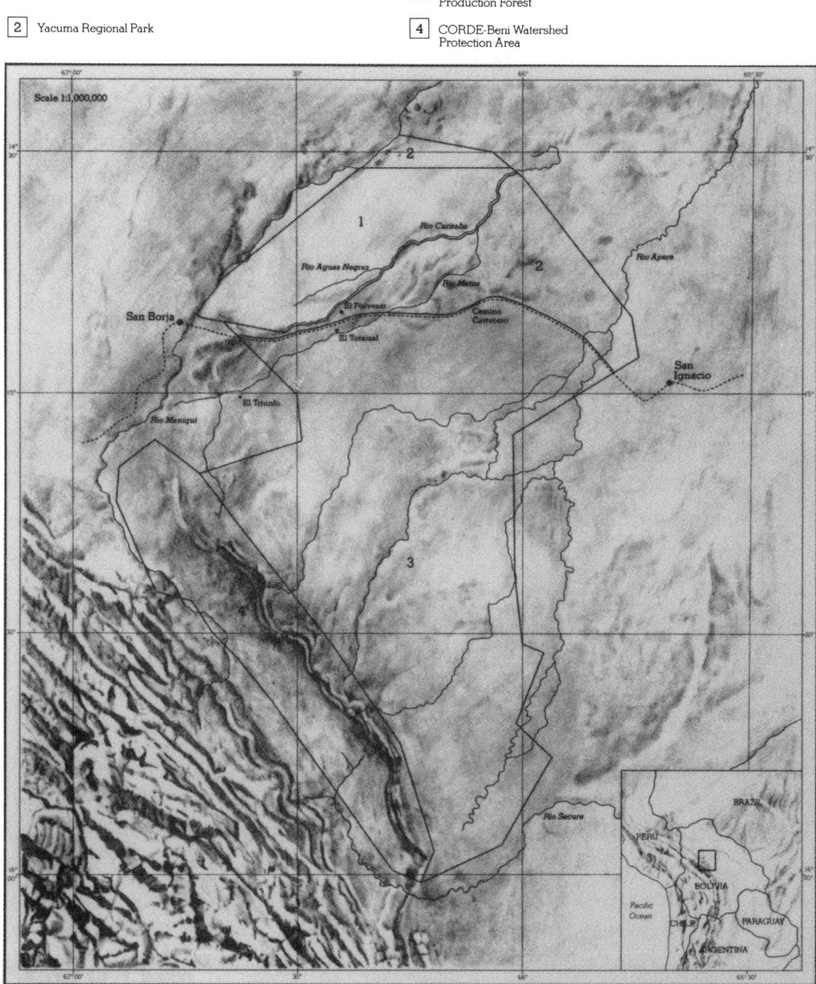

Figure 12.1

cal desire for the conservation of their natural resources. The Beni Regional Forest Service (CDF-RN), the principal government agency in charge of the administration of natural resources, invited the Beni Biosphere Reserve and Conservation International to assist them in their effort to create a sound forestry policy for the exploitation of the Chimane Permanent Forest Reserve. The three institutions then became the center of an effort aimed at the development of a management plan and the necessary institutional structure for the sustainable use of the resources within the 1,000,000 hectares surrounding the Beni Biosphere Reserve.

To increase local commitment and available resources, Conservation International signed a collaboration agreement with the Bolivian government—an agreement that included the first debt-for-nature swap in history—and an integrated and multisectorial regional environmental management effort began.

The opportunity to protect a major regional nature reserve, at the same time initiating serious environmental planning in Bolivia, brought together a strong institutional consortium incorporating local, regional, and national, as well as international organizations. Such organizations were part of the critical support system that would ensure the continuity of the program.

Planning was initiated through an Interinstitutional Commission (CTI) supported by the CDF-RN, Conservation International, and the local Association of Timber Companies. Serving as an advisory committee to the CDF-RN, the CTI's main task was to implement in the Chimane Forest a regional example of development based on conservation principles. This example would be used in the elaboration of the regional conservation strategy for Beni Province. The commission included representatives from the CDF-RN, Conservation International, the BBR, the Bolivian Institute of Ecology, the National Conservation League (LIDEMA), and the local Association of Timber Companies. All of these organizations volunteered their services to assist in a regional planning effort led by the regional government.

The participants in the CTI strongly believed that the participation of the peoples of the Beni in the decision-making process regarding the Chimane Forest was to be encouraged by creating open forums of discussion where minority groups' concerns would be heard. Thus CTI sponsored a series of regional forums, public decision-making seminars, and meetings.

Institution Building and Capacity Development

Activities undertaken by the CTI had a two-pronged approach. First, with the assistance of Bolivian and international expertise, a multiproject integrated program (the Chimane Conservation Program) was developed. The program responded to initial population and resources overviews (*diagnósticos*) and disciplinary concerns.

Second, the CTI drafted an action plan to stimulate the institutional development of the CDF-RN. The technical and administrative capacity of the CDF-RN to manage, regulate, and monitor the sustained use of the Beni's forest resources would be strengthened by incorporating operation policies, personnel training, and program planning.

To comply with the first task, guidelines for basic biological and social research were drafted and urgent projects implemented by participating institutions. Among these projects were an action plan for immediate attention to the Chimane Permanent Production Forest (CPPF); a preliminary management plan for the sustainable use of forest resources in the CPPF; a biological overview of the CPPF; a vegetation map of the CPPF; and a socioeconomic overview of the CPPF.

The CPPF socioeconomic analysis identified human settlements, colonization trends, and present and future resource use. It also provided solid data that would allow inclusion of the interests of local peoples in the planning process and encourage long-term participatory management of the Chimane Forest. Ultimately, it became a vital instrument in encouraging historical changes in the Chimane Forest—changes concerning indigenous territories, which will have an important impact on the future of indigenous peoples of eastern Bolivia.

By the end of 1988, the basic organizational and physical infrastructure for an integrated program was ready and the Interinstitutional Commission, or CTI, was dissolved. The Chimane Conservation Program was officially created. The program, overseen by a multisectorial board of directors, consisted of four subprograms: forestry, biology, socioeconomic affairs, and training. The proposed board of directors initially included all institutional members of the CTI, as well as representatives of pertinent governmental organizations, production groups, and social sectors (indigenous peoples, town committees, provincial authorities).

Strengthening local, regional, and national support systems (institutions) that will not only continue to oversee the projects, but will replicate functional examples within the country, was also considered to be of great importance. Conservation International has ongoing cooperative projects with six Bolivian institutions, including all the members of the Interinstitutional Commission.

Finally, training and public awareness is a vital component of the Chimane Conservation Program. Sustainable production in the Chimane Forest demands diversification, long-term planning, and a radical change in methodologies, strategies, and attitudes toward the use of the forest and its resources. It is therefore essential to design and implement ad-hoc programs to train technical personnel within governmental agencies and pertinent institutions, as well as to motivate the private sector and to gain the support of local inhabitants. Training is perhaps most critical in forestry, since forest products will continue to be a major part of the region's economic base.

Resource Protection and Management

The BBR Management Plan was the first management tool in the area. In the Chimane Forest, urgency moved the forestry subprogram ahead. The second major area undergoing long-term management planning is the Chimane Permanent Production Forest (CPPF)—590,000 hectares of forest initially zoned for exploitation by seven timber companies. Strong local support for a policy of sustainable forest management exists in the Beni, but up to the present management resources have been absent and timber harvesting has been unplanned and unregulated. The process for granting long-term timber concessions in the area has begun; meanwhile, companies continue to receive annual permits to extract timber in their areas.

Technical assessment of the CPPF was initiated by a team of foresters that included Dr. Gerardo Budowski, from the Universidad para la Paz in Costa Rica; Ing. Leonidas Vega, FAO and USAID consultant based in Surinam; and Ing. Luis Goitia, a Bolivian forester. Subsequently, a Preliminary Forestry Management Plan for the CPPF was developed by a team of Bolivian foresters and approved by the CDF-RN. Finally, the Chimane forestry subprogram developed a project titled "Conservation, Management, Utilization and Integrated and Sustained Use of the Forest in the Chimane Region, Beni Department, Bolivia," which through the Bolivian government was granted three years' support by the International Timber Trade Organization (ITTO).

Information Management

The Chimane Ecosystem Conservation Program must become a demonstration area for the sustainable use of resources. Therefore, studies were initiated to collect the baseline information needed to identify resource utilization criteria.

The Bolivian Conservation Data Center (CDC), assisted by Conservation International and in conjunction with the Beni Biosphere Reserve, the Bolivian Academy of Sciences, and the Institute of Ecology at the National University, will complete a "Bio-Resources Overview of the Chimane Area." The overview assesses existing biological information for the region and puts this information in the context of current land use and other social pressures to determine conservation priorities and gaps in knowledge. The final document will be used by the Chimane Ecosystem Conservation Program board of directors as a planning tool. The program needs the support of a local, responsive, and comprehensive information center. The flow of information within and among the subprograms is of vital importance.

The Challenge: Participatory Planning and Management

The concept of a Regional Ecosystem Conservation Program is a viable alternative for developing countries, where the pressure to increase economic productivity and fulfill human needs competes with the necessity to set aside large tracts of land for strict protection. The immediate success of the Chimane Ecosystem Conservation Program is that it has bought ten years for the resources and the peoples of the area. To secure effective participation and integration of all sectors involved is still the greatest challenge.

The most recent development in the program is that the indigenous peoples of the area have requested the central government of Bolivia to declare part of the Chimane Permanent Production Forest an "indigenous territory." An ad-hoc commission has been created to respond to the request. Can traditional land-use practices be successfully integrated within the management plans for the region? Can the active participation of indigenous peoples in the planning process be achieved? Can imminent social change effectively incorporate technical recommendations? The institutions involved in the Chimane Ecosystem Conservation Program strongly believe so, and the real job of making the program work has just begun.

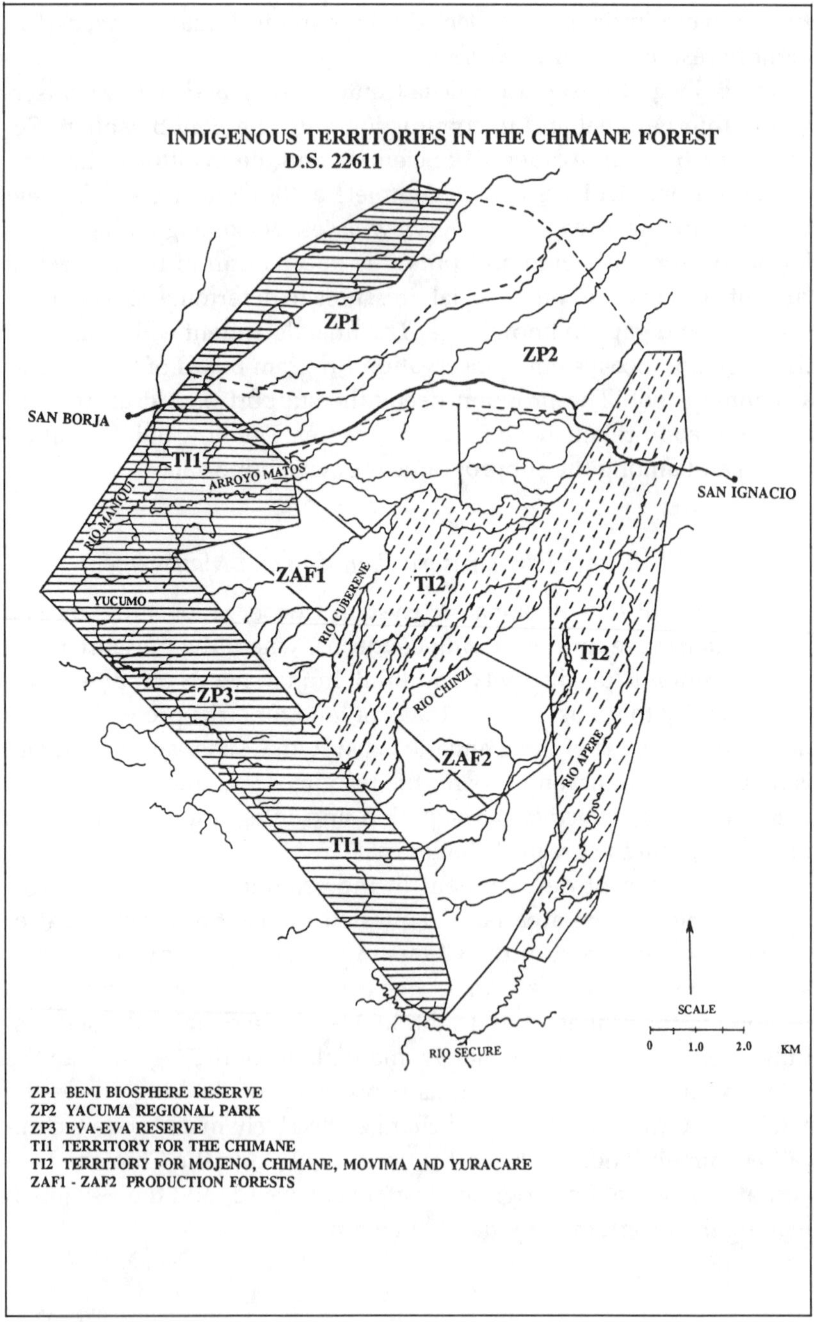
INDIGENOUS TERRITORIES IN THE CHIMANE FOREST
D.S. 22611
ZP1
ZP2
SAN BORJA
TI1
ARROYO MATOS
RIO MANIQUI
SAN IGNACIO
ZAF1
RIO CUBERENE
TI2
YUCUMO
TI2
RIO CHINZI
ZP3
ZAF2
RIO APERE
TI1
SCALE
0 1.0 2.0 KM
RIO SECURE
ZP1 BENI BIOSPHERE RESERVE
ZP2 YACUMA REGIONAL PARK
ZP3 EVA-EVA RESERVE
TI1 TERRITORY FOR THE CHIMANE
TI2 TERRITORY FOR MOJENO, CHIMANE, MOVIMA AND YURACARE
ZAF1 - ZAF2 PRODUCTION FORESTS

Figure 12.2

Postscript

Following the presentation of this paper at the 1989 Man and Biosphere Workshop, vital changes have taken place in the Chimane Forest. The paper has not been completely updated, since it was intended to present the program's initial philosophy and project development. However, a more complete chronology of events (Appendix 12.1) and a recent map(figure 12.2) have been included.

Historical events such as the March for Dignity and Territory have encouraged changes in Bolivia's five-hundred-year heritage of colonialism. During the month of October 1990, more than three hundred Indians from the Beni and their supporters marched four hundred miles from Trinidad in Beni to La Paz to demand a hearing by the federal government. The president of Bolivia, Jaime Paz Zamora, granted their requests. Institutions involved in the Chimane Ecosystem Conservation Program feel extremely proud and pleased to have been tools of lasting change.

Thus the Chimane Ecosystem Conservation Program and the institutions involved in the area are readjusting their plans and projects. Conservation International is now concentrating its efforts on strengthening the Great Chimane Council and Chimane representation in the Beni Biosphere Reserve, and cooperating with the council in resource-use projects.

ACKNOWLEDGMENTS

This paper is based on material collected by the Conservation International staff during its years of involvement in Bolivia's conservation movement. I am grateful to Maria Teresa Ortiz, Guillermo Mann, and Ken Margolis, who provided me with important insights and information in formulating my vision of the program, and to many Bolivian colleagues for their experiences and long sessions of participation and cooperation. I also would like to thank Dr. Kent Redford and Amanda Stonza for their patience and editorial assistance. For any errors in interpretation, I alone remain responsible.

LITERATURE CITED

Conservation International. 1988. *The Chimane Ecosystem Program, Bolivia.* Washington, D.C.: Conservation International.

Dahlberg, Kenneth A. 1977. Ecological effects of current development processes in less developed countries. In T. C. Emmel, ed., *Global Perspectives on Ecology,* Palo Alto, Calif.: Mayfield. 135–147.

Denevan, William M. 1963. Additional comments on the earthworks of Mojos in northeastern Bolivia. *American Antiquity* 28:540–544.

—— 1966. The aboriginal cultural geography of the Llanos de Mojos of Bolivia. *Ibero-Americana* No. 48. Berkeley, Calif.: University of California Press.

Mann, Guillermo 1987. *The Information Continuum for Conservation of Biological Diversity in the tropics.* Washington, D.C.: Conservation International.

Important Events That Took Place After the Presentation of This Paper

February 1989: Conservation International composes a position paper advocating inclusion of Indians in Beni conservation planning process.

1989: An independent Bolivian Interministerial Commission is formed to study the issue of Indian territories in the Chimane Forest. The Interministerial Commission recommends that lands (*polígonos*) on the forest fringe be given to the Indians, leaving timber concessionaires in the center of the CPPF.

November 1989: Central de Pueblos Indígenas del Beni (CPIB) is created (from the Central de Cabildos Indígenales Moxeños) to represent all indigenous groups in the Beni. CPIB rejects the *polígonos* decision as biased and inaccurate.

March 1990: Conservation International signs a collaboration agreement with CPIB and Consejo Chimane to provide technical and financial support for marketing and commercialization of *jatata* (a palm whose leaves are used in thatching). Conservation International and LIDEMA coproduce a series of public-awareness radio programs focusing on indigenous communities of the Chimane Forest (traditions and culture, resource management practices, territorial claims). Members of Chimane Council participate in script writing and production.

August 1990: Government inaction and Indians' rejection of *polígonos* leads to historic March for Dignity and Territory from Trinidad, Beni, to La Paz. For one entire month, more than three hundred

Indians from the Beni march four hundred miles to make themselves heard by the federal government.

October 1990: Territory in Chimane Forest is granted to Indians (see figure 12.1); Paz Zamora establishes commission to draft Indigenous Law to give legal standing to traditional leaders and governing bodies. Timber companies in the CPPF are required to cease cutting by the end of October 1990 and remove all felled timber by the end of the year. Companies would be reassigned concessions elsewhere in the forest.

THIRTEEN

The Cuyabeno Wildlife Production Reserve: Human Needs and Natural Resource Conservation in the Ecuadorean Amazon

FLAVIO COELLO HINOJOSA

The protection of Amazonian lands in Ecuador was given major impetus in 1974 when consultants for the Food and Agriculture Organization (FAO) of the United Nations were invited to help evaluate the important renewable natural resources of the region and to train Ecuadorean technicians. The final result of their work was the production of a document, "The National Conservation Strategy," which established conservation priorities for key areas. The zones designated are now being managed by the Department of Natural Areas and Forest Resources (Departamento de Areas Naturales y Recursos Silvestres) of the Ministry of Agriculture and Livestock (Ministerio de Agricultura y Ganadería).

During the 1970s, Napo Province in northeastern Ecuador became the center of the oil industry in the Amazon. The community of Lago Agrio marks the location of the first productive well of the region. As a result of the oil discoveries, the province was subject to unplanned development, which resulted in the colonization of extensive areas by spontaneous settlers from various parts of the country and displacement of native groups from their ancestral lands. Oil exploitation in the Amazon continues, as does the arrival of new settlers, who follow the roadways opened by concessionaires. Land speculation, logging, and the growth of illegal activities, including the drug trade, have also followed.

Management and conservation in the Amazonian region have

proved to be very difficult. Even zones that have full legal standing as protected areas have not escaped the development pressure of the oil, mining, and logging industries. The lack of adequate settlement programs (as well as political and private interests) has weakened the conservation process in the country.

Several important indigenous groups that inhabit reserve lands and surrounding areas have traditionally hunted and gathered products of the forest. Unfortunately, the majority of these communities are now being very rapidly acculturated. This process has resulted in a dramatic change in their felt needs as well as an increasing dependence on foods, medicines, education, and other products from outside their own environment. Moreover, indigenous traditions and knowledge of the forest are not being passed on to younger generations and will eventually be lost.

Economic dependence has also led these indigenous communities to exploit natural resources in imprudent ways. Wildlife and plant products are gathered for use in the manufacture of handicrafts that are in turn sold for prices so low that they neither reflect the value of the resources nor do they adequately compensate the collectors for the effort and time expended.

Aware of the complex situation in the region, the Departamento de Areas Naturales y Recursos Silvestres, with the support of the World Wildlife Fund (WWF), designed a management plan that lays out basic norms and principles for both conservation and planned development in the area. Taking into account that this is a zone dedicated to the "production" of wildlife, the plan promotes the establishment of new programs that offer alternative subsistence practices to the area's inhabitants. The programs try to incorporate the rich indigenous traditions of the region and also reflect current social, economic, and political realities.

The Cuyabeno Wildlife Production Reserve

The Ecuadorean conservation system counts among its legally constituted areas several Amazonian reserves that are particularly noted for their great biological and ecological diversity and other qualities that make them unique in both Ecuador and the world.

The Cuyabeno Wildlife Production Reserve, located in Napo Province, forms part of the Amazonian Biotic Province. It is characterized by humid tropical forest, maximum and minimum temperatures of 26°C and 18°C, and mean annual rainfall of 3,500 to 4,000 millimeters. Monthly rainfall varies considerably, but seasonal patterns are not regular. Supporting the exuberant vegetation of the region are soils that are fragile, very limited for agriculture, and suited only for the growth of forest.

The reserve was created by the Acuerdo Ministerial No. 0322, 07/16/79; its boundaries were ratified and published in the Registro Oficial No. 69, 11/20/79. It encompasses an area of 254,760 hectares, largely lands subject to annual flooding. The animal and plant species of the reserve are exploited by indigenous Siona-Secoya who inhabit the reserve, as well as by Quichua, Secoya, Cofane, and Shuar groups (from the southeast of Ecuadorean Amazonia) who live in surrounding areas along the Aguarico River. There are also groups of new colonists living along roads constructed by the oil industry, both within and near the reserve.

As mentioned above, the opening of roads has prompted massive immigration into the western part of the reserve. This invasion led to a carving up of the reserve's territory, and a substantial area was turned over to new settlers. Fortunately, since tracts of pristine forest are still to be found on the eastern edge, the reserve was expanded to make up for the land lost to colonists. Despite preservation efforts, social and economic pressures continue to push settlers into protected areas.

The reserve offers multiple alternatives for the wise and sustainable use of resources to benefit both the various native groups and communities of spontaneous immigrants. According to a census conducted in 1985, there were 1,000 families from various parts of the country living within the Cuyabeno Reserve. These settlers were principally engaged in subsistence agriculture, land speculation, and logging.

The Management Plan

The management plan of the Cuyabeno Wildlife Production Reserve seeks interinstitutional and community cooperation to conserve and develop this protected area. Particular emphasis is being placed on finding ways of reconciling the wise use of natural resources with the needs of the local population. In arriving at the plan, many factors have been considered. Among important issues were the challenge of preserving the high biodiversity of both *terra firme* and inundated forests in the reserve, the particular ecological fragility of some areas, and the injunction that specific areas of forest must remain protected forests. Great importance in shaping the reserve was also given to human use issues, particularly the presence of native communities, within and outside the reserve, and of settlers, largely in surrounding areas.

In attempting to make realistic plans, the Departamento de Areas Naturales y Recursos Silvestres and the WWF had to take into account problems stemming from the oil industry's generally unplanned growth and the ongoing influx of immigrants to the area. Immigrants continue to bring with them problems of deforestation and conversion of forest to impermanent cropping, as well as expansion of human settlements into the protected area. Buying and selling of land by speculators posing as settlers is also a problem. Finally, internal migrations of indigenous communities located in areas surrounding the reserve has complicated planning.

Objectives, Policies, and Strategies

A series of objectives, policies, and strategies that allow for the integration of conservation and development were formulated. Both their planning and implementation involve the active participation of local population groups. The overall objectives of the management plan are to promote conservation of the reserve's vulnerable riches while fostering the appropriate use of its natural resources in order to achieve balanced, sustainable development for local populations.

In order to achieve these complex objectives, general policies were developed that stress the need to minimize conflicts between ethnic groups and between indigenous and immigrant populations, at the same time ensuring full participation of local populations in conservation and development activities. Another policy priority has been the integration of socioeconomic, cultural, and environmental concerns. To maximize benefits of the development plan, it was considered most important that the plan include provisions for education, training, and extension at all educational levels and to all groups.

The strategies that were developed to implement these policies are both multifaceted and far-reaching. On the conceptual level, the planners determined that priority must also be given to the development of methodological and conceptual approaches that allow for a clearer understanding of the relationship between the environment and the development of human communities. The formulation of new models of environmental recovery on degraded lands was also considered a priority. These are needed if the rehabilitation of areas within and outside the reserve that have been degraded by excessive exploitation is to proceed effectively.

High priority was also given to developing effective means of working cooperatively with indigenous peoples and colonists, as well as with the development agencies present in the area. The organization of local groups was considered very important. Thus plans were made to both strengthen current community organization and facilitate the formation of self-governing and self-reliant forms of social organization at the local, regional, and national levels.

The establishment of monitoring capabilities and improved information exchange is another important strategy. Plans were laid for an innovative natural resource monitoring system that would involve the active participation of the community. Precedence was also given to the formulation of a strategy to guide interinstitutional and community activities, and to identify and tap internal and external sources of funds and technical support to carry out plans, programs, and projects needed in the area.

Functions of Local Administration

The duties of the reserve's administrators are complex and combine the functions of developing the management plan and designing and coordinating programs in a variety of areas including environmental education, tourism, silviculture, scientific research, wildlife management, and specific projects. Among the aims of these programs is the establishment of a relationship of trust with the local population and, by making local participation in research a reality, also to provide a mechanism to facilitate data collection. Local involvement will also help to gain community support for proposed activities and aid acceptance of decisions to protect natural resources.

Resource Use and Development of Alternatives

Forests occupy 121,876 square kilometers of the Ecuadorean Amazon. It is estimated that 1,200 square kilometers of that forest are destroyed each year (CONADE/PNUMA 1987). Deforestation radically changes the tropical environment and alters the traditional way of life of indigenous forest groups such as the Siona-Secoya, Quichua, and Cofane. Changes and adaptations that indigenous peoples undergo can have both positive and negative aspects. The development of beneficial resource use and employment alternatives for indigenous peoples is one of the priorities of the Cuyabeno administration.

The most common problem characterizing the use of natural resources in the area is the lack of adequate settlement policies. It is unknown what parameters were used to determine the standard family unit or farm for immigrants to be 50 hectares, that is, a piece of land 250 meters wide and 2,000 meters long. What is certain is that easy access to land encourages other settlers to move into the area, in direct conflict with conservation goals. These colonists continue to clear forest for pastures and plantations of coffee, corn, rice, and African oil palm. As more settlers enter, the vegetation of the area is rapidly destroyed, and with it habitats of the Amazonian fauna (CONADE/PNUMA 1987:32–63).

The 2,550 square kilometers of the reserve not only protect outstanding natural wonders, but also contain natural resources of great economic value. An important objective in protecting this area as a reserve is to allow wildlife populations to recover from depletion due to illegal hunting and fishing. Special attention is being given to campaigns to increase respect for the natural riches that are the patrimony of the Ecuadorean nation. Control and protection of critical areas where colonists live is the most challenging task.

The formidable problems that the area presents have led to the development of a multipronged program that combines protection of flora, fauna, and traditional territories of the indigenous communities with the management of scientific research, tourism, and wildlife; restoration of degraded environments; and monitoring of the environmental impacts of colonization, oil extraction, and economic development.

Because the Cuyabeno Wildlife Production Reserve is located in one of the most biologically diverse areas of the world (Mittermeier and Plotkin, pers. comm.), it offers unique terrestrial and aquatic ecosystems for scientific research, tourism, and experiments in wildlife management for the benefit of the varied human populations of the region.

Considering the priorities of the reserve and the complex mandate to pursue both development and conservation while keeping in mind the economic needs of both indigenous and colonist groups within and outside the reserve, certain procedures are being followed. Permanent contact with community leaders is maintained through site visits, and representatives of the reserve's administrative staff participate in meetings to provide information and discuss specific subjects. Communities are also involved in wildlife research activities, and active participation of the population is sought in the evaluation and implementation of all activities called for in the management plan.

Major Development Programs

Conservation Program

Plans for conservation and protection of natural resources have not been neglected. The cooperation of indigenous and settler communities is the cornerstone of the effort to maintain boundaries between indigenous territories, colonization zones, and the reserve.

Park guards, assisted by technical and professional personnel, keep in close communication with indigenous communities; they cooperate in establishing management priorities and in dispelling doubts that might lead to disputes among the indigenous groups, the colonists, and the reserve administration. There have been instances in which new settlers have ignored the law and invaded territories they claim are empty and owned by no one. They often base their claims on an interpretation of colonization laws, namely that land belongs to "him who works it." Faced with such problems, representatives of community organizations have helped by condemning land seizures that threaten local stability. These leaders have actively supported the agreements that were made to protect and maintain reserved areas and to stop the formation of new settlements.

Community Training Program

Education programs include training on the community level. These projects have multiple objectives: to give the indigenous community a broader perspective on the effects their actions have on the environment, to help them set their own conservation priorities, and to give them guidance in forming opinions on environmental matters.

Training objectives for inhabitants of the recently settled zones located close to the reserve are somewhat different. Education programs in these communities seek to help the settlers acquire a sense of their identity and place within an environment foreign to them, and to help them find alternative ways of using the tropical environment to improve their lives. Through extension activities, an attempt is made to convince these new arrivals of the benefits that wise use of resources will yield.

Forestry Development Program

We believe that to solve some of the problems of the reserve, we must begin from outside the park and work inward. The relationship of the colonists to their surroundings is very different from that of indigenous peoples. A general objective of the program is to implement among recent settlers who live in communities surrounding the reserve a series of practices based on the traditional patterns of native communities. To accomplish these changes, the following actions have been proposed:

1. Design research projects to evaluate the relative advantages of introducing exotic species of economic plants and the production of species native to the reserve, and of harvesting methods based on indigenous patterns.
2. Study traditional agroforestry and forest management patterns of the Siona-Secoya, Quichua, and Cofane communities that could be incorporated into production models for colonist communities.
3. Establish sustainable agriculture systems based on agroforestry and fruit-tree production.
4. Develop agricultural methods that would foster the rehabilitation of degraded soils employing agricultural rotation practices used by indigenous farmers and adapting experimental agro-silvi-pastoral models to local conditions.
5. In addition to allowing for natural habitat recovery, reforest critical areas, particularly recently colonized zones within and on the outskirts of the reserve, using appropriate methods and species.

Ethnobotany Training Program

Although both the Siona-Secoya and the Quichua have a rich tradition of using plant medicines, the Cofane community inhabiting areas contiguous to the reserve possesses the broadest ethnobotanical knowledge (Plotkin, pers. comm.). However, this knowledge is being lost daily. Today only the older community members know the secrets of the forest. As a result, a program to promote the use of natural, traditional remedies that have been shown to be effective has been proposed. We hope that this program will help inculcate into the local community an attitude of self-reliance and desire for improvement in health matters. We also hope to involve the national and international scientific communities in compiling this knowledge with a view toward developing appropriate health technologies based on forest products and to promote the creation of indigenous "medicine banks," as has been done successfully by native communities of the Ecuadorean Andes.

Tourism Management Program

Development of tourism is one of the important economic options for settlers and indigenous peoples of the region, especially for the Siona within the Cuyabeno Reserve. Tourism generates numerous economic benefits for the local population and provides incentives for conservation.

Members of the Siona community were the first to insist that the forest be maintained in "as natural a state as possible." This attitude was later adopted by communities peripheral to the reserve, including the Quichua, Cofane, and Secoya, as well as the Shuar of the extreme southwestern edge.

Visits are arranged either by tourism companies in the main cities or by small groups that arrive independently. Visitors travel along the rivers of the reserve in canoes handled by native people. Members of indigenous communities also work as occasional tour guides and cooks, thus earning an income. Prices are agreed upon between the companies and the native community and total about $25 per day, plus fuel expenses. Nonnative tour guides offer tours for groups that are not organized and that are unfamiliar with the area's administration. As a result, the Departamento de Areas Naturales y Recursos Silvestres is establishing several legal requirements and licenses to work in protected areas. This will facilitate both control of tourism activities and law enforcement.

Each visitor to the reserve must pay an entrance fee to the Departamento de Parques Nacionales; foreigners pay $10, while Ecuadorean nationals pay 500 sucres (US $1). These funds are paid into a special account of the Dirección Nacional Forestal for reinvestment in the reserve.

Wildlife Harvesting Program

Members of the indigenous communities within the Cuyabeno Reserve, as well as a considerable number of colonists, exploit the natural resources of the reserve, especially its fauna, for their subsistence needs. Hunting and fishing provide approximately 25 percent of the calories consumed by Siona-Secoya, Quichua, and Cofane families (Vickers 1976; Vickers 1983:456; Vickers and Plowman 1984) living in and around the protected area.

It is difficult to determine exactly the income that natives and nonnatives derive from hunting and fishing. However, based on a study conducted in the Brazilian Amazon, Paucar and Gardner (1981) report that five of the six largest mammals—the peccary, deer, tapir, paca, and agouti—can produce an average of 240 kilograms of meat per square kilometer (equivalent to US$500 per square kilometer) without negatively affecting the natural reproduction of these wildlife species.

Paucar and Gardner (1981:3) also estimate that if nonmammals, including fish, birds, and turtles—all species consumed by Cuyabeno communities—are included, the value of meat produced can reach up to US$4,000 per square kilometer per year.

Hunting, fishing, and gathering of forest products must be regulated in the park, since nonnative people often exploit the flora and fauna using firearms and other materials that are prohibited. The management plan includes special hunting and fishing provisions for indigenous peoples, and there are agreements between these communities and management officials to ensure that wildlife harvesting is done in a wise fashion. However, some members of indigenous communities, particularly the Cofane, also sometimes engage in illegal activities.

Even though the Cofane, an indigenous community adjacent to the wildlife reserve, live within a substantial territory, they are looking for new areas in which to pursue subsistence hunting and fishing activities. The Cofane have become more acculturated than their neighbors and are exploiting wildlife for commercial purposes. They sell forest products to colonists and tourists, and despite attempts at control, their actions, together with those of some colonists, are having serious effects on wildlife in native territories, as well as in the wildlife reserve and buffer zones. Management officials seek to prevent the kind of commercial exploitation of wildlife that is now being carried out by Cofane Indians without forbidding the exploitation of wildlife altogether. The goal of regulation is to ensure that populations recover and that animals are available for exploitation by human populations in the future.

Wildlife Production Program

With the assistance of the World Wildlife Fund, US, the Ministerio de Agricultura y Ganadería, through its Departamento de Areas Naturales y Recursos Silvestres, plans in 1989 to begin experimental management of wildlife in captivity and semicaptivity. The goal of this program is to find viable economic and subsistence alternatives for the human communities of the region by producing meat for sale and for local consumption. Experiments in raising animals in corrals and nurseries will be started with a view to using other methods later.

Currently, populations of the giant river turtle (*Podocnemys expansa),* collared pecarry (*Tayassu tajacu),* and white-lipped peccary (*T. pecari*) are under study. In the next few months we expect to start studies on *Agouti paca, Dasyprocta fulginosa,* and *Hydrochaeris hydrochaeris.*

The ethnozoological and ethnobiological knowledge of the Siona-Secoya, Quichua, and Cofane is being used in these studies. The colonists too have shown increasing interest in managing wildlife in a manner that fosters self-sufficiency and self-reliance, and that eases pressures on the reserve's resources.

The WWF and the Departamento de Areas Naturales estimate that by managing populations of *Podocnemys expansa* in small pools, approximately 22,000 kilograms of meat per hectare per year can be produced (Paucar and Gardner 1981:14). Data collected from indigenous communities of the area show that a kilogram of smoked turtle meat is sold for 300 sucres ($1 = 530 sucres). Thus calculating conservatively, a value of $20,000 per hectare per year could be attained. This estimate does not include the thousands of turtle eggs collected along river beaches during the breeding season, each of which is sold for 10 sucres.*

*Each turtle lays approximately thirty to ninety eggs per year.

Several other animals also are being considered. For instance, under optimal conditions such as those found on the savannas of Venezuela, capybaras grow rapidly. However, in the humid tropics where they must forage for hours or days to obtain their food, they do not exhibit this characteristic. Experiments in management in captivity and semicaptivity are necessary to determine whether the capybara, as well as other animals, ought to be added to the list of selected species.

Research Program

Scientific research is an important component of the management plan. Various research projects are being carried out in the reserve. All research activity focuses on the relationship between people and their environment, with a view toward finding the most acceptable resource use alternatives. One of the most important plans is for a program of research in biology and ecology being arranged with the Universidad Católica de Ecuador. The diverse projects that will be subsumed under this program will contribute to the development of wildlife management. The program also will include training in Amazon ecology for students and professionals.

Foreign researchers also are involved in tropical studies at the reserve; among institutions active at Cuyabeno are the universities of Aarhus (Denmark) and Wageningen (The Netherlands), the Missouri Botanical Garden, and possibly the Jersey Wildlife Preservation Trust and the New York Zoological Society.

In the area of plant resources, it is assumed that much potential economic value is being lost because of inadequate knowledge, as well as lack of appropriate management models. Important research directions being planned are an inventory of both species and uses of forest flora and fauna and research aimed at the development and promotion of new and effective natural products.

Physical infrastructure for research in the reserve has been provided largely by the Departamento de Areas Naturales with funding from the state; however, financial and training support from other institutions has been indispensable.

Even though we do not know the exact income local residents derive from the exploitation of wildlife in the region, it is obvious that they maintain a high degree of economic dependence on such activities.

No effective experimental trials in wildlife management have been reported for Ecuador. However, we firmly believe that adequate management of plant and animal species is more profitable than clear-cutting the forest for short-term agriculture and the creation of pastures.

Settlers living within and outside the reserve have the alternative of participating in sustainable methods of wildlife production that can produce an adequate income without destroying the resource base. The alternatives proposed in the management plan derive from an appreciation for the human needs and the economic, social, and political character of the region and the country. These alternatives also support the production, conservation, and development objectives of the Departamento de Areas Naturales y Recursos Silvestres.

Apart from the programs of wildlife management, we believe that tourism offers the best potential for satisfying the needs of the native communities of the region, and we hope to expand these activities to the settler community, taking into account, of course, their different cultural and economic needs.

Scientific research must continue to provide answers to the questions of man-environment relationships based on long-term observations in native communities.

Finally, it is necessary that there be support and agreement at high levels of government on the importance of conservation and development of the natural patrimony of the country. Such commitment will also aid in fostering harmonious coexistence and mutual respect among indigenous peoples, recent settlers, and the government.

ACKNOWLEDGMENTS

The author would like to thank Carlos Villarreal, economist and director of Proyecto AWA, Ecuador, for his valuable contributions to and comments on this manuscript.

LITERATURE CITED

Consejo Nacional de Desarrollo (CONADE/PNUMA). 1987. *Proyecto propuesta de manejo de recursos naturales en la región amazónica ecuatoriana.* Quito, Ecuador: Consejo Nacional de Desarrollo.

Hiraoka, Mario and Shozo Yamamoto. 1989. Agriculture and development in the upper Amazon of Ecuador. *Geographical Review* 7094:423–445.

International Union for the Conservation of Nature and the United Nations Environmental Program (IUCN/UNEP). 1986. *Managing protected areas in the tropics.*

Paucar, Angel, and Alfred Gardner. 1981. *Establecimiento de una estación de investigaciones científicas en el Parque Nacional Yasuní de la República del Ecuador.* Quito, Ecuador: Ministerio de Agricultura, Programa Nacional Forestal.

Vickers, William T. 1976. Cultural adaptation to the Amazonian habitats: The Siona Secoya of eastern Ecuador. Ph.D. Thesis, University of Gainesville, Florida.

—— 1983. The territorial dimensions of Siona-Secoya and Encabellado adaptation. In R. B. Hames and W. T. Vickers, eds., *Adaptive Responses of Native Amazonians* pp. 451–478. New York: Academic Press.

Vickers, W. T. and T. Plowman 1984. Useful plants of the Siona and Secoya Indians of Eastern Ecuador. *Fieldiana Botany,* no. 15. Chicago: Field Museum of Natural History.

FOURTEEN

The Wildlands and Human Needs Program: Putting Rural Development to Work for Conservation

DENNIS GLICK AND
MICHAEL WRIGHT

Land titling, agricultural credit, artisan cooperatives, ecotourism, appropriate technologies, sustainable agriculture, rural community leadership training, and a host of other rural development tools are rapidly becoming an integral component of wildland management strategies for the Neotropics (WWF 1988). In fact, at this time in the history of conservation in the American tropics, the initiation of low-impact rural development practices may be as important as the establishment of new parks and reserves, or at least an integral complement to these stricter protection-oriented activities. Tremendous human pressures on natural resources are rapidly converting many of the most diverse ecosystems on the planet to biological deserts or geographically isolated patches of natural vegetation (Soulé and Wilcox 1980). These may or may not remain ecologically viable, depending upon their size, shape, and other factors, and will most certainly be further degraded as surrounding resource "buffers" are whittled away by rural populations. Putting into place the rural development strategies and practices that will stabilize these populations, reduce their impact on wildland resources, actively involve them in resource-protection decisions and activities, and improve their quality of life appears to be essential if the long-term survival of protected parks and reserves of tropical America is to be achieved (WWF 1987).

Recognizing this premise, the World Wildlife Fund (WWF) has initiated an ambitious effort to support field projects attempting to link the preservation of biological resources with the needs of the rural poor. The underlying premise of its Wildlands and Human Needs Program (WHNP) is that the biological resources of concern to conservationists are threatened by the inability of the rural poor to sustainably meet their development needs from these and adjacent lands (WWF 1988). Another premise of the program is that the destruction of natural resources will, in the long run, worsen the plight of those who depend on them. The program was initiated in 1985 and has thus far supported more than twenty projects in Latin America, the Caribbean, and Africa. Similar integrated conservation and development projects, though not officially part of the program, are also under way in Asia, and, in fact, many project proposals now submitted to WWF for support reflect this new orientation.

Background and Context of the Wildlands and Human Needs Program

The World Wildlife Fund, an independent nonprofit organization which since 1961 has been active in international conservation of wildlife, natural habitats, and natural resources, has become increasingly aware of the conservation-development link. Initially WWF focused primarily on individual species protection. However, early in its history, this strategy expanded to include habitat preservation, and later entire ecosystems became the target of WWF conservation efforts. With the publication of the *World Conservation Strategy* (WCS) in 1980, the organization formally declared as one of its primary objectives "the maintenance of species and ecosystems for their sustainable utilization by human populations" (WWF, IUCN, and United Nations Environmental Program 1980). Increasingly, WWF has supported projects that reflect this approach, although the organization was generally reactive rather than proactive in its interest in more development-oriented conservation.

In 1985, the United States Agency for International Development (USAID) and WWF initiated negotiations to design and implement a matching grant program for conservation work in tropical nations where USAID is active. After a series of discussions, the project proposal for the WHNP was submitted to the Office of Private Voluntary Cooperation (PVC) of USAID.

While some international organizations support both development and conservation efforts in Third World nations, the WHNP grant represents one of the first joint ventures between a development-oriented agency and a nature conservation group that focuses specifically on the integration of these two fields. The grant provides financial support for the widespread and multifaceted activities needed to effectively implement an "on the ground" program in Wildlands and Human Needs. In addition, these funds allowed WWF to institutionalize its development-oriented activities and establish a separate program with its own criteria and staff.

During the program's first three years, WWF has established close ties with the United States Private Voluntary Organization (PVO) development community through the organization of an advisory committee. The committee meets twice a year and includes representatives of Coordination in Development (CODEL), Catholic Relief Services, Save the Children, Pan American Development Foundation, International Institute for Environment and Development, the PVO office of USAID, Christian Children's Fund, Africare, International Center for Research on Women, Interamerican Foundation, Cultural Survival, CARE, and the Organization of American States. With the assistance of the advisory committee, WWF has approached the population issue indirectly in the recognition that increasing the quality of life of rural peoples (their health and that of their children, increased opportunities for women, and assurance of resource ownership) is perhaps the most fundamental precondition to the population stabilization necessary to relieve pressure on the environment.

Many biologically important wildlands in the developing world have survived as a result of their distance from centers of economic activity. However, another result of this isolation is that the rural communities living in or adjacent to these areas often represent society's destitute, left behind in the overall development thrust of government programs. The growth in numbers and the fundamental needs of these people are commonly seen as a major threat to the long-term viability of wildland areas. In contrast, the premise of the Wildlands and Human Needs Program is that only through a focus on these populations will we find the key to the sustainable management of wildlands and the preservation of biological diversity (WWF 1988).

The basic challenge for the Wildlands and Human Needs Program is to design projects which maintain the ability of natural ecosystems to produce an array of goods and services for human benefit without depleting their natural values. Many of the biologically diverse project sites face severe limitations on their capacity to sustain greatly increased levels of exploitation. These areas lack immediate potential for more intensive development due to their soil classification, topography, or natural characteristics. These are often the only areas for the rural poor, and their further deterioration will increase poverty. On such marginal lands, management of the area in its relatively natural state may offer the best option for the extraction of a sustainable flow of goods and services for meeting local needs.

The Wildlands and Human Needs Program begins with the identification of a critical biological resource such as the extensive tropical forests of the Pacaya Samiria National Reserve in Peru. It is at the frontier where man presses against these unique natural resources that some of the poorest of the poor cling tenaciously to a marginal existence. The isolation of these rural populations (in the case of Pacaya Samiria, some fifty thousand people that live on the borders of the reserve) makes their access to government resources limited. Providing "extension services" to assist such communities in the implementation of ecologically sound land-use practices is one benefit that may be supported by the program.

The Wildlands Program does not seek to identify a single solution to the problems of poverty and biological destruction. Interaction with the local community to identify impediments to development is the necessary first step of each project if it is to address specific local needs. In many rural communities, wildland resources provide one element in a diverse approach to subsistence. Wildland resources also provide fundamental survival insurance. The Wildlands Program is attempting to strengthen diverse survival strategies that have low ecological impacts and, where possible, encourage additional enterprises that improve the overall quality of life in rural communities.

Program Methodology

The strategy of promoting conservation through development recognizes that responsibility for conservation lies primarily with governments, whereas the use of natural resources rests largely on private initiative (WWF 1988). Conservation and development are not inherently incompatible but often fail to occur harmoniously for lack of communication, conflicting goals, wrong incentives, and other reasons. In recent years the governments of many developing countries, especially in Latin America, have become more aware of their shortcomings in achieving sound natural resource management at the local level. At the same time, these governments and the general public have acknowledged how important and urgent sound resource use has become.

This evolving political and social reality has provided a favorable context within which integrated conservation and development projects can be initiated. In order to provide some structure for the WHNP, WWF has established criteria for selecting project sites in Latin America and Africa. Most of the sites exhibit the following characteristics (WWF 1987):

- An area of high biological significance, often, but not always, with an officially declared protected area.
- A peripheral zone that buffers the biologically important area.
- People living in the area who use natural resources from the biologically significant area.
- Generally the projects include a local nongovernmental organization whose principal interest is conservation of the protected area and believes that sound development of natural resources by and for the people who use them will help conserve protected areas.
- Government involvement in or approval of the project.

The strategy of attaining conservation through development assumes several characteristics of the site and the resource users (McCaffrey and Landazuri 1987):

- The site has high conservation value; it contains species or habitats of high scientific or social interest at the international level or of special interest locally.
- The site is subject to strong development pressures; current resource use is near, or even beyond capacity or will be so in the near future.
- In many instances outside interests threaten to usurp resources from local peoples.
- The site exhibits some degree of development opportunity; it has underutilized resources, resource use could improve, or benefits from existing use could be redistributed.

- Data exist or can be obtained to determine the resource capacity and development capacity of the site.
- The local population engages in diverse economic activities and uses several resources—as opposed to depending on a single resource—which provide flexibility, spread risks, and increase receptivity to change.
- Local people have secure rights to natural resources, or, where rights are lacking, a major project goal is to secure such rights to enable people to make long-term investments in conservation because they feel secure in the likelihood that they will realize the benefits of the investments.
- A limiting factor prevents people from increasing or improving resource use; removal of the limiting factor will promote development, and hence conservation. The limiting factor can be technical, financial, political, legal, or institutional.
- Appropriate technology or institutional or social innovation can overcome the limiting factors; the technology meets people's needs and is within their means. Usually it will build on existing practices and skills and use local materials.
- Local populations can make decisions and carry them out; they can assess their needs and choose means to satisfy them that rely primarily on existing community resources.

WWF offers financial support, training, technical assistance, administrative and managerial expertise, and, when appropriate, the publicity and endorsement of the World Wildlife Fund. WWF seeks to build a partnership with each nongovernmental organization (NGO) or government agency that strengthens it and makes it increasingly self-sufficient.

To catalyze appropriate rural development requires close interaction with local people to determine what stands between them and higher or more sustainable yields, i.e., identifying limiting factors for development. Sometimes the solution is technical but requires action in political and social areas as well. Once a solution is in place, it must be sustained. This means people must be convinced that it works, and they must be able to apply it or gain access to it themselves or via institutions they trust, hence the significance of appropriate technology, training, and institution-building to the success of the projects.

The ultimate goal of the World Wildlife Fund, the conservation of protected areas, rests on the support and collaboration of the local people. It also requires effective management by local and governmental institutions. When resource users and institutional interest groups act on the belief that conservation furthers their interests, conservation will result.

Status of the Program

As of December 1988, the WHNP has been in existence for three years. Twenty projects have been officially supported by the program in Latin America, the Caribbean, and Africa. Perhaps more significant, however, is the fact that there are now dozens of other projects in these regions, and in Asia too, that are virtually identical in their orientation to the projects supported by the WHNP. In fact, most of the project proposals being received by WWF at this time include a rural development component and conservationists throughout the developing world are embracing this strategy in their wildland protection efforts. The WHNP certainly did not invent this approach to preserving biodiversity, but the implementation of this program has encouraged many conservationists in Latin America, Africa, and Asia to continue and expand their pioneering efforts to integrate conservation and development.

Wildlands and Human Needs projects in the Neotropics are as diverse as the natural and cultural resources they seek to protect. The sites include several biosphere reserves such as the Sian Ka'an and Sierra de Manantlan reserves in Mexico, the Río Platano Reserve in Honduras, and the Beni Reserve in Bolivia. They include aquatic ecosystems such as the mangrove swamps of the Monterrico Biotope Reserve of Guatemala and the coral reefs of the Southeast Coast Project in Saint Lucia. Several of the sites are occupied or utilized by indigenous people such as the Awa Indians involved in the Awa Ethnic Forest Reserve project in Ecuador and the Uros Indians in the Lake Titicaca National Reserve in Peru. Colonists and loggers, two groups often considered as principal culprits of tropical deforestation, are also target groups for several projects such as the Cuyabeno Wildlife Production Reserve in Ecuador and the Small-Scale Sawyers Project in Dominica.

At all of these sites and the others supported by the WHNP, a variety of rural-development-related activities based on the local socioeconomic, political, institutional, and ecological conditions are being carried out. Land titling, agricultural extension, organization of cooperatives, community organization, development of community-managed ecotourism and artisanship, establishment of rural health services and appropriate education programs, securing of revolving loans for rural development projects, installation of appropriate technologies, agroforestry, intensive forage production for cattle and other animals, aquaculture, captive propagation of wildlife and participatory management of reserve wildlife, and community forestry management are some of the rural development activities that have been employed by Wildlands Program–supported projects.

All of these projects are in various stages of development. While some, such as the Southeast Coast Project of Saint Lucia, the Awa Ethnic Forest Reserve in Ecuador, and the Gandoca Integrated Land Use Project in Costa Rica, have made impressive advances; others such as the Río Platano Biosphere Reserve in Honduras and the Cuyabeno Reserve in Ecuador are seriously threatened. Even the above mentioned Awa, Gandoca, and Saint Lucia projects are still in their early stages of development, and it would be premature to declare them a successful integration of wildland conservation and sustainable rural development. But steady progress is being made, and it appears that this new conservation strategy may prove appropriate for many Latin American parks and reserves. Once again, these activities will not supersede the idea of completely protected conservation units such as national parks, but rather will complement and improve the prospects for these core natural areas' survival in the long term.

Not only have conservationists generally supported this approach, but governmental and nongovernmental development agencies increasingly have realized the need to incorporate resource protection activities into their rural development projects. NGO development and humanitarian aid groups such as CARE and Save the Children are initiating grassroots conservation activities, and several of the governmental agencies such as the overseas development organizations of the United States, Canadian, British, Dutch, Scandinavians, and Swiss governments are all supporting environmental protection efforts, including the protection of wildlands and wildlife. Perhaps even more significant has been the interest and participation of multilateral banks such as the World Bank and the Interamerican Development Bank.

The project matching was renewed by USAID in October of 1988, which will continue the grant through 1993. In addition, WWF has pledged to more than match this amount. Unfortunately, local financial support for these projects has not increased. In fact, in some cases it has been reduced. This, to a certain degree, reflects the serious economic crisis facing many of these beleaguered nations.

In November of 1988 representatives from all of the Latin American WHNP projects met at one of the project sites in Colombia for a two-week workshop on rural development and wildland management. Development philosophy, concepts, and techniques were discussed, and project managers exchanged information. In addition, these individuals were able to reflect on their problems and progress, evaluate their programs, and plan for the future.

An important result of this workshop was the sobering realization that no patented models exist for realizing the successful integration of conservation and sustainable development. There are some basic guidelines for working with local populations and for increasing chances for sustainable development activities, but the local natural and cultural resource characteristics, the institutional structures and human resources available, and a host of other details necessitate that integrated conservation and development projects be custom-designed for each particular circumstance. The conservation projects supported by the WHNP in Latin American and Caribbean forests (WWF 1988) give an idea of the range of possible strategies.

In the Manantlan Biosphere Reserve of Mexico, resource managers from the University of Guadalajara have initiated a wide array of resource investigation and management activities that are providing the information base for an expanding ecodevelopment program with the surrounding *ejidos*. The fifteen hundred or so inhabitants of the region (many of them indigenous peoples), regularly harvest resources such as firewood, timber, and wildlife from the protected area. The reserve has established an ecodevelopment unit that is beginning to initiate activities such as aquaculture and appropriate agriculture and cattle-raising projects, as well as health and education programs.

At the twelve-million hectare Sian Ka'an Biosphere Reserve in Mexico, the WHNP has supported the conservation development work carried out by an NGO known as the Amigos de Sian Ka'an. This has included support of a lobster management project with the local fishermen living within the reserve. These fishermen and other reserve neighbors also regularly harvest several species of palm, which they use for a variety of purposes, including the construction of lobster traps. The Wildlands Program supported research on these palms and the development of management programs and educational materials. The program also has supported the establishment of an experimental sustainable agricultural project and an intensive forage production study. A nature tourism program is also in the initial stages of development. Interinstitutional problems have threatened the survival of this project. But these administrative setbacks are being resolved.

In Guatemala, the Wildlands Program is assisting the University of San Carlos in the management of the Monterrico Biotope Reserve. This coastal site is occupied by fishermen, and its beaches are visited by Guatemalan tourists. An overall management plan is being developed for the reserve, and some ecodevelopment projects have been initiated; these include the establishment of an iguana-raising facility and a turtle-protection project. While this reserve has existed for several years, lack of continuity in management activities has slowed the progress of this important project. Increased funding and technical support should help to stabilize management efforts.

The Río Platano Biosphere Reserve is an important Central American wildland suffering from an array of development pressures. The WHNP sponsored an emergency planning workshop with more than forty participants from a dozen different agencies. The plan is now being implemented, and a management team has established a base of operations in the problematic upper end of the reserve watershed. The focus of its work is with recently arrived colonists, but attempts are also being made to more fully involve the original reserve inhabitants, primarily Miskito and Paya Indians, in the protection and management of the area.

At the Gandoca Integrated Development Project in Costa Rica, the local project managers from the NGO The New Alchemy Institute (ANAI) have initiated a variety of interrelated resource management activities ranging from land titling (according to Costa Rican law, untitled land must be "improved," i.e., deforested, to show ownership) to the establishment of more than twenty-five communally operated tree nurseries. In addition, they have initiated iguana-raising projects, reef fish management activities, and agricultural extension and diversification activities. Plans are being developed for ecotourism programs, and a national wildlife refuge has also been established in the center of the project area. Deforestation in this area continues, but it appears that the cumulative impact of the many ANAI projects is beginning to arrest uncontrolled forest cutting.

In the La Amistad Biosphere Reserve, also in Costa Rica, the WHNP supported the work of an anthropologist studying the land and resource utilization practices of the Cabecar Indians. This information was used in the development of the management plan for the reserve. Problems in fully integrating the reserve's indigenous population into administrative programs still plague this effort. But plans are being formulated to carry out a "south-south" exchange of Indian leaders from this reserve to other protected sites in Latin America such as the Awa project in Ecuador, where natives themselves are carrying out management activities.

In Ecuador, the Awa Indians in the northeastern mountains bordering Colombia are in the process of establishing their own forested reserve. These people live in scattered settlements, and their ecologically benign development practices have left the forests largely intact. With the help of the ministry of foreign affairs, the Awa have secured title to this land and demarcated the reserve boundaries, and are currently writing a management plan for the entire 110,000 hectare unit. In addition, extensive efforts have also been carried out to organize their communities and assist them in securing legal rights to their resources. The Ecuadoran Awa success has served as a catalyst for Awa Indians living on the Colombian side of the border, and efforts have now been initiated to establish a similar reserve there that would encompass nearly 400,000 hectares of this incredibly diverse Choco forest (see Orejuela, this volume).

Also in Ecuador, the WHNP has supported the development and implementation of a management plan for the 255,000 hectare Cuyabeno Wildlife Production Reserve in the Amazon region. This area is threatened severely by oil exploration and development and colonization, which has followed in the wake of the petroleum companies. In addition, tourism is increasing in the area and is now an important source of income for the Siona-Secoya Indians living within the reserve. The management plan calls for agricultural and wildlife management extension activities to be initiated with the local populations. Work on the initiation of these programs is just getting under way, and illegal forest cutting is still widespread. In addition, nature tourism operators have caused some problems with unauthorized construction of facilities (see Coello Hinojosa, this volume).

The giant Pacaya Samiria National Reserve in Peru (2 million hectares), is the site of several Wildland Program–supported resource protection and ecodevelopment activities. Based on the reserve's management and operational plans which were developed by The Committee for the Pacaya Samiria Reserve (COREPASA), an interinstitutional management agency, rural development projects are being carried out in many of the isolated communities scattered around the reserve. Considerable wildlife is hunted from the site, and emphasis is being placed on developing plans for sustainable harvesting of wildlife and the development of alternatives to hunting and other resource-damaging activities.

At the Beni Biosphere Reserve in Bolivia (see Campos, this volume), the Center for Interdisciplinary Community Studies (CIES) has received financial support for the development and initiation of their education and extension program.

In Dominica the WHNP is supporting the work of the Eastern Caribbean Natural Area Management Program (ECNAMP), which is assisting small-scale sawyers in establishing labor intensive forest harvesting techniques that minimize ecological damage and maximize profits. This project includes the development of a furniture-making and handicrafts industry, which could improve the value of the raw materials. The premise behind this project is that increased value of resources and increased involvement in resource management by those utilizing these resources should result in more sustainable forestry practices.

For two years the WHNP also supported the University of the West Indies' Center for Resource Management and Environmental Studies in Barbados. This is one of the most important centers for training resource managers in the Caribbean. Students from this program carry out their field studies at some of the WHNP-supported projects such as the above-mentioned Southeast Coast Project.

In addition to these WHNP-supported projects in the Neotropics, the program also supports activities in six African countries including sites in Sierra Leone, Cameroon, Central African Republic, Zambia, Malawi, and Madagascar. These projects, much like those described above, employ a wide variety of rural development tools.

Lessons Learned

1. Costs and Benefits

In some instances, the major benefit of the WHNP is not the identification of sustainable uses of wildland resources but rather the mitigation or provision of some alternative benefit such as a health center or a land title. This, for example, could be the case in an area with extremely fragile resources. It would be erroneous to think that in all instances the interests of conservation and local needs can be harmonized easily or completely. In contrast, some projects are succeeding principally because the potential for increased benefits does exist and the program removes obstacles to realization of these improvements by local people.

Strict benefit analysis of the projects is difficult because of the nature of wildland resource benefits—often intangible or difficult to measure. While dollar values can be given to some wildland-derived goods such as firewood and lumber, other utilitarian benefits are harder to assess. This is perhaps preferable though, lest these resources be deemed less "economically valuable," justifying some other lucrative but ultimately destructive use.

2. Institution Building

The greatest successes of the program have been achieved thus far where an existing counterpart institution, already familiar with and involved in conservation or development activities, has been the implementing organization. Several of the projects, in fact, had already been initiated at the time that WWF began providing support. However, specialized training for project managers, increased awareness and promotion of rural development, and increased attention to planning, monitoring, and evaluation has further strengthened even the relatively well-established organizations.

3. Sustainability

The WWF strategy of addressing the issue of sustainability includes use of local personnel and indigenous NGOs, keeping recurrent costs low, stressing self-reliance and avoiding dependency on foreign funds and exotic technologies, and the use of local technical knowledge. Sustainability can be addressed through participation of beneficiaries and inventories to assure project design is geared to local circumstances. Rural communities can be tenacious in their support for projects that are seen to address their priorities.

Local resource users are often unorganized and vulnerable. With little margin for error, potential beneficiaries are reluctant to be exposed to the risks inherent in new technologies and have difficulty sustaining them if payouts are not immediate or if success is episodic. Demonstration projects, using local resource users, are used whenever possible to allow the communities to make their own value judgments on the merits of new approaches. Project flexibility is important, and adaptation over the life of the project is encouraged in light of new information and local reaction. And, of course, the sustainability of projects is greatest where the implementing organization is a local NGO with broad-based grassroots participation.

4. Benefit Distribution

In all projects in which the focus is geographically based, the benefits are designed to enhance the quality of life for the residents living in or near the parks, buffer zones, or other regions of concern. The extent to which the benefits are received by the community rather than the household or individual is determined by the particular orientation of the project. These are determined based on the critical needs defined by the project staff and local communities. Projects strive for benefits that are virtually guaranteed in order to establish credibility. Similarly, project managers must be careful and realistic regarding what is promised to local populations, understating possible benefits and letting results, not promises, drive the program.

5. Local Participation

Participation is defined as local involvement in project design and sharing decision-making with area inhabitants during actual implementation. WHNP's experience is that projects are more likely to be successful when they involve local people from the start and are designed to meet a specific need identified by the community. Success can be measured in a variety of ways; however, elements such as sustainability, leadership development, and local control are more likely when grassroots participation is encouraged.

6. Leadership Development

In addition to institutional strengthening, a critical component of many projects is leadership. Strong leaders, however, do not always allow broad-based participation. Encouraging leadership in such a way that it supports the development of long-term, open participation is a difficult challenge.

7. Innovation and Technology Transfer

The goal of sustainability requires the withdrawal of most external funding; however, initially such funds may be critical to absorb some of the risks involved when the rural poor undertake new activities. The WHNP has taken the view that previously existing, locally tested techniques are more likely to succeed than major, complex, and imported solutions. Thus the program does not stress radically innovative technical solutions, although several of the projects do include new, though tested, technologies. Most effective technology transfer occurs when peers exchange information. In other words, local people are trained as trainers. Hands-on and on-site training is the most effective.

Another approach is for projects to have an "extensionist," whether explicitly defined as such or not. Extension techniques have been successful in increasing general understanding of project objectives and methods. However, extension and education should not be confused with participation. Improving or modifying existing technologies or projects, based on the recommendations of the local community, has been the most effective means of improving production or quality of life.

8. Replication of Approach

A variety of elements in many of the projects are replicable. However, it would be a serious mistake to think that any of the projects could be transferred to other areas. The specific conditions and elements which are important vary from site to site and community to community. Each project must be, as mentioned before, custom tailored to the national, cultural, economic, natural, and institutional/political idiosyncrasies of its location. What can be transferred, though, is an understanding of the approach and philosophy, the processes, techniques and materials used, and what has worked and what hasn't and why (WWF 1988).

The Future

What is the future of conservation in Latin America and the developing tropical world in general? How far will this integrated approach to natural resource protection carry conservation organizations such as the World Wildlife Fund or the Amigos de Sian Ka'an? Will we soon be indistinguishable from the development organizations? Will our underlying goal of preserving of biological diversity be lost in the rush to embrace this currently popular approach to conservation?

Predicting what the future holds is difficult at this time. But rather than tout the Wildlands and Human Needs Program as the only way to protect the wildlands and wildlife of the developing world, WWF proposes that it is best viewed as yet another tool for realizing our principal goal, which is and will continue to be the preservation of biological diversity. We approach the use of this tool with caution, but because of the severity and velocity of the deterioration of tropical wildlands, we are willing to take the risks involved in implementing unproven strategies.

Traditional conservation methods such as more academic wildlife research, the establishment of national parks, and nature education are not dead, but rather are being supplemented by pragmatic, development-related activities that, in most instances, complement these methods. There will be conflicts, however, and environmental degradation will be unavoidable in some cases, even when so-called "sustainable" development practices are employed. A refusal to expand conservation options and continuing reliance on strictly "protectionist" strategies furthers the gap between conservationists and an increasingly desperate rural population. But with thoughtful and patient collaboration, mist nets and machetes can coexist.

LITERATURE CITED

McCaffrey, D. and H. Landazuri. 1987. *Wildlands and Human Needs: A Program Evaluation*. Washington, D.C.: United States Agency for International Development/World Wildlife Fund.

Soulé, M. and Bruce Wilcox. 1980. *Conservation Biology, A Evolutionary-Ecological Perspective*. Sunderland, Mass.: Sinauer.

World Wildlife Fund (WWF). 1987. *First Year Report on the Matching Grant for a Program in Wildlands and Human Needs*. Washington, D.C.: World Wildlife Fund.

—— 1988. *Mid-Course Report on the Matching Grant for a Program in Wildlands and Human Needs*. Washington, D.C.: World Wildlife Fund.

—— 1988. *Third Year Report on the Matching Grant for a Program in Wildlands and Human Needs*. Washington, D.C.: World Wildlife Fund.

World Wildlife Fund, International Union for the Conservation of Natural Resources, United Nations Environmental Program. 1980. *The World Conservation Strategy*. Gland, Switzerland:TK

FIFTEEN

Building Institutions for Sustainable Development in Acre, Brazil

MARIANNE SCHMINK

In 1986 the Federal University of Acre (UFAC) and the University of Florida (UF) began a program of technical cooperation focusing on ecological, social, and economic aspects of forest extraction, agroforestry, and agricultural systems used by rubber tappers and settlers in the state of Acre. The collaborative project sought to strengthen the technical capacity of UF, UFAC, and other local and national institutions through a research and training program based on the methodology of "adaptive research." The core objective of the program was to explore means to increase the income of local small producers and to reduce pressures on the resource base and on biological diversity. The multidisciplinary, interinstitutional project sought to contribute to the formulation of sustainable policies for agroforestry and agricultural development, emphasizing the rational use and conservation of natural resources.

The concept of sustainability has the virtue of drawing attention to the complex conditions under which socioeconomic and environmental change take place over the long term. Beyond these general notions, however, there is little consensus regarding the precise time frame, measures, and practical requirements for sustainability. Many definitions focus on technical production criteria for sustained yield harvests or on economic sustainability as measured by supply-demand and market projections. Broader questions of social and political sustainability, and of strategies to institutionalize sustainable development alternatives, are also essential determinants of the long-term success of technical and economic planning. Addressing these institutional issues requires the integration of insights from the natural and social sciences with the technical focus that has dominated the development field for the last few decades. The UF-UFAC project is designed to build a coherent institutional strategy within which technical, social, and political issues can be addressed over the long term.

Sustainable development requires a favorable social, political, and economic environment, including the organizational mechanisms to support the design, experimentation, and implementation of appropriate practices. Building these institutions is a time-consuming and elusive process, and there are few models. Just as technical solutions must be adapted to each specific site, strategies for institutional change are necessarily unique to local settings. One of the key potential contributions from social science to the conservation and development field is in the analysis of local responses (by specific social groups in particular ecosystems) to macro-level economic and political changes.

The experience of the UF-UFAC project in Acre provides lessons in institution-building strategies that may be useful elsewhere. First, the Acre project is grounded in an ongoing analysis of how political and economic processes influence resource use. This "political ecology" approach provided a framework for adapting institutional strategies to changing political and economic circumstances (Schmink and Wood 1986).

Second, the project promoted an integrated approach to development and conservation problems. The project experimented with mechanisms to bring together the resources of local institutions and with interdisciplinary, participatory methods of research and extension. It drew on the expertise of a network of specialists from Brazil and the United States. The "adaptive research" methodology (described below) brought together researchers and extensionists from different disciplines in a formal, structured dialogue with producers that provided a common reference point for reaching a consensus about priority problems and potential solutions. Technicians who participated in the training program formed an interinstitutional group capable of coordinating a common research, action, and policy agenda. This capacity was enhanced as several project participants moved into key decision-making positions in local governmental and nongovernmental agencies.

Definition of a common set of goals and an integrated, participatory work philosophy shared by representatives of a dozen organizations involved in development and conservation work in Acre are key accomplishments of the project so far. Another is the newly created interest group composed of project participants who can orient their various organizations in the direction of more effective collaboration in work with small producers in Acre. This group, representing public, private, and research sectors in Acre and their collaborators elsewhere, is in a position to adapt to changing opportunities and limitations and to build a sustained institutional effort on behalf of Acre's natural resource base and the small producers that depend on it.

The Political Ecology of Acre

Many of the factors that influence land-use practices lie outside the control, or even the knowledge, of local producers who are part of an increasingly integrated global system. The analysis of the interface between the macro-level political and economic context and specific ecological systems and local populations is therefore an essential complement to the focus on individual producers. Even apparently well-conceived plans for sustainable development often go wrong for reasons related to macro-level political and economic realities and their penetration at the local level. The goals of resource conservation and sustainable development are fundamentally at odds with most postwar nationalistic development projects oriented to expanded production and short-term capital accumulation. Ecological disasters also are often political and economic successes for those in power and their constituencies, which helps to clarify why apparently "irrational" policies continue to be implemented (Schmink 1987).

Understanding the political and economic context of ecological change requires going beyond the technocratic model often used in development and conservation circles, or a pluralistic conception of the state as neutral arbiter of competing interests. Actions by the state (including planning and administrative agencies, legislative bodies, the courts, the military, and police) tend to favor the interests of the economically dominant social classes. But different agencies and levels of the state apparatus have distinct agendas and constituencies. The analysis of such political dynamics, and of the changing configuration of state policies and practices, provide a stronger basis for pragmatic approaches to institutional change. Such approaches can seek to shift the balance of power in the direction of relatively disadvantaged groups within society (Esman and Uphoff 1984).

Brazil stands out as a country whose impressive postwar economic growth was accompanied by environmental degradation and persistent maldistribution of wealth. Capitalist investors were favored over small producers, who typically bore the brunt of development policies. Amazon development initiatives such as expansion of cattle ranching, hydroelectric power plants, and charcoal-based processing of minerals typically benefitted investors from outside the Amazon region at the expense of local economies and ecosystems. At times, particularly under the military dictatorship that held power from 1964 to 1985, the armed forces pursued their own agenda, even at the expense of the economic elites who were their main source of political support. But since a civilian government took over in 1985, pressures for reform have mounted, and a more open political system has forced the state to respond to broader constituencies.

One of those new constituencies is the population of rubber tappers who inhabit the forested areas of the western Amazon state of Acre. Their remarkable political mobilization emerged in response to the local effects of federal Amazon development programs. After a decade of local organizing efforts, in the mid-1980s, the movement gained the support of national and international nongovernmental organizations, and the rubber tappers embarked on a successful campaign to convince the federal government to create "extractive reserves," areas set aside for continued small-scale extraction. Small producers, especially the organized rubber tappers, have also gained greater access to resources and the decision-making process in Acre's state government. Analysis of the historical factors leading to this favorable institutional environment highlights the interactions between world markets, federal policies, migration, political organization, and local resource use (Allegretti 1979; CEDEPLAR 1979; Martinello 1988; Oliveira 1985; Schwartzman 1989).

Located on Brazil's western border with Bolivia and Peru, the settlement of Acre by nonindigenous peoples began a century ago, stimulated by the boom in extraction of natural rubber for world markets between 1850 and 1920. The region's indigenous groups also were absorbed into the rubber trade. Tens of thousands of migrants entered Acre as rubber tappers. At the time, the territory belonged to Bolivia, but after a series of conflicts it was purchased by Brazil in 1903. A second wave of migrants was recruited during World War II, mostly from northeastern Brazil, with United States financing. Acre's culture is strongly dominated by the northeastern origins of the majority of its population.

The rubber tappers traditionally lived an isolated existence dispersed among the distant *colocações*, homesteads where each tapper extracted latex from trees situated along trails through the forest. At the end of each day the tappers smoked the latex until it formed large balls that were sold at the trading post. The economic system was dominated by the supply and credit system known as *aviamento*, which maintained the rubber tappers in constant debt to the trading posts, in their turn controlled by the owners of the large rubber estates known as *seringais*. The tappers also paid rent to the owners for the right to tap the latex along the rubber trails on their land. The rubber economy was entirely oriented to export, and the large export houses in the cities of Belém and Manaus captured the profits from the rubber trade and the *aviamento* system. The producers themselves were often prohibited from engaging in subsistence activities, so as not to distract them from rubber tapping and to maintain their dependence on the trading posts.

The success of Asian rubber plantations beginning in 1911 spelled the end of the boom in Amazonian natural rubber, with the exception of the short-lived war effort financed by the United States when Japan blocked access to Asian rubber sources during World War II. With the decline of rubber, Acre's economy gradually diversified to include collection of Brazil nuts and subsistence agriculture, fishing, and hunting. After World War II, many rubber estate owners found themselves unable to repay the credit they had received under the special wartime programs. As a further threat to the traditional system, rising inflation in the 1950s and 1960s undermined the viability of the *aviamento* system, which depended on credit extended over long periods. During this period, rubber tappers in many areas became more autonomous as their erstwhile patrons went bankrupt or moved on to other activities.

The territory of Acre was elevated to statehood in 1962. The beginnings of federal investment in the region led to the growth of towns and new economic opportunities that attracted rural elites away from the declining rubber estates. More drastic changes began to take place in the 1970s, stimulated by ambitious federal programs in the Amazon under the military government that took power in Brazil in 1964. A road link was established between Acre and the capital in Brasília in the late 1960s, and the land along its borders was transferred to federal jurisdiction in 1971. During the early 1970s, Acre's governor took advantage of the euphoria surrounding the Trans-Amazon Highway colonization effort and other federal initiatives to open up Acre's lands for sale to southern investors. The newly created Amazon Bank (BASA) provided easy credit for new investments in agriculture and ranching, displacing previous support for rubber and other forest products. Land turned over rapidly as indebted owners of rubber estates sold their rights to outsiders. As elsewhere in the Amazon, land sales soon led to legal confusion, since titles had been issued in Acre by many different authorities throughout the region's history: the state of Amazonas; Bolivia; the short-lived independent republic of Acre under Plácido de Castro; and the federal land agency INCRA. Conflicting claims to title covered an area larger than the whole state by half.

While the state's lands changed hands, however, the rubber tappers who occupied them remained in the forests, where they continued to tap rubber, collect Brazil nuts, hunt, fish, and cultivate their subsistence gardens. Most of Acre's new landowners were interested in establishing cattle ranches, an enterprise that requires relatively little productive investment and for which credit was readily available. Their first task was to rid the land of the rubber tappers that occupied it, either by buying the rights to their modest homesteads, by threats, or by outright violent expulsion. Many tappers who had little experience with money and few employment alternatives were tempted to sell and move to town, where they quickly found they had few skills with which to make a living.

By the mid-1970s, many tappers had drifted back to the rural areas and, along with those who remained in the forest, began to organize through local unions. Their movement sought to defend the legal rights they had acquired, through long-term residence and use, to occupy the rubber trails. They developed nonviolent tactics to prevent the new landowners from clearing land, such as *empates,* or stand-offs, in which whole communities camped in front of bulldozers. They also sought to provide literacy and basic arithmetic skills, health care, and marketing alternatives that would increase the rubber tappers' autonomy from merchant intermediaries. The success of the tappers' resistance movement led to violence against its leaders; among others, Wilson Pinheiro (head of the union in Brasileia) was killed in 1980, and Chico Mendes (head of the union in Xapuri) was murdered in 1988.

By the late 1980s, the rubber tappers movement began to receive international attention due to links between local, national, and international nongovernmental organizations concerned with environmental protection and human rights (Schwartzman 1989). Environmental groups in the United States and Europe had begun to pressure the multilateral development banks to pay serious attention to the environmental impact of the projects they financed. One of the key targets of criticism was World Bank-financed colonization in the neighboring state of Rondônia. The existence of an organized local constituency provided an important argument against the extension of road-paving and colonization efforts beyond Rondônia's border into Acre. The movement gained in strength with this international support. The formation of the National Council of Rubber Tappers broadened the movement beyond Acre's borders, and links were formed with organizations of indigenous peoples. The prizes awarded to Chico Mendes in 1987 by the United Nations and the Future Worlds Society lent legitimacy and visibility to their struggle.

A new governor of Acre, elected in 1987, responded to this changing political scenario by declaring his intention to pursue "forest-based development" for the state. He created new state agencies concerned with environmental protection and with alternative technologies oriented to forest products. Their efforts range from large-scale, sustained-yield commercial harvesting of tropical hardwoods to modernizing the logging industry to implementing extractive reserves in the state.

The Challenge for Acre's Institutions

Grassroots pressure linked to local and international support networks has helped to counterbalance the pressures leading to further forest conversion for less sustainable and equitable land uses (such as cattle ranching) in Acre. Their impact on both federal and state policies has created a more favorable institutional setting for experimentation with sustainable resource-use systems appropriate for small producers. As dramatic political and economic changes continue in Brazil, putting these alternatives into practice still poses enormous organizational, technical, and political challenges. Not the least of these pressures is the continuing opposition by the regional landholding elites to the extractive reserves and other alternative land-use proposals (Schwartzman 1989).

The tremendous technical problems are most evident in Acre's newly created extractive reserves. Although rubber tappers have occupied the state for decades, current pressures on their production systems have led to a greater tendency toward forest clearing for agricultural and animal production. The tappers lack schools and basic health services, as well as technical assistance. The challenge is to increase and diversify forest productivity in sustainable ways, and to improve the marketing and processing systems so that forest producers are able to earn more and raise their standard of living.

The success of the extractive reserves in Acre will depend not only on resolving these internal problems, but also on the changes beyond their borders. The paving of the BR-364 highway from Pôrto Velho (Rondônia) to Rio Branco (the capital of Acre), now underway, will increase pressure on land and other natural resources. Development initiatives oriented to road-building and occupation by ranchers and farmers from outside the state threaten to increase deforestation, with ominous consequences for the local flora and fauna in an area of high biological diversity. This change in the state's productive base undermines the principal sources of survival for the rubber tappers (both inside and outside the reserves), traditional agricultural colonists, and indigenous communities. The whole range of land uses, competing interests, and multiple goals will affect the future of forest-based alternatives now being considered.

The diversity and complexity of these production systems, and the pressures they suffer, call for a comprehensive institutional approach that addresses the whole context. The implementation of both ecologically and socioeconomically sustainable development alternatives adapted to the situation in Acre will depend in part on the long-term capability of local organizations to monitor and influence the direction of development efforts. This means addressing both technical and political issues and strengthening the potential collaboration between grassroots organizations, nongovernmental support groups, research and teaching institutions, and government planning and extension agencies.

The UF-UFAC collaborative project sought to build on the underutilized resources of the universities to coordinate interinstitutional efforts to address the technical problems of implementing sustainable development for Acre's small producers. Like other organizations devoted to research and teaching, universities have participated little in the planning and implementation of Amazonian development projects. Yet because of their academic function, universities are an appropriate forum for discussion and experimentation with alternative technologies. Their quasi-governmental status also makes them more stable than most local nongovernmental organizations, yet relatively removed from local political factors that constrain state agencies.

Historically, Amazonian institutions have had a relatively weak role in shaping regional development processes. Most of the important economic decisions concerning Acre and other Amazonian areas have been made by national or international agents whose interests often run counter to the needs of the local population. The current institutional environment in Acre offers some promise of changing this pattern.

A network of local, national, and international nongovernmental organizations currently provides technical assistance and support to the movement organizations such as local unions and the National Council of Rubber Tappers. These NGOs also represent the interests of local small producers in negotiations with government and donor agencies. They are a crucial vehicle for participation by local populations in decisions about resource use.

The continued mobilization of local populations, and of adequate resources to support their efforts, is the most urgent element in assuring continued support for alternative development strategies now being proposed in Acre. Grassroots political pressure is key to a continuing dialogue between government and local populations, especially in a country where environmental politics are not yet institutionalized.

The responsiveness of public-sector agencies is not only a political but also a technical issue, since Brazil's agriculture and forestry research and extension agencies have not developed programs to respond to the needs of small producers. The UF-UFAC Collaborative project sought to build that technical capacity among researchers and extensionists from Acre's agrarian-sector agencies. Their collective work would contribute to the prospects for success in addressing the state's development problems over the long run.

The UF-UFAC Collaborative Project

During the initial two-year period (1986–1987) of reciprocal visits by UF and UFAC faculty and administrators, the two universities formed multidisciplinary groups of faculty and (in the case of UF) graduate students and defined a set of common goals. These were to

1. analyze small-scale family systems of production in Acre, the limitations they confront, and the interrelations among members of the household, crops, animals, forest products, the market, and the institutional environment;
2. combine the methods of social, biophysical, and agroeconomic sciences to identify and solve urgent problems of small-scale production under conditions of limited resources;
3. improve institutional and administrative relations and their potential impact on agrarian policies, infrastructure, and sustainable development in Acre;
4. apply the adaptive research approach to the creation and development of a program to generate, validate and disseminate scale-specific technology for long-term resource management by small farmers and rubber tappers; and
5. synthesize and disseminate the results of the program to a broader audience of researchers, policymakers, and the public in Acre, in the rest of Brazil, and in other tropical countries.

Once the interuniversity linkage was cemented through substantive discussions and a formal exchange agreement, a planning grant from the Ford Foundation permitted the UFAC and the UF to broaden their ties to other institutions in the United States and Brazil, especially in Acre. The UF provided a channel for exchanges with institutions and funding agencies outside Brazil. In Acre, governmental and nongovernmental agencies that work with small farmers, rubber tappers, and Indians participated in a two-day seminar to discuss the proposals of the two universities. This meeting laid the foundation for the participation by representatives of a dozen federal, state, and private agencies in the training-and-research phase of the project.

At the outset, project participants focused on two principal groups of small producers in Acre. The large number of rubber tappers who inhabit the region, with their profound knowledge of the Amazonian forest, are central to the project. Rubber tappers exploit rubber (*Hevea brasiliensis*) and Brazil nuts (*Bertholetia excelsa*), both products of great importance to Acre's economy. They also exploit many other forest products. It is the contention of this work and the object of UF-UFAC collaboration that with proper technical assistance, the economic viability of their traditional activities can improve without irreversibly altering the environment.

A second target of interest to the project is small agricultural producers, a group comprising former rubber tappers from Acre and migrants from other regions of the country. With the paving of the BR-364 highway from Pôrto Velho to Rio Branco, the number of small farmers is sure to increase. Population growth creates an urgent need to identify and encourage sustainable and environmentally sound agricultural practices. Such methods must be developed not with ideal conditions in mind, but rather based on a realistic assessment of the constraints small farmers confront in the region.

The Training Strategy

The primary activity of the UFAC-UF collaborative project was training of the interinstitutional group (later named PESACRE) during two short courses in 1988 and 1989, with the support of grants from the Ford Foundation. Participants included specialists in agronomy, soils, forestry, biology, economics, sociology, anthropology, geography, education, statistics, chemistry, and physics. They represented seventeen institutions, including universities in Acre, Amazonas, and Florida; nongovernmental organizations; state agriculture, forestry, and environmental agencies; and local branches of federal or regional agricultural research and extension agencies; and those in charge of colonization and environmental protection. The participating institutions have mandates in the areas of teaching, research, extension, agricultural settlement, community development, environmental protection, and Indian affairs. The UF trainers (an anthropologist, an agricultural economist, and an agronomist), developed the courses with the support of UF's International Training Division and produced a Portuguese-language manual based on farming systems research and extension training materials (Hildebrand, Poats, and Waleka 1989). Several members of the PESACRE group participated as trainers during the second course. The UFAC handled local arrangements.

The adaptive research approach presented in the courses was a modified version of farming systems research and extension (FSR/E), with a strong emphasis on environmental protection and natural resource management. FSR/E is a multidisciplinary method that joins research and extension in a common effort to develop appropriate technologies (Hildebrand 1986). Producers are integrated with researchers and extension agents in a systematic procedure to identify and resolve problems. Multidisciplinary teams are composed of personnel from different research and extension institutions, who work with producers to identify problems and limitations and then to create, adapt, and test specially designed solutions.

The teams use rapid surveys (called *sondeios* in Portuguese) to identify existing problems and limitations on small farms or rubber trails. The producers are incorporated into a sequence of activities that formulate and test solutions in the producers' own fields, as well as on experiment stations. Close contact is maintained with producers, and the activities of research and extension are integrated instead of separated. As a result, communication problems are reduced and the gap between problem identification and technology adoption is minimized. In this way, the method incorporates the adaptation, learning, and diffusion by the "client" population, steps that have always been very important for innovation and change in agricultural technology. Participation by representatives from key institutions also provides a direct link to planning and infrastructural development.

The two-week course in 1988 covered FSR/E diagnostic methods, whereas the second three-week course provided a complete sequence including diagnosis, design, and analysis of on-farm or on-site experimentation. The thirty-four course participants in 1988 and the forty-one in 1989 included four UF graduate students who developed their own research projects based on the course. Other research and extension project proposals emerged from the discussions of priorities by the PESACRE group during and after the two training courses.

Interviews with producers using the *sondeio* method provided the material for the diagnosis of key problems. Seven producers, including two tribal Indians, were invited to the training site in 1988 to provide information about their production systems as the basis for training in interviewing and modeling systems. During both years, course participants carried out three-day *sondeios* with agricultural settlers and rubber tappers. A total of 112 settlers were interviewed in two colonization areas near Rio Branco. The diagnosis of settler systems indicated that colonists who had previously lived in Acre as rubber tappers used the forest resources on their lots more intensively than did the migrants.

During the same period the group interviewed thirty rubber tappers dispersed in five different areas near the towns of Plácido de Castro and Xapuri. The rubber tapper diagnosis yielded significant insights into the differences between these two regions. In both 1988 and 1989, course participants produced written diagnostic reports with conclusions and recommendations for both groups of producers (PESACRE 1988; PESACRE 1989a; PESACRE 1989b).

Diagnosis

Settlers Interviews with settlers showed that planning and implementation of the colonization projects were carried out without due consideration of the area's natural resource potential. The project failed to take advantage of the empirical knowledge of the rubber tappers who were already living in the colonization areas. In contrast to the rubber tappers, the majority of the settlers came from other regions of Brazil and lacked knowledge of the potential of the natural resources and how to use them rationally. Moreover, technical knowledge to indicate areas appropriate for colonization was not derived from field studies. This led to deforestation of areas rich in Brazil nut trees and rubber trees for agricultural purposes, despite their low soil fertility and poor water resources.

The production systems used by these settlers are based on cutting and burning the forest and planting rice, corn, beans, cassava, and different fruit trees. However, many of these settlers do not remain in the areas they initially colonized, since their production systems are not sustainable. The instability of the production systems is usually a result of inadequate use and management of natural resources, leading to general environmental degradation.

A host of problems undermines the economic potential of colonists' agricultural production. Most of the agricultural products are harvested during and at the end of the rainy season, when unpaved roads do not permit access to markets. The lack of on-site storage facilities reduces the quality and value of the products. In many instances, production costs and the end price of the products place these small farmers at a distinct economic disadvantage. Yield declines, either as a consequence of changes in soil fertility or of increased weed infestation and the lack of technology, extension, and inputs (seeds, fertilizers, pesticides, and equipment), increase the vulnerability of these small farmers to the prevailing economic forces. Other factors limiting the activities that could be developed by the settlers include the lack of water sources or its poor distribution in the property, and the poor quality of the public health services.

The small farmers understand the fragility of their production systems and hope to base their activities on crops (such as coffee, cocoa, black pepper, and sugarcane) and cattle (as a source of protein and as savings for emergencies), which demand less labor once established and have more value in the market. Once accomplished, this would result in more stable production systems that could guarantee enough income.

The expansion of the areas with these crops and with pastures is a reality in the colonization projects in Acre. Pasture establishment in areas formerly used for agricultural purposes is also a means of increasing the sales value of the land. The migration of the small farmers in search of new forest areas or to the urban areas (where they live in slums) once the natural resources are exhausted, and the high rate of land turnover, reflect the results of this sequence of conversion from forest to agriculture to pasture.

Despite the problems mentioned above, agricultural production increases in the colonization projects, and some small farmers have managed to consolidate their holdings. Among the settlers who still remain in their areas, property is seen as a means to guarantee the family's future. They believe that despite their difficult living conditions, they would not be able to guarantee the same standard of living in the city. There, they would necessarily depend on a salary which, besides being difficult to obtain, is not enough for the family's survival. On their properties, even under the worst of conditions, they are able to feed their families.

Rubber Tappers The extraction of rubber and Brazil nuts from the natural forest provides most of the income for rubber tappers in this region of Acre. In addition, they make use of a variety of other natural forest products, cultivate small swidden subsistence plots, and raise domestic animals for their own use. Some rubber tappers have begun to expand their cattle herds and pastures.

The productive unit of the native rubber tapping system, the *colocação,* is measured not by land area but in accordance with the distribution of the looping trails through the forest along which the rubber trees are distributed. A tapper family exploits an average of three trails, with about one hundred fifty to two hundred rubber trees each, from which liquid latex is extracted on a daily basis during the drier season lasting from April through December. Tapping and processing rubber requires nine to ten hours per day of the adult male members of the family. During January and February, men devote most of their time to the collection of Brazil nuts if they are abundant within their *colocação.* These two products provide nearly all of the family's cash income. Women and children are responsible for the care of domestic animals and help with most phases of production and processing of maize, rice, beans, cassava, and fruits used by the family.

Compared with the small farm systems of the settlers, the rubber tapping systems of production provide a relatively stable subsistence and have a far less drastic impact on the natural environment. However, rubber tapping systems are also undergoing change due to development pressures. Common problems face the rubber tapper families in the five areas where fieldwork was carried out, but the potential for resolving these problems depends to a great extent on institutional factors.

The isolation of the tappers settled in small clearings several hours distant from one another presents formidable problems of transportation, marketing, and access to services. Growing pressures on the natural resource base and uncertainty regarding land tenure have tended to undermine the stability of the natural forest extraction practiced over decades. Finally, the dependence of most rubber tapper families on the sale of rubber and Brazil nuts as their sole source of monetary income will become increasingly untenable as new plantations of both products within Brazil enter the market in the near future.

Despite these difficulties, the rubber tapper system of production is remarkably stable compared to other land uses in the state. Its impact on forest resources has been minimal and selective. Rubber tapper families are closely tied through informal kin and nonkin cooperation and show little disposition to move on to other activities. Instead, they seek to diversify their production systems and improve their productivity in order to remain as forest dwellers without undue dependence on the market for their subsistence needs. Their possibilities for achieving these goals depend on the potential for resolving the problems of land tenure, marketing, and access to technical assistance and other services through increased autonomy and internal organization. The degree of autonomy from landowners and intermediaries, and the extent of political organization, determine significant differences among current systems of natural forest rubber tapping.

Tappers located near Xapuri, where most tappers have been free of domination by patrons for some time and where political organization is particularly strong, have begun to address some of the most important constraints to improving productivity and sustainability. With the support of nongovernmental organizations they have constructed a number of schools in remote areas, which provide basic literacy and arithmetic skills that allow the tappers and their families to defend themselves in market transactions and other interactions (Campbell 1990). The schools also provide an important focus for community activity and a meeting place for political discussions. Their presence helps to stabilize a population of young people who would otherwise move elsewhere because of the lack of schools. In contrast, there are no schools in the rubber areas close to Plácido de Castro, where political organization is weaker, and tapper families must send their children to nearby towns to study.

Near Plácido de Castro, the rubber tappers still depend to varying degrees on trading post owners or other intermediaries, sometimes referred to as patrons, for transport of their rubber and other goods and for provisions. Only those fortunate enough to be located close to a road can choose with whom to trade and purchase what they need in the nearby town. Because the traders control transportation and prices, the producers have responded by expanding their subsistence activities in order to avoid market dependence. By contrast, the rubber tapper movement near Xapuri recently founded a cooperative that purchases rubber, Brazil nuts, and other products from producers at a better price, and provides transport at cost between central warehousing points and isolated *colocações*. The cooperative also sells basic provisions at a lower cost to its members. As a result, other traders in this area have adjusted their own prices in order to compete.

The sustainability of the rubber tapper system depends on finding ways to improve productivity while maintaining a functioning natural ecosystem. Development of technologies and market strategies to exploit a wider range of forest products on a sustained-yield basis requires a long-term research and extension effort in collaboration with the people who know and live in the forest. Assurance of current land rights and of those of future generations are a prerequisite for such a project. In response to immediate market pressures and to the uncertainty of their future as rubber tappers, producers sometimes knowingly overexploit rubber trees or produce lower-quality rubber through less time-consuming processing methods. In politically organized areas, especially those set aside as extractive settlements, however, the population has begun to impose its own regulations on rubber extraction and processing; clearing for pasture and along watercourses; extraction of timber for sale or cutting of fruit trees for fruit; and hunting with dogs or for more than subsistence needs. Tappers also protect useful trees in the forest, along rubber trails, and near clearings and experiment with transplanting local species in abandoned fallows. These innovations, and the local organizations that have supported them, provide the potential for future research and extension efforts.

The Research and Extension Strategy

The diagnosis carried out by the PESACRE group indicated the dynamic character of local small production systems. The evident diversity and change underscored the need to build long-term problem-solving capabilities. The research revealed a lack of cooperation among local institutions as one of the main factors limiting the capacity to develop a process of planning rational policies for agricultural and agroforestry development in the state of Acre. This suggested the need not only to strengthen the institutions, but also to intensify the cooperative nature of activities in search of sustainable agroforestry and agricultural systems adapted to the soil and climatic conditions of Acre.

Priorities for these cooperative actions included a) study of sustainable agroforestry and agricultural systems aimed at recovering degraded areas; b) development of rational methods of use and management of natural resources; c) study of methods of pest and disease control for the main annual, perennial, and horticultural crops in the field and during storage; d) study of the economic potential of native plants and animals; e) development of mechanisms aimed at strengthening rural communities; f) adaptation of education techniques and school calendars according to the activities of small farmers and rubber tappers; and g) recuperation and study of the culture of rubber tappers.

As a result of the diagnostic reports, several research and extension projects were initiated in 1988 and 1989 by members of the PESACRE group and by graduate students from the UF. A small "seed money" fund from the Ford Foundation supported preliminary research, and local institutions contributed their human resources and infrastructures. Some studies have already been concluded and the results made available for future applied work (Campbell 1989; Campbell 1990; Pinard 1991).

In 1990, PESACRE was consolidated as an independent nongovernmental organization dedicated to research, training, and dissemination of information related to sustainable development practices for small-scale producers in Acre. In partnership with the University of Florida, the group was awarded a three-year grant of nearly $1 million from the United States Agency for International Development to support research, extension, and training activities in agroforestry and natural forest management. These funds permit the group to consolidate its role in facilitating interinstitutional cooperation to address the complex problems facing the state of Acre.

PESACRE group members also began to integrate the goals and philosophy of the project's work with low-resource producers into their own organizations' programs. By the end of 1989, group members had taken positions as president of the Workers' Center (CTA—a key NGO supporting the rubber tappers' movement) and as chairman of the agrarian sciences department at the UFAC. Others later were named to responsible positions in local offices of national, regional, and state organizations including Institute for Technical Assistance and Rural Extension (EMATER), Brazilian Institute for Agricultural Research (EMBRAPA), Technological Foundation of Acre (FUNTAC), Environmental Institute of Acre (IMAC), and National Institute for Amazonian Research (INPA). In 1990, UFAC project coordinator Mâncio Lima Cordeiro was named Secretary for Agrarian Development of the state of Acre.

Novel Strategies for Institution Building

The UF-UFAC Collaborative Project experimented with a novel combination of strategies to build institutional capability and to develop a sound methodological basis for research and action. Participants included researchers and extensionists selected because of their common interest in or ongoing work with small producers within their own organizations. The project provided these technicians with the skills to be responsive to emerging problems and with the opportunity to work with those from other agencies with different points of view. The links between research, extension, and policy help to mold an integrated, not a sectoral, perspective. The multidisciplinary, interinstitutional group called PESACRE is a key product of the project. The group neither substitutes for nor competes with existing organizations, but rather provides integration and a forum in which group members can work and discuss common issues outside the constraints of their own institutions. The group functions not as a political interest group but as a professional collective, a cross-institutional vehicle for debate, discussion, sharing of information, and linking of research, policy, and extension.

Experiences elsewhere have shown that issue-oriented working groups can be an effective way to draw together the resources and knowledge of researchers, planners, and extensionists who are otherwise dispersed (Schmink 1986; Schmink et al. 1986). More efficient use of existing information provides a firmer basis for innovation. The ascension of several Acre project participants to decision-making positions was an unforeseen opportunity to strengthen the institutional bases for the project. Other group members have shifted from one agency or job to another during the course of the project. Over time, these connections help to consolidate an "institutional field" to meet the challenge of implementing sustainable development programs now being proposed for Acre.

The UF-UFAC training strategy was the key to establishing a strong basis for the PESACRE group in a short time. The FSR/E methodology, especially the *sondeio* and discussion of its findings, trained participants to listen to producers and understand their logic and priorities and to observe whole systems and their interrelationships rather than focus only on their particular professional interests. Preparation of reports forced participants to listen to one another, reach a consensus on findings, set priorities, and develop recommendations. Literally every word of the *sondeio* reports was discussed by the group, and the reports were finished within a week after the end of the fieldwork. The collective research and training experience provided in a matter of weeks a vehicle for agenda-setting that took more than a year of monthly meetings for working groups elsewhere to achieve (Schmink 1986).

A relatively favorable political climate for the implementation of alternative technologies for small producers exists in Acre, at least for the short term. The continued strength of grassroots political movements will help to assure the possibility of continuing these programs. The state of Acre is small enough to make a statewide impact possible under current conditions. The state agency FUNTAC is coordinating a special, intensive effort to implement Acre's extractive reserves. The work thrust of the Secretariat for Agrarian Development has complemented that effort with a statewide extension program. The university and other research agencies will provide the long-term research and training support. It is impossible to predict Acre's political future, which is subject to change with each gubernatorial election. But the interinstitutional character of PESACRE gives the group the flexibility to adjust the overall work strategy to future constraints and opportunities.

The UF-UFAC collaborative project is oriented to the long-term process of building the strength of existing local institutions by making more efficient use of scarce and dispersed human and material resources. The role of outside "experts" was primarily as facilitators and as equal collaborators in this process. The project philosophy was not to introduce exotic technical packages or solutions, nor to produce world-class research based on external technical assistance.

Building up from existing human and institutional resources takes longer but has important multiplier effects that improve the long-term capability to work with outside experts, to develop appropriate technical solutions, and to sustain political support for their implementation. If the program is successful in achieving its goals, the ultimate beneficiaries will be the limited-resource producers in Acre and the natural systems on which they depend for their livelihood.

LITERATURE CITED

Allegretti, M. H. 1979. Os Seringueiros—Um estudo de caso em um Seringal Nativo no Acre. Master's thesis, University of Brasília.

Campbell, C. 1989. Community mobilization and education for conservation: a case study of the rubber tappers in Acre, Brazil. *Latinamericanist* 24:2 (May):2–7.

—— 1990. The role of popular education in the mobilization of a rural community: A case study of rubber tappers in Acre, Brazil. Master's thesis, University of Florida.

Centro de Desenvolvimento e Planejamento Regional (CEDEPLAR). 1979. Migrações Internas na Região Norte: O caso do Acre. Belo Horizonte, Brazil: Universidade Federal Minas Gerais/SUDAM.

Esman, M. and N. Uphoff. 1984. *Local Organizations.* Ithaca, N.Y.: Cornell University Press.

Hadley, M. and K. Schreckenberg. 1989. *Contributing to Sustained Resource Use in the Humid and Sub-humid Tropics: Some research approaches and insights.* Paris: UNESCO, MAB Digest 3.

Hildebrand, P., ed. 1986. *Perspectives on Farming Systems Research and Extension.* Boulder, Colo.: Lynne Reinner Publishers.

Hildebrand, P. S. Poats, and L. Waleka. 1989. *Introdução a pesquisa e extensão em sistemas agrícolas e florestais.* (Translated and revised by M. Proença and C. Rancy). Gainesville, Fla.: University of Florida.

Martinello, P. 1988. A "Batalha da Borracha" na Segunda Guerra Mundial e suas conseqüências para o Vale Amazónico. Cadernos UFAC 1, Série C, Estudos e Pesquisas. Brazil: Universidade Federal do Acre.

Oliveira, L. A. P. de. 1985. *O Sertanejo, o brabo, e o posseiro: Os cem anos de andanças da população Acreana.* Rio Branco, Brazil: Fundação Cultural do Acre.

PESACRE. 1988. *Curso sintese de pesquisa e extensão em sistemas agroflorestais (PESA), relatório final.* Rio Branco, Brazil: Universidade Federal do Acre/ University of Florida (UFAC/UF).

—— 1989a. *Método de pesquisa e extensão em sistemas agricolas e florestais (PESA), relatório do 2O. treinamento.* Rio Branco, Brazil: Universidade Federal do Acre/University of Florida (UFAC/UF).

—— 1989b. Sustainability of agroforestry and agricultural systems used by rubber tappers and settlers in the state of Acre, Brazilian Amazonia. Paper presented at the 9th Annual Farming Systems Research/Extension Symposium, October, Fayetteville, Arkansas.

Pinard, M. 1991. Impacts of stem harvesting on populations of *Iriartea deltoidea* (Palmae) in an extractive reserve in Acre, Brazil. Master's thesis, University of Florida.

Schmink, M. 1985. The "working group" approach to women and urban services. *Ekistics* (January 1985), 310:76–83.

—— 1987. The rationality of tropical forest destruction. In J. C. Figueroa Colón, F. H. Wadsworth, and S. Branham, eds., *Management of Forests in Tropical America: Prospects and Technologies*, pp. 11–30. Rio Piedras, Puerto Rico: United States Department of Agriculture Forest Service.

Schmink, M., J. Bruce, and M. Kohn. 1986. *Learning About Urban Services for Women in Latin America and the Caribbean.* New York: The Population Council.

Schmink, M. and C. H. Wood. 1987. The "political ecology" of Amazonia. In P. D. Little and M. M. Horowitz, eds., *Lands at Risk in the Third World: Local Level Perspectives*, pp. 38–57. Boulder, Colo.: Westview.

Schwartzman, S. 1989. Deforestation and popular resistance in Acre: From local movement to global network. Paper presented at the meeting of the American Anthropological Association, November, Washington, D.C.

PART IV

New Directions in Research and Action

Introduction

KENT H. REDFORD AND
CHRISTINE PADOCH

The task of integrating the goals of biological conservation with the rights of traditional peoples falls into the chasms that separate disciplines; anthropologists are not versed in biology, biologists are not taught economics, and foresters do not learn political science. Yet these and many other disciplinary perspectives and tools are needed to address the problems facing managers and politicians. In part 4, "New Directions in Research," the authors provide examples of novel ways of both doing and thinking designed to integrate the needs of forest peoples with biological conservation.

Forest products, both timber and nontimber, are among the most heavily exploited natural resources in tropical forests. The first two essays in part 4, by Jan Salick and by Kent H. Redford, Carolina Murcia and Bert Klein, discuss new ways in which such products can be sustainably harvested. Salick worked with the Amuesha Indians in the Peruvian Amazon, where the well-known Palcazu Project is based. It was here that experts developed the "strip-cut" method, a form of natural forest management based on the dynamics of forest gap regeneration. Strip-cutting should simultaneously allow for the extraction of timber, the conservation of nontimber forest products, and the preservation of biodiversity.

To the Amuesha Indians, nontimber products are extremely important, and conservation of them is of the utmost importance. Salick discusses the distribution and uses of useful plants in different forest types in her essay, showing that all forest types are used intensively by the Amuesha and that each area has a unique suite of species of useful plants. Then, through the use of native informants, Salick analyzes the impact of the strip-cut method on the abundance of useful plants. She concludes that, at least in the short term, natural forest management is compatible with the management of nontimber forest products crucial to Amuesha subsistence.

In a similar light, Redford et al. propose the incorporation of wild game animals into tropical agroforestry systems. Virtually all attention and work on agroforestry has been devoted to the botanical side of research, and if animals are incorporated, they are virtually always domesticated species. Presenting data and a set of recommended species, the authors present a methodology for use in areas of low human population density and large tracts of relatively undisturbed habitat. These conditions are necessary because, if violated, the agroforestry-game system would serve only to concentrate game animals, making them easier to hunt rather than increasing game populations. This is a phenomenon frequently unrecognized by those advocating habitat modification as a way of increasing hunting.

The next four papers address the discrepancies between the ways in which tropical forests are used by their human inhabitants and the values placed on these forests by the larger economic and political systems.

In the first of these essays, Peter May analyzes the social equity effects of changes in property regimes resulting from permanent land use conversion and technological innovation. Making the case for the importance of forest extraction to the well-being of many local peoples, he points out that many extractive schemes have been managed for sustained yield under common property management regimes. Using case studies of rubber tapping and *babaçu* nut harvesting, he documents how agrarian change, associated deforestation, and technical innovations are hastening the demise of extractivism.

In order to redress this situation May proposes institutional shifts in property rights systems such as those championed by rubber tappers in the Brazilian state of Acre. He concludes with a list of suggested areas in which funding organizations could invest in order to promote institutional modifications and support refinement of common property management practices in use by Neotropical forest dwellers. It is clear that without action of the sort suggested by May, the rural poor will either become poorer or migrate and add to the already overwhelming masses of urban poor.

The question of the value of different land use systems is picked up in essay 19 by Susanna Hecht. Starting from the premise that the rate of deforestation in the Amazon must be slowed, and the fact that much of the clearing is being done as a means of controlling land and consolidating access to the economic benefits of landed property, she compares the three most common land use systems in Amazonia: colonist agriculture, cattle production, and forest extraction.

With the provocative title "The Good, the Bad, and the Ugly," Hecht constructs a model that analyzes the costs and returns of these systems, comparing only the aggregated gross returns. The conclusion reached by Hecht is that even though extraction as a land use is not very lucrative, its returns from productive activities based on actual use of land resources is better than either cattle production or colonist agriculture. She ends with the important statement that throughout the Amazon—and, for that matter, throughout much of Latin America—rural households engage in forms of extraction for both subsistence and cash. These activities are everywhere undervalued by economists and development planners. Hecht ends with a set of recommendations that would begin to ameliorate this by allowing for a more precise analysis of the economics of forest extraction.

In the next essay, Jason Clay discusses novel efforts to improve the economic situation of forest extractors. During the last few years, Cultural Survival, a indigenous-rights nongovernmental organization, decided that the best way to protect the future viability of human communities, while at the same time ensuring the conservation of the forests that they depend upon, was to expand the market for sustainably harvested forest commodities. This organization is working to increase the world market for selected tropical forest products, and helping to establish cooperatives among traditional forest dwellers who collect these forest products. In so doing, Cultural Survival is hoping to guarantee producers higher and more stable prices for their commodities and generate income for appropriate and sustainable development projects in tropical forest areas. So far the most successful product produced through this scheme is Rainforest Crunch, a confectionery made with products produced in the Brazilian rain forest.

In this effort, Cultural Survival has blazed a new trail for organizations concerned with the plight of the rural poor. Clay is open in his discussions of the potential drawbacks of this type of action, pointing out the biological, social, and economic dangers that must be avoided if large-scale marketing of "minor forest products" is to be sustainable.

Last, R. J. A. Goodland provides a provocative discussion of what he deems to be the research and action priorities for the conservation of tropical forests and the traditional peoples relying on them. He lays out four areas that must be addressed in order to stop the loss of Neotropical moist forest: population management, economic incentives, education/political pressure, and site-specific alternatives.

Goodland does not agree with the views of some of the other authors who contributed to this volume. In particular, he feels that in most cases there is no need for more research, that what is needed is implementation of the research that has already been done. With his broad overview, Goodland's paper challenges many of our societies' practices and lays out a blueprint for achieving sustainable development, or a reasonable and equitably distributed level of economic well-being that can be perpetuated for many human generations. It is clear that only with this set of goals can we hope to combine the conservation of tropical forests with the challenge of ameliorating the situation of the tropical forest peoples.

SIXTEEN

Amuesha Forest Use and Management: An Integration of Indigenous Use and Natural Forest Management

JAN SALICK

Worlds are to be learned and applied from indigenous management of tropical forests. Conservation and management of tropical ecosystems are challenges to developers, conservationists, and indigenous populations alike. In the Amazon basin, as in many other areas of the tropics, biodiversity must be conserved for the future use of humankind, resources should be developed and managed rationally for the national good, and indigenous peoples must be assured of continuing access to and availability of natural resources that sustain their livelihood. Although these goals appear divergent, their proponents are beginning to bridge their differences and integrate their efforts. In the long run, no one group or goal can be accommodated without recognizing the rights and rationales of the others.

In one effort to coordinate the development of forest resources with the conservation of tropical rain forest biodiversity and the management of forest products used by an indigenous group of Amazonian people, the Amuesha of Peru, the system of natural forest management allows for the regeneration of nontimber vegetation along with the timber products for which the system was devised. The regeneration of these nontimber products helps to conserve the subsistence and forest-use practices of the native people who derive an income from timber products. Only through natural forest management can the demand for both timber and cash income be met and nontimber forest products and biodiversity be conserved.

Indigenous plant specialists have pointed out some problems with the system and some anomalies in the regeneration of harvested forest stands, but these do not seem insurmountable. The problems are the exceptions that give us hope for integrating disparate goals. The problems also define new areas of conservation biology that need to be explored.

The People and the Setting

The Amuesha people (or Yanesha, as they refer to themselves) are a small, indigenous group of about five thousand individuals who inhabit the tropical rain forests of east-central Peru (see figure 16.1). Linguistically, the Amuesha are classified as Arawakan and are rec-

Figure 16.1 Center of the Amuesha People

ognized as being very distinct from their nearest neighbors, the Campa (Wise 1976). Historically, larger populations of Amuesha probably inhabited the upper drainages of the rivers Palcazu, Pichis, Pozuzo and Perene, although now the group is centered in the Palcazu Valley, where the study that is the focus of this paper was conducted (Smith 1977). Nine native communities have received land titles in the valley (Smith 1983), and groups and cooperatives for the production and marketing of agriculture, pottery, and forest products (Stocks and Hartshorn) and the management of communal reserves and the Yanachaga-Chemillen National Park (Brack 1987) have been formed. The basis of the Amuesha economy is subsistence agriculture (Salick 1986a; Salick 1986b; Salick 1989b; Salick and Lundberg 1990) supplemented by fishing, gathering, off-farm labor, and, to a lesser extent, hunting (Smith 1977; Barclay 1985). Here, the Amuesha use and management of forest products and the integration of these activities with commercial forestry is explored.

Commercial forestry in the Palcazu Valley has become an important concern to the Amuesha and other inhabitants since the recent construction of a road into the area. The Carretera Marginal (Perimeter Road) was envisioned by the previous president of Peru, Fernando Belaunde Terry, to connect the valleys of upper Amazon tributaries along the eastern edge of the Andes. Sections of the road, including a dirt road into the Palcazu Valley connecting the valley with the Central Highway and the markets of Lima, were partially completed during his tenure.

President Belaunde also instigated development projects along the Carretera Marginal funded by numerous national and international donor agencies including The United States Agency for International Development (USAID) in the Palcazu Valley. During the early 1980s USAID in Washington was experiencing pressure from groups concerned both with tropical deforestation and with indigenous peoples' rights. Specific mention was made of the Palcazu Valley and the Amuesha. USAID responded to this pressure by forming an environmental Central Selva Natural Resource Management Project, which included anthropologists devoted to furthering native rights and practices, foresters and agronomists intent on conserving and managing tropical biodiversity, and myself as an ethnobotanist studying indigenous natural resource management and conservation. A central concern of the project was that land uses should reflect environmental capabilities. Thus a national park was formed along the Yanachaga-Chemillen Valley crest, protected forests were designated along the San Matias crest and elsewhere (totaling 60 percent of the valley area), 23 percent of the total area was given to managed forests, and agricultural zones were planned for permanent and annual crops. The forestry sector dominated; a native forestry cooperative was formed (Stocks and Hartshorn in press) and a system for natural forest management designed (Hartshorn 1989). Agricultural projects, which took up a minor proportion of the land area, were based on principles of sustainable agriculture and maintenance of indigenous farming systems (Staver 1984).

Natural Forest Management

Natural forest management is an attempt to work with the diversity of a natural forest habitat rather than replace it with one or a few timber species grown in plantations or in enrichment plantings. In the tropical rain forest, where the conservation of biodiversity is a priority, one of the major advantages of natural forest management is that with the regeneration of timber comes a wealth of other plants not included in reforestation or enrichment planting schemes. Only through natural forest management is there hope to meet simultaneously the demand for both timber and cash income and to conserve nontimber forest products and biodiversity for present and future use.

The natural forest management scheme developed within the Central Selva Natural Resource Management Project has several distinguishing features (Hartshorn 1989). The method is based on an understanding of the dynamics of natural forest gap regeneration (Hartshorn 1978; Special Feature 1989) and the principles of strip cutting (Parker 1986). Low- or medium-level technology is used. Long, thin strips of rain forest are clear-cut across the slope. Oxen then extract the logs and poles, which are milled for timber, chemically preserved for posts, or baked into charcoal by an Amuesha cooperative devoted to the harvesting and processing of forest products. Natural forest regeneration resulting from coppicing, seed banks, and seed rain then begins; a rotational cycle estimated at thirty to fifty years is planned. Little manipulation of the regeneration or successional stages is projected. The lack of deliberate manipulation allows for the regeneration of nontimber vegetation, albeit inadvertently, along with the timber products for which the system was proposed.

Ethnobotanical studies have shown the importance of nontimber products to the Amuesha (Salick 1986c). Several aspects of Amuesha resource use, including agroforestry systems(Salick 1986a, Salick 1986b, Salick 1989b, Salick 1989c, Salick and Lundberg 1990 and Salick and Lundberg in press), and methods of germplasm management (Salick 1986d, Salick 1986e, Salick 1992, and Salick and Merrick 1990), have been discussed elsewhere; their use of forest products will be discussed here (see also Salick 1986c).

Nontimber Forest Products

The usefulness of nontimber forest products in the tropics has been long recognized (Burkill 1935), and forests have been managed for the production of a few commercially important products (rubber, tannins, rattan, damar, medicinals, etc.). Indigenous groups have intensively used (Boom 1985; Boom 1986; Balée 1986; Prance et al. 1987) and managed their forest resources for centuries (Posey 1983; Posey 1984; Balée and Gély 1986). Extractive reserves dedicated to nontimber production are being set up in Brazil (Schwartzman 1986), but little attempt has been made to manage forests for nontimber products within commercial forestry ventures. The natural forest management portion of the Central Selva Natural Resource Management Project offered an opportunity to initiate management of forests for useful products along with the development of timber resources.

Two studies on Amuesha use and management of forest plants were carried out: 1) a quantitative comparison among major forest types of plants used by the Amuesha and 2) an attempt to integrate indigenous use with natural forest management.

Comparisons of Useful Plants Among Forest Types

The five lowland forest types most common on the Palcazu Valley floor were sampled for useful plants. Three were undisturbed types: forests on alluvial soils, forests on weathered clay soils of terraces and hills, and forests on white sand outcrops. These divisions are recognized both by Amuesha indigenous taxonomy (Salick 1989b, c) and by academically trained ecologists (Foster 1981). Two were disturbed

Table 16.1
Comparison of Number of Useful Plant Species Per 10m² in Five Lowland Forest Types

Forest Type	Number of 25 m² Plots Sampled	Number of Plant Species Used Per 10m²	Number of Significant Differences Among Forest Types
Primary Lowland Forest Types			
Weathered clay soils	9	5.1	n.s.
Alluvial soils	3	4.0	n.s.
White sand soils	2	3.4	n.s.
Secondary Forests			
Slash-and-burn fallows	4	5.6	n.s.
Selectively logged	4	3.6	n.s.

forest types: fallows resulting from Amuesha swidden agriculture and selectively logged forest.

In these five forest types, randomly located plots measuring 5 × 5 meters (25 square meters) were inventoried for useful plant species employing an Amuesha herbalist as an informant (table 16.1). Useful nontimber plants found within the plots included species producing edibles, medicinals, resins, oils, ritual objects, culinary herbs, drinks, dyes, poisons, materials for containers, crafts, and construction, and more (see appendix). These products are used by the Amuesha much more frequently and intensively than is wood, which is primarily important as a source of firewood, posts, canoes, and paddles, rather than lumber.

The major findings of the study were that *all* forest types, both primary and secondary, are intensively used by the Amuesha and that no one forest type has significantly more useful plants.* Different for-

*It is possible that with further sampling, quantitative differences in the number of useful plants among forest types would be detected, but this would not detract from the overall importance of habitat diversity.

est types, however, provide different useful plant species. This finding is important, since some forest types are disappearing, especially primary forest on alluvial soils suitable for agriculture. The herbalist informant said that he is already limited in his practice of traditional medicine by the lack of key species found only in that habitat. He also said that overexploitation of some species in other habitats has made them extremely scarce. Thus if a forest subsistence base is to be maintained, it is important to conserve both species and habitat diversity within a native people's home range.

Management of Nontimber Forest Products Within Natural Forest Management

An "experimental strip" 20 meters wide and 75 meters long, facing north-south to minimize light entering the forest gap, was clear-cut in April 1985 at the Central Selva project site. A shelter belt was left around the strip to provide seed trees (figure 16.2). Plants in the strip that are useful to the Amuesha were inventoried before felling and a year after felling.* These inventories included the great majority of species present because of the Amuesha's broad-ranging use of forest species (cf. studies of other Amazonian groups—Balée 1986; Boom 1986; Prance et al. 1987). These results were corroborated by dendrological inventories (Aspinal 1985; Ramirez 1986; Hartshorn 1986).

Overall, the number of useful plants and useful plant species increased dramatically with regeneration. Relative importance or percentages of useful species and individuals in each plant-form category are more meaningful, since the total numbers of species and individuals increased during regeneration due to their small size. Comparative statistics on nontimber products within the experimental strip before harvest and one year after cutting are differentiated by growth form (table 16.2). Often, results show the same trends for both numbers and percentages of species or individuals.

The regenerating forest includes fewer species of palms and fewer individual ferns than the original vegetation, whereas there are more species and many more individual herbs and grasses. These trends may reflect the very early successional nature of the strip, although some of the herbs and grasses collected have been collected rarely and

*Data from a third inventory done three years after felling (1988) are not yet published, but trends are congruent.

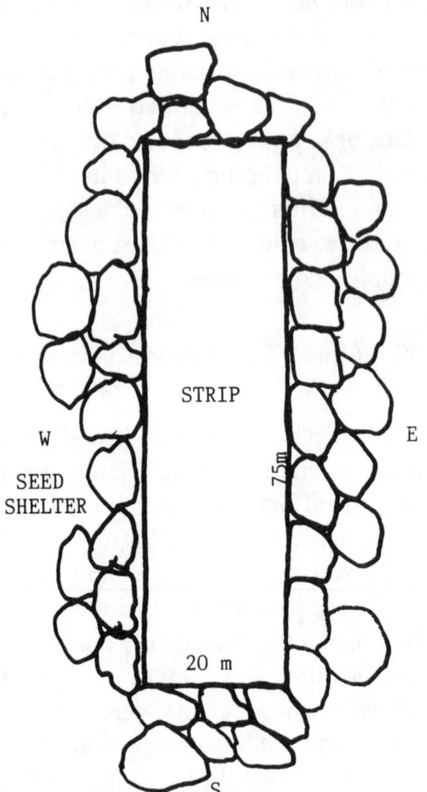

Figure 16.2 The "Experimental Strip"

may have unusual forest-gap niches. The increase in trees useful for nontimber products is due to the small size of individuals, as well as changes in species composition.

The increased dominance of vines in the strip is of concern, since they directly compete with tree seedlings. Vines can affect tropical forest succession greatly (Putz 1984). The Amuesha recognize several species of vines present as weeds and find them of no use whatsoever (not a common Amuesha category for plant species). The postharvest treatment of cutting and leveling slash apparently encouraged vines to sprout throughout the strip.

Regeneration from cut plants resprouting or coppicing is very important; 40 percent of all useful plants reestablish themselves in this manner (table 16.3). Dendrological inventories confirm this tendency

Table 16.2

Comparative Statistics on Useful Plants (Nonwood Uses) in 20m × 75m Strip Before Harvest (Mature Forest on Clay Soil) and Regeneration One Year After Cutting

	Totals	Trees	Vines	Herbs	Palms	Grasses	Ferns	Tree Ferns	Epiphytes	Shrubs
Before Strip-Cut (1985)—Mature Forest on Clay Soils										
Number of useful plant species	64	25	17	5	7	0	2	2	5	1
% of useful plant species	100	39	27	8	11	0	3	3	39	2
Number of useful plant individuals	441	92	118	22	109	0	69	21	6	4
% of useful plant individuals	100	21	27	5	25	0	16	5	1	1
One Year After Strip-Cut (1986)—Regeneration										
Number of useful plant species	142	55	26	37	4	9	2	2	4	3
% of useful plant species	100	39	18	26	3	6	1	1	3	2
Number of useful plant individuals	2,302	1,007	511	419	192	81	48	15	25	4
% of useful plant individuals	100	44	22	18	8	4	2	1	1	0

Table 16.3
Regeneration of Useful Plants (Nonwood) by Sprouts or Seeds, One Year After Harvest of 75m × 20m Strip

	Total	Sprouts	Seeds
Species of Useful Plants	116	50	66
% of Species	—	43	57
Individuals of Useful Plants	2,160	862	1,298
% of Individuals	—	40	60

(Hartshorn 1989) for trees as well. This type of regeneration is different from natural gap regeneration (Special Feature 1989) and similar to regeneration in Amuesha-managed fallows (Staver 1990) as well as indigenous fallow management in other parts of the world (Mackie 1986). Using indigenous fallow and forest management as the theoretical underpinning to natural forest management and applying described techniques would be productive. For example, to maximize regeneration of desired species, in addition to managing seed sources, emphasis might be placed on management practices that encourage healthy coppicing.

Strip regeneration is also different from natural gap regeneration and more like that of swidden fallows in the proportion of secondary plant species encountered (table 16.4). Before harvesting, 6 percent of useful species were secondary, compared to 24 percent during regeneration of the strip. Undoubtedly, the strip is a greater disturbance to the rain forest than is a gap resulting from a few fallen trees, and light-demanding species (e.g., grasses, herbs, *Cecropia* spp., etc.) can more easily enter. Parker (1986) points out that even if secondary forest succession is not being obviated with strip cutting, there are many other advantages to strip cutting and few tested alternatives. To the Amuesha, management of forested areas undergoing secondary succession is analogous to their management of utilized fallows. There should be ample opportunities to integrate Amuesha indigenous management practices and their use of secondary-forest products.

Nonetheless, some problems with the regeneration of nontimber products within the cut strip must be addressed before this form of natural forest management can be deemed a success. Some species important to Amuesha subsistence were lost, and there is no evidence

Table 16.4

Habitat Characterization of Useful Plant Species Found in 75m × 20m Strip Before Harvest (Mature Forest) and During Regeneration (One Year After Harvest)

	Before Harvest		After Harvest	
Habitat	Number of Species	% of Species	Number of Species	% of Species
Forest	60	94	90	76
Regrowth	4	6	28	24
Totals	64		118	

If a species is most commonly found in mature forests, it is characterized as a forest species; if it is more common in early secondary forest regrowth or in pastures, it is characterized as a regrowth species.

that they will regenerate in the near future. Most important among these is *tamshi* (*Heteropsis* and *Marcqravia* spp.), which is used to weave Amuesha baskets and mats and to construct everything from houses (nails are costly) to fish traps and baby hammocks. This important forest product is becoming increasingly scarce near Amuesha settlements and must be collected as much as a day's walk away. If a forest management scheme further reduces *tamshi* distribution, the Amuesha will feel the effect quickly. Other species not present in the regenerating plot include thatch (e.g., palms San Capilla, *Hiospathium* sp.), construction materials (e.g., *camona* and *camonilla, Irartella* spp.), arrow poison (e.g., *pumahuasca* or *curari, Curarea* sp.), and some medicinals (e.g., *puntasangre,* Gesnereaceae, *Columnia* sp.). Whether these species reestablish themselves will only be seen with time. The potential for management of key species such as *tamshi* within the forest management systems could be investigated using indigenous management methods.

Integration of Indigenous Forest Use with Commercial Management

The results of the study on the integration of indigenous forest use with commercial management suggest a broad approach to conservation; clearly, conservation of useful forest products falls within conservation biology. Numerous natural products from the tropical rain forest are used by the Amuesha of central eastern Peru. All forest types sampled in this study, both primary and secondary, are intensively used by the Amuesha, and yet some forest types are disappearing. The practice of indigenous medicine is already limited by the lack of key species due to habitat destruction, while overexploitation of other useful species has made them extremely scarce. These problems of indigenous use and management of forest resources have not been addressed within conservation biology. The study of population biology and habitat conservation needs to be integrated with research into resource use and economics.

Natural forest management is compatible with management of nontimber forest products crucial to Amuesha subsistence. Natural forest management is certainly preferable to the establishment of timber estates, plantations, or enrichment planting, since it allows for the regeneration of nontimber forest products essential for indigenous subsistence. There are, however, some specific problems which need further research. The increased dominance of secondary forest species and vines suggests that primary forest species may be at a disadvantage with strip cutting. With the importance of plant resprouting or coppicing in regeneration, the management of coppicing used in indigenous management of fallows should be studied to enhance growth of timber and other useful plants. Key species important to indigenous resource use that are missing from regenerating natural forest need to be studied and individually managed. In addressing all these problems it would be worthwhile to refer to indigenous practices of fallow and forest management (Staver 1990; Salick 1989b; Salick and Lundberg 1990, Denevan and Padoch 1987). Despite these problems, there are other advantages to strip cutting (Parker 1986) and limited tested alternatives that depend on natural forest management (Mergen and Vincent 1987). The techniques of natural forest management are compatible with indigenous peoples' need for and management of useful forest plants. This integration is not only technologically advantageous but beneficial in maintaining the people's stake in management and in assuring that their lives are enhanced by technological development rather than merely disrupted by it.

To reach the broader goal of preserving, conserving, managing, and using plants for people's subsistence and profit, it is useful to listen closely both to those demonstrating what indigenous cultures have to offer technologically and to those working in development efforts in both conservation and natural resource management. May our interactions bear fruit . . . and medicines, resins, oils, ritual objects, culinary herbs, edibles, drinks, dyes, poisons, materials for containers, crafts, and construction, and much more.

LITERATURE CITED

Aspinal, W. 1985. *Inventario forestal: Faja de investigación I.* Iscuzacin, Peru: Proyecto Especial Pichis-Palcazu.

Balée, W. 1986. Ethnobotanical studies among Tupi-Guarani speaking Indians in Amazonia. In *IEB*, ed., Discoveries in Economic Botany, New York: New York Botanical Garden.

Balée, W. and A. Gély. 1989. Managed Forest Succession in Amazonia: The Ka'apor case. *Advances in Economic Botany:* 129–158.

Barclay, F. 1985. Analisis de la division de trabajo y de la economia domestica entre los Amuesha de la selva central. USAID, Lima, Peru.

Boom, B. M. 1985. Amazonian Indians and the forest environment. *Nature* 314:324–00.

—— 1986. Ethnobotanical studies of the Chacabo and Parane Indians. In *IEB*, ed., *Discoveries in Economic Botany: Research and Results at the Institute of Economic Botany, 1981–1986.* New York: New York Botanical Garden.

Brack, A. J. 1987. *Plan maestro: Parque Nacional Yanachaga Chemillen.* Lima, Peru: The Nature Conservancy.

Burkill, I. H. 1935. *A Dictionary of the Economic Products of the Malay Peninsula.* London: Crown Agents.

Denevan, W. M. and C. Padoch, eds., 1987. *Swidden-Fallow Agroforestry in the Peruvian Amazon. Advances in Economic Botany* 5. New York: New York Botanical Garden.

Foster, R. 1981. *Brief Inventory of Plant Communities and Plant Resources of the Palcazu Valley.* McLean, Virginia: JRB Associates.

Foster, R. and N. V. L. Brokaw. 1982. Structure and history of the vegetation of Barro Colorado Island. In E. G. Leigh, Jr., A. S. Rand, and D. M. Windsor, eds., *The Ecology of a Tropical Forest: Seasonal Rhythms and Long Term Changes,* pp. 67–81. Washington, D.C.: Smithsonian Press.

Hartshorn, G. S. 1978. Treefalls and tropical forest dynamics. In P. B. Tomlinson and M. H. Zimmermann, eds., *Tropical Trees as Living Systems.* New York: Cambridge University Press.

—— 1986. Tree regeneration on first demonstration strip. Dendrology consultants fifth report (March), Palcazu Valley, Peru. Tropical Science Center, San Jose, Costa Rica.

—— 1989. Application of gap theory to tropical forest management: Natural regeneration on strip clear-cuts in the Peruvian Amazon. *Ecology* 70:567–569.

Institute of Economic Botany (IEB). 1986. *Discoveries in Economic Botany: Re-*

search and Results at the Institute of Economic Botany 1981–1986. New York: New York Botanical Garden.

Mackie, C. 1986. The landscape ecology of traditional shifting cultivation in an upland Bornean Rainforest. In *Proceedings of 1986 Conference on Impact of Man's Activities on Tropical Forest Ecosystems*, 425–464. University Pertanian Malaysia, Serdang.

Mergen, F. and J. R. Vincent. 1987. *Natural Management of Tropical Moist Forests: Silvicultural and Management Prospects of Sustained Utilization*. New Haven: Yale University Press.

Parker, G. G. 1986. *Analysis and Recommendations concerning the Natural Forest Management System used in the Central Selva Project, Peru*. Peru: United States Agency for International Development.

Posey, D. A. 1983. Indigenous ecological knowledge and development of the Amazon. In E. F. Moran, ed., *The Dilemma of Amazonian Development*, 225–258. Boulder, Colo.: Westview Press.

—— 1984. A preliminary report on diversified management of tropical forests by the Kayapo Indians of the Brazilian Amazon. In G. T. Prance and J. A. Kallunki, eds., *Ethnobotany in the Neotropics. Advances in Economic Botany*, 1:112–126. New York: New York Botanical Garden.

Prance, G. T., W. Balée, B. M. Boom, and R. L. Carneiro. 1987. Quantitative ethnobotany and the case for conservation in Amazonia. *Conservation Biology* 1:296–310.

Putz, F. E. 1984. The natural history of liana on Barro Colorado Island, Panama. *Ecology* 65:1713–1724.

Ramirez, W. 1986. *Informe de subunidad de investigación forestal: Inventario-manejo*. Iscozacin, Peru: Proyecto Especial Pichis-Palcazu.

Salick, J. 1985. Subsistencia y mujeres solas entre los Amuesha, Shupihui: Centro de Estudios Teologicos de la Amazonia 10:323–333.

—— 1986a. Ethnobotany of Amuesha. Ecological basis of Amuesha agricultural systems, Peruvian upper Amazon. USAID/Peru, Lima.

—— 1986b. Ethnobotany of Amuesha. Variation and nascent changes in indigenous agricultural systems of the Amuesha, Peruvian upper Amazon. USAID/Peru, Lima.

—— 1986c. Ethnobotany of Amuesha. Use and management potential of forest plant resources: An integration of indigenous use and natural forest management in commercial forestry. USAID/Peru, Lima.

—— 1986d. Ethnobotany of Amuesha. Cocona (*Solanum sessiliflorum*). USAID/Peru, Lima.

—— 1986e. Ethnobotany of Amuesha. A Cocona Cookbook. USAID/Peru, Lima.

—— 1989a. Cocona (*Solanum sessiliflorum*): Production and breeding potentials of the peach-tomato. In G. E. Wickens, N. Haq, and P. Day, eds., *New Crops for Food and Industry*. London: Chapman and Hall.

—— 1989b. Ecological basis of Amuesha agriculture. In D. A. Posey and W. Balée, eds. *Resource Management in Amazonia. Advances in Economic Botany* 7:189–212. New York: New York Botanical Garden.

—— 1989c. Bases ecologicas de los sistemas agricolas Amuesha. *Amazonía Indigena* 9:3–16.

—— 1992. Crop domestication and the evolutionary ecology of cocona (*Solanum sessiliflorum* Dunal). *Evolutionary Biology,* vol. 26.

—— In press. Subsistence and the single woman among the Amuesha, upper Peruvian Amazon. *Soc. and Nat. Resources.*

Salick, J. and M. Lundberg, 1990. Variation and change in Amuesha agriculture. In G. T. Prance and M. J. Balick, eds., *New Directions in the Study of Plants and People. Advances in Economic Botany.* 8:199–223. New York: New York Botanical Garden.

—— In press. Indigenous agroforestry in times of change: Amuesha agriculture of the upper Peruvian Amazon. ICRAF, Nairobi.

Smith, R. C. 1977. Deliverance from chaos for a song: A social and religious interpretation of the ritual performance of Amuesha music. Ph.D. dissertation, Cornell University, Ithaca, N.Y.

—— 1983. Las comunidades nativas y el mito del gran vacio Amazonico. AIDESEP, Lima, Peru.

Special Feature. 1989. Treefall gaps and forest dynamics. *Ecology* 70:535–576.

Staver, C. P. 1984. Agroforestry approaches to the sustained development of the Central Peruvian Amazon. Paper presented at the First Symposium on Humid Tropics, Belém, Brazil.

—— 1990. Why farmers rotate fields in maize-cassava-plantain bush fallow agriculture in the wet Peruvian tropics. *Human Ecology* 17:401–426.

Stocks, A. and G. Hartshorn, in press. The Palcazu Project: Forest management and native Amuesha communities. In J. Nations and S. Hecht, eds., *The Social Causes of Deforestation in Latin America.* Ithaca, N.Y.: Cornell University Press.

Wise, M. R. 1976. Apuntes sobre la influencia Inca entre los Amuesha. Factor que oscurece la clasificacion de su idioma. Lima, Peru: *Revista del Museo Nacional.*

Nontimber Forest Products in 75 × 20m Strip of Lowland Rain Forest on Clay Soils, Palcazu Valley, Peru

Part 1 (1985), Mature Forest Before Natural Forest Management Cut

Annonaceae			
	Annona	*cherimoya del monte*	Comestible (fruit).
	Annona	*anonilla blanca*	Tiestuff.
	Unonopsis	*icoja blanca*	Medicament for colds. Drink infusion from bark.
Apocynaceae			
7263	*Bonafousia undulata*	*senango*	Parasiticide. Antianemic. Medicament for rheumatism (root and resin).
7275	*Bonafousia undulata*	*senangillo*	Medicament for *choke de aire* chills. Inhale vapor.
Araceae			
7265	*Anthurium tessmannii*	*muela de Chacarero*	Antidote for snakebite. Drink infusion and apply as plaster.
7256	*Heteropsis* sp.	*tamshi*	Tiestuff (aerial roots). Element for weaving.
7249	*Philodendron cf. krukovii*	*hierba loca*	Element for divining (leaves). Leaves dance when put into water. Analgesic for labor pains. Drink infusion.
7244	*Philodendron cf. linnea*	*orejas de venado*	Antihemorrahagic/antimenorrhagic. Drink infusion mixture with Salick 007243 *Philodendron* sp. and Salick 007245 *Columnea tessmanii.*

7243	*Philodendron deltoideum*	*curada de scopeta*	Antihemorrahagic/antimenorrhagic. Drink infusion mixture of this with Salick 007244 *Philodendron linnea,* 007245 *Columnea tessmanii.*
7269	*Philodendron* sp.	*Hierba corazon trepadora* muela de Jergon	Antihemorrhagic/antimenorrhagic. Drink infusion. Antidote for snake bite. Apply as plaster.
Bignoniaceae			
7245	*Martinella obvata*	*yucilla*	Anti-infective for eye infections. Grate and squeeze juice into eye.
Boraginaceae			
7276	*Cordia nodosa* Lam.	*uvilla blanca*	Comestible (fruit).
Burseraceae			
	Protium sp.	*copal blanco*	Luminary. Element for caulk (resin). Used for boats. Antidote for snakebite (resin). Eat.
	Unidentified	*corteza*	Comestible (fruit). Dyestuff.
Caryocaraceae			
	Caryocar glabrum	*almendra colorada*	Comestible (nut and oil).
Celastraceae			
7261	*Cheiloclinium cognatum*	*chuchuhuasillo amarillo*	Medicament for rheumatism. Drink infusion. Tonic for blood. Drink infusion.
7264	*Cheiloclinium cognatum*	*chuchuhuasillo rojo*	Parasiticide. Drink strong infusion.
7259	*Cheiloclinium* sp.	*chuchuhuasillo de hoja ancho*	Analgesic for stomachache. Drink infusion.
Cyclanthaceae			
7241	*Thoracocarpus bissectus*	*tamshi macho*	Element for fiber used in weaving or for rope.
Dilleniaceae			
7252	*Tetracera*	*bejuco colorado*	Medicament for liver and kidneys. Potable (sap).
Elaeocarpaceae			
7274	*Sloanea* sp.	*caimitillo amarillo*	Comestible (fruit).

No.	Species	Common name	Use
Fabaceae			
7242	Unidentified	*santa rosa*	Medicament for liver ailments (sap). Drink.
Gesneriaceae			
7245	*Columnea tessmanii*	*punta sangre colorada*	Antihemorrhagic/antimenorrhagic. Drink infusion mixture with Salick 007243 *Philodendron* sp. and Salick 007244 *Philodendron linnea.*
Guttiferae			
7266	*Rheedia* sp.	*duraznillo amarillo*	Comestible (fruit).
7278	*Rheedia* sp.	*duraznillo calato*	Comestible (fruit).
Lauraceae			
	Aniba sp.	*sacha canela*	Medicament for stomachache and colic. Drink infusion from bark.
7271	Unidentified	*plano blanco*	Element for splints.
	Unidentified	*palo plano*	Element for splints.
Loganiaceae			
7240	*Potalia amara*	*caspisenango*	Sedative for crying babies. Bathe baby in infusion.
7268	*Strychnos jobertiana*	*chuchurasco colorado*	Tonic for strength and vigor.
Meliaceae			
	Guarea sp.	*barbasco del monte*	Poison used in fishing (root).
Menispermaceae			
7277	*Abuta grandifolia*	*chuchuhuasillo amargo*	Parasiticide.
7253	*Curarea toxicofera*	*pumahuasca*	Analgesic for stomachache, parasites. Anticarcinogen. Drink bitter fusion.
Mimosaceae			
	Inga sp.	*pacae blanco*	Comestible (fruit).
Moraceae			
	Pourouma guianensis	*tacona bacalau*	Medicament for cough and bronchitis. Drink infusion from bark.
	Pourouma sp.	*uvilla blanca de monte real*	Comestible (fruit).
7281	*Pseudolmedia laevis*	*pama calata colorada*	Comestible (fruit).

Monimiaceae			
7248	*Siparuna* sp.	*achimosa de arbol*	Antiviral. Inhale vapor.
Palmae			
	Iriartea	*camona*	Element for canals and flooring.
	Iriartella	*camonilla*	Element for canals.
	Bactris	*chontilla*	Comestible (fruit and heart).
	Euterpe	*huasai*	Comestible (fruit and heart). Medicament for convulsive coughing.
		palma San Capilla	Element for roofing.
		palmiche blanca	Element for roofing, weaving, and wrapping food.
Piperaceae			
7250	*Piper obligum*	*matico del monte real*	Medicament for rheumatism and colds. Inhale vapors.
	Piper sp.	*matico del monte real calata*	Medicament for wounds, especiallly gunshot wounds. Bathe wound in infusion (leaves).
Pterophytes			
7258	*Cyclodium meniscioides*	*helecho blanco*	Analgesic for toothache (sap). Apply directly to tooth.
7272	*Lomariopsis japurensis*	*macurilla trepadora*	Tonic for general good health. Inhale vapor.
7257	*Lomariopsis japurensis*	*helecho trepadora*	Medicament for witchcraft or *choke de aire*.
	Trichipteris sp.	*helecho de arbol* (tree fern)	Medicament for snakebites and tarantula, scorpion, and ant bites.
Rubiaceae			
7255	*Geophila cordifolia*	*corazoncito*	Medicament for fungi. Bathe in infusion (leaves).
7270	*Pentagonia gigantifolia*	*cascadilla amarilla*	Medicament for diarrhea.
Sapotaceae			
7273	Unidentified	*lucma blanca*	Comestible (fruit).
Sterculiaceae			
	Theobroma obovatum	*cacao del monte*	Comestible (fruit).

Verbenaceae			
7251	*Petrea* sp.	*hierba aspera*	Cleaner for guns that no longer shoot straight. Knuckle-duster. If you hit someone, they will turn black.
Zingiberaceae			
7246	*Costus* sp. nov.	*cana cana*	Medicament for liver and stomach ailments (stem and leaves).
Unidentified			
—		*ashumin*	Comestible (fruit).
—		*flecha del monte*	Element for arrows.
—		*hierba balsamo*	Medicament for infection. Apply as plaster (leaves).

Nontimber Forest Products in 75 × 20m Strip of Lowland Rain Forest on Clay Soils, Palcazu Valley, Peru

Part 2 (1986), Regenerating Forest One Year After Strip Clear-Cut

Acanthaceae			
7317	*Pseudoeranthemum lanceolatum*	*yuca yuca*	Growth enhancer. Plant with *yuca* (*Manihot esculenta*) to encourage large root formation.
7385	*Duguetia* cf. *flagellaris*	*cherimoya del monte*	Comestible (fruit).
7292	*Guatteria megalophylla*	*huanganahuasca*	Tiestuff.
7350	*Guatteria* sp.	*chumairo caspi*	
7326	*Rollinia* sp.	*anonilla blanca*	Tiestuff.
Apocynaceae			
7372	*Bonafousia undulata*	*senango*	Parasiticide, antianemic.
7291			Medicament for rheumatism (root and resin).
7300	*Lacmellea* sp.	*chocoque amarillo*	Comestible (fruit).
Araceae			
7339	*Alocasia macorrhiza*	*sheshake*	
7305	*Anthurium clavigerum*	*costilla de Adan*	Medicament for aging, witchcraft, and illness. Inhale vapor.
7314	*Dieffenbachia* cf. *parvifolia*	*pasalla blanca*	Tiestuff.
7370	*Dracontium* cf. *loretense*	*hierba jergon*	Antidote for snakebite (root), grind and apply plaster. Used especially for bothrops

7279	*Monstera cf. spruceana*	*bejuco trepadora corazon*	Antimenorrhagic. Drink decoction.
7349	*Philodendron cf. deltoideum*	*bejuco culata rastrera*	Antihemorrhagic for miscarriage or postpartum bleeding. Drink infusion.
7332	*Philodendron cf. linnea*	*orejas de venado*	Antihemorrhagic. Drink infusion.
7386	*Philodendron insigne* Schott	*ajo del monte trepadora*	Medicament for cancer and rheumatism. Cook and eat stem.
Asclepiadaceae			
7348	*Tassadia obovata*	*bejuco lechera de hoja fina*	
Bignoniaceae			
7334	*Martinella obovata*	*yucilla*	Anti-infective for eye infections. Grate and squeeze juice into eye.
7374	*Martinella* sp.	*yucillo macho*	Anti-infective for eye infections. Grate and squeeze juice into eye.
7351	*Pyrostegia dichotoma*	*sachagrandilla*	Comestible (fruit).
Boraginaceae			
7357	*Cordia nodosa*	*uvilla*	Comestible (fruit).
Burseraceae			
7293	*Protium apiculatum*	*copal blanco*	Luminary. Element for caulk (resin). Used for boats. Antidote for snakebite (resin). Eat.
7336	*Protium odulosum*	*copal amarillo*	Luminary. Element for caulk (resin). Used for boats. Antidote for snakebite (resin). Eat.
Caryocaraceae			
7290	*Caryocar glabrum*	*almendra colorada*	Comestible (nut and oil).
Cecropiaceae			
7391	*Cecropia sciadophylla*	*tacona ojo de venado*	Seed eaten by animals.
7388	*Cecropia* sp.	*tacona pata colorada*	Seed eaten by animals.
7392	*Cecropia* sp.	*uvilla blanca de monte real*	Seed eaten by animals.

7387	*Cecropia* sp.	*uvilla colorada*	Seed eaten by animals.
7390	*Cecropia* sp.	*uvilla blanca*	Seed eaten by animals.
7361	*Pourouma* sp.	*tacona* sp.	Comestible (fruit).
7352	*Pourouma* sp.	*uvilla*	Comestible (fruit).
7360	*Pourouma* sp.	*uvilla blanca*	Comestible (fruit).
Celastraceae			
7280	*Cheiloclinium (?) cognatum*	*chuchuhuasilo amarillo*	Medicament for rheumatism and colds (leaves and bark). Drink infusion from bark or inhale vapor from leaf decoction.
Compositae			
7353	*Erechtites valerianaefolii*	*yauyanka morada*	Anti-infective/vulnerary (leaves). Squeeze juice into wound.
7307	*Pseudoelephantopus spiralis*	*lechuguillo*	Pasture.
Connaraceae			
7367	*Connarus*	*barbasco del monte*	Element for fish poison (roots).
Cyclanthareae			
7356	*Cyclanthus bipartitus*	*cola de golonodrina*	Element for food wrap.
Cyperaceae			
7311	*Calyptrocarya glomenrulata*	*colcha macho*	Pasture.
7363	*Cyperus diffusa*	*sachacolcha*	Pasture.
7376	*Pleurostachys* sp.	*cortadilla macha*	Component of charm to combat witchcraft. Bathe in infusion.
7321	*Scleria aff. secans*	*cortadilla rastrera de tres esquinas.*	Cleaner for guns (leaves). Component of curtain for injuring bats (leaves).
Dilleniaceae			
7323	*Tetracera* sp.	*bejuco colorado*	Potable (sap). Depurative for kidney and liver (sap).
7375	*Tetracera* sp.	*bejuco cortadilla colorado*	Cleaner for guns.
Elaeocarpaceae			
7320	*Sloanea* sp.	*caimito tushma*	Comestible (fruit).
Euphorbiaceae			
7354	*Alchornea glandulosa*	*huampillo blanco*	Parturifacient (bark). Drink infusion.

7347	*Sapium marmieri*	*cortapercha colorada*	Anti-infective (latex). Apply as plaster.
Gentianaceae			
7306	*Irlbachia alata*	*sachatabaco*	Medicament for chicken lice (leaves). Line bottom of chicken coop.
Gesneriaceae			
7313	*Besleria aggregata*	*Santa Maria blanca*	Antiviral (leaves). Used especially for grippe.
Gramineae			
7310	*Ichnanthus panicoides*	*hierba balsamo*	Anti-infective (leaves and resin). Apply directly.
7319	*Lasiacis sorghoidea*	*caricillo morado*	Feed for guinea pigs. Medicament for venereal disease (root). Drink infusion.
7382	*Orthoclada laxa*	*colcha macho*	Pasture.
7345	*Paspalum cf. conjugatum*	*colcha*	Pasture.
7377	Unidentified	*colcha del monte*	Pasture.
Guttifereae			
7344	*Marila* sp.	*pama calata colorada*	Comestible (fruit).
7342	*Rheedia* sp.	*duraznillo amarillo*	Comestible (fruit).
7299	*Rheedia* sp.	*duraznillo calato*	Comestible (fruit).
7358	*Rheedia* sp.	*durazno calato*	Comestible (fruit).
7285	*Vismia* sp.	*achimosa calata*	Antiviral. Inhale vapor.
7298	*Vismia* sp.	*café de hoja ancha*	Anti-inflammatory for boils (resin).
7322	*Vismia* sp.	*café del monte*	Anti-inflammatory for boils (leaves). Grind and apply as plaster.
Heliconiaceae			
7282	*Heliconia velutina*	*platanillo del monte*	Element for wrapping (leaves). Used in cooking.
Labiatae			
7303	*Hyptis* sp.	*sheshtaigra blanca de pastales*	Medicament for wounds and infections (leaves).
Lauraceae			
7362	*Aniba*	*sacha canela*	Analgesic for stomachache and colic. Drink infusion.
Leguminosae			
7346	*Acacia* sp.	*pashakillo blanco*	

7378	*Fabaceae*	*santa rosa*	Depurative for liver.
7379	*Inga cf. macrophylla*	*pacaecillo colorado*	Comestible (fruit).
7327	*Inga* sp.	*pacae colorado*	Comestible (fruit).
7368	*Inga* sp.	*pacae coto*	Comestible (fruit).
7295	*Inga* sp.	*pacaecillo blanco*	Comestible (fruit).
Lepanianthes			
7343	*Pothomorphe umbellata*	*hierba raya*	Anti-infective (leaves). Heat and apply as plaster. Antidote for sting of sting-ray (leaves). Heat and apply as plaster.
Loganiaceae			
7371	*Potalia amara*	*chirosenango*	Parasiticide, antianemic, medicament for rheumatism (roots and resins).
7240	*Potalia amara*	*caspisenango*	Sedative for crying babies. Bathe baby in infusion.
Lycopodiaceae			
7365	*Lycopodium cernuum*	*hierba del pensamiento*	Medicament for insomnia and worries. Bathe head in infusion (leaves).
Marantaceae			
7340	*Calathea wallisii*	*achira blanca*	Element for food wrap.
7316	*Calathea wallisii*	*achira morada*	Element for food wrap.
7393	*Monotagma juruanum*	*platanillo del monte*	Element for food wrap (leaves). Used in cooking.
7359	*Monotagma* sp.	*bijanillo de bara*	Element for food wrap (leaves). Used in cooking.
Marcgraviaceae			
7312	*Norantea guianensis*	*mata palo*	Medicament for burns and lesions. Drink (sap).
Melastomataceae			
7329	*Bellucia* or *Loreya* sp.	*palo estrella Romana*	Comestible (fruit).
7338	*Miconia serrulata*	*palo estrella blanco*	Comestible (fruit).
Menispermaceae			
7380	*Curarea toxicofera*	*bejuau amargo*	Parasiticide (stem). Drink infusion.
Monimiaceae			
7355	*Siparuna* sp.	*achimosa de arbol*	Antiviral. Inhale vapor.

Moraceae			
7297	*Pseudolmedia macrophylla*	*pama morongke*	Comestible (fruit).
7366	*Pseudolmedia laevigata*	*pama amarillo*	Comestible (fruit).
7281	*Pseudolmedia laevis*	*pama calata colorada*	Comestible (fruit).
7331	*Trymatococcus amazonicus*	*mayan blanco*	Comestible (fruit).
Passifloraceae			
7308	*Passiflora coccinea*	*granadilla sachavaca*	Anti-infective. Grind and apply as plaster.
7389	*Passiflora cf. laurifolia*	*granadilla amarilla*	Comestible (fruit).
Phytolacceceae			
7309	*Phytolacca rivinoides*	*huanturmo*	Comestible (leaves). Used in salads and soups. Element for laundry soap (fruit).
Piperaceae			
7330	*Piper hispidum*	*matico soldado de monte real*	Anti-infective/vulnerary (leaves). Used especially for gunshot wounds. Wash area with decoction.
7286	*Piper obliqum*	*matico del monte real calata*	Anti-infective/vulnerary (leaves). Used especially for gunshot wounds. Wash area with decoction.
7288	*Piper obliqum*	*matico del monte real*	Antidote for snakebite (leaves). Grind and apply as plaster.
7384	*Piper* sp.	*matico colorado*	Vulnerary (leaves). Boil and apply as plaster. Medicament for rheumatism (leaves). Boil and apply as plaster.
7383	*Piper* sp.	*maticillo*	Vulnerary (leaves). Boil and apply as plaster medicament for rheumatism (leaves).
Pterophytes			
7287	*Cyclodium meniscioides*	*helecho blanco*	Analgesic for toothache. Chew directly.
7373	*Lindsaea* sp.	*helecho negro*	Analgesic for toothache. Chew directly.

7284	*Trichipteris* sp.	*helecho blanco del arbol*	Antidote for snake, scorpion, and ant bites (sap).
7294	*Trichipteris* sp.	*helecho negro de arbol*	Antidote for snake, scorpion, and ant bites (sap).
Rubiaceae			
7301	*Hamelia patens*	*arco sacha-morada*	Medicament for *granos* and "ill effects of the rainbow". Decoction (leaves).
7304	*Geophila orbicularis*	*corazoncito*	Medicament for fungus infection. Bath with infusion (leaves).
7333	*Palicourea* sp.	*hierba de tres esquinas*	Anti-infective/vulnerary (leaves). Boil and apply as plaster.
7341	*Pentagonia gigaut folia*	*cascadilla amarilla*	Medicament for liver ailments (leaves). Boil with honey, strain, and drink decoction cold.
7337	*Pentagonia macrophylla*	*cascadilla blanca*	Medicament for hepatitis (bark). Drink infusion.
7289	*Pentagonia macrophylla*	*cascadilla colorada*	Medicament for hepatitis (bark). Drink infusion.
7302	*Sipania hispida*	*amorcillo amarillo*	Anti-infective. Bathe wound in decoction (leaves).
Sapotaceae			
7335	*Pouteria(?) durlandii*	*caimito blanco*	Comestible (fruit).
Solanaceae			
7325	*Cyphomandra* sp.	*tomate de arbol de monte*	Comestible (fruit). Anti-infective for eye infections. Grate stem and put juice into eye.
Sterculiaceae			
7296	*Theobroma subincanum*	*cacao del monte*	Comestible (fruit).
Ulmaceae			
7314a	*Trema micrantha*	*pasalla blanca*	Tiestuff (inner bark).
Verbenaceae			
7283	*Aegiphila integrifolia*	*san Pablo*	Anti-infective. Grind and apply juice on infection (leaves).
7328	*Aegiphila* sp.	*ocuera*	Intensifier for cocaine quids. Burn stem and use

			scrapings as a substitute for calcium.
7364	*Petrea* sp.	*hierba aspera*	Cleaner for guns.
Zingiberareae			
7324	*Costus* sp.	*cana cana*	Medicament for liver and stomach ailments (stem). Peel and chew.
Undetermined			
7315	—	*ajo del monte*	Medicament for rheumatism and cancer (stem). Drink decoction.
7318	—	*bejuca lechero*	
7381	—	*bijau trepadora*	Element for food wrap.
7369	—	*espantaje del diablo*	Component of charm to combat illness or witchcraft. Burn and vaporize house.

SEVENTEEN

Incorporation of Game Animals into Small-Scale Agroforestry Systems in the Neotropics

KENT H. REDFORD, BERT KLEIN, AND CAROLINA MURCIA

Agroforestry is defined by the International Council for Research in Agroforestry (ICRAF) as "all land-use practices and systems in which woody perennials are deliberately grown on the same land management unit as annual crops and/or animals" (Gholz 1987). Throughout the world researchers are attempting to develop, adapt, and refine such agroforestry systems to fit local conditions. Virtually all this attention has been devoted to the botanical component of agroforestry systems, and if animals are incorporated, they are virtually always domesticated species (Gholz 1987). Yet in many parts of the world wild animals serve as important sources of meat and as such would make suitable candidates for inclusion in agroforestry systems.

Game animals, particularly frugivorous animals that feed on perennial fruit trees, already play a role in small-scale agroforestry systems in the Neotropics. In areas of low human population density with large tracts of relatively undisturbed habitat such as many parts of the western Amazon, fruit and frugivory are important aspects of the Neotropical forest. The ways in which traditional peoples living in these areas have harvested frugivorous game species and incorporated such harvesting into their food gathering systems present intriguing ideas for incorporating such systems into planning for new agroforestry projects.

Fruit and Frugivory

A large proportion (variously estimated at 50 to 90 percent) of the trees of Neotropical, humid lowland forests produce fleshy fruits (Smythe 1986). The quantity of fruit produced by these plants can be very high, with one estimate of between .85 and 1.3 metric tons per hectare per year (Fittkau and Klinge 1973).

Most tropical fruits are rich in water and carbohydrates, contain a relatively large proportion of indigestible seeds, and are poor in fats and protein. What are termed "high quality" fruits—those rich in fats and proteins—are much less common and tend to occur in certain palms and in a restricted number of large-seeded canopy tree families. The "high quality" nutritional aspects of these fruits make them attractive to dispersers, frequently large vertebrates (Fleming, Breitwisch, and Whitesides 1987). The production of fruits attractive to vertebrates is common: for example, on Barro Colorado Island, Panama, 78 percent of 291 canopy tree species and 87 percent of 131 subcanopy tree species produce such fleshy fruits (Howe 1984). A more general estimate has been provided by Fleming, Breitwisch, and Whitesides (1987), who state that from 50 to 90 percent of the species of tropical shrubs and trees, depending on habitat, rely on fruit-eating vertebrates to disperse their seeds. In a survey of one hectare in the western Amazon, Balée (1987) noted, based on informant responses, that 86 percent of the 138 tree species produced fruit eaten by game animals.

These fruit-eating game animals include a significant proportion of the total vertebrate fauna. Many species of birds, mammals, and even a few reptiles eat fruit. Not only is the number of species high but their biomass is also: in Peruvian tropical forests frugivorous species make up 80 percent of the mammalian and avian biomass (Terborgh 1986).

Humans, for whom fruit has long served as an important source of nutrition, also are among the direct consumers of tropical forest fruits. This reliance is reflected in the diversity of trees whose fruits humans eat. Boom (1987) records 36 percent of the 94 species of trees found in one hectare of Bolivian tropical forest produced fruit eaten by the Chácobo Indians. In a similar study in the eastern Amazon, Balée (1987) recorded 25 percent of 138 tree species as producing fruit used by humans.

Frugivores as Game

Fruit-eating animals are an important component in the animal communities of tropical forests and as such would be expected to be important game animals as well. Not all frugivores are hunted; for example, hundreds of small passerine birds, which rely heavily on fruit, are rarely killed. Based on the data presented in Redford and Robinson (1987) and Robinson and Redford (1986), however, it is clear that most important game species are frugivores. Nine of the top eleven mammalian game taxa and all seven of the top avian game taxa eat fruit.

Terborgh (1986) supports this conclusion with data from an unhunted site in Peru where over half of the mammalian biomass is contained in populations of six frugivorous species: howler monkeys (*Alouatta seniculus*), spider monkeys (*Ateles paniscus*), tapirs (*Tapirus terrestris*), pacas (*Agouti paca*), and the two species of peccaries. Birds show a similar situation, with six frugivorous species making up more than 60 percent of the total avian biomass: two tinamous, three cracids, and a toucan. All of these frugivorous taxa, both mammalian and avian, are important game animals.

Hunting of Frugivorous Game

The characteristics that make a fruit attractive to potential vertebrate dispersers also make it attractive to humans. This human attraction takes two forms: first, as a food source and second, as an attractor of game. The attraction of game animals to fruiting trees has been exploited by hunters in several ways. First, hunters know that some species of wild fruiting trees attract more game than others. During fruiting season, it is frequently to these forest trees that hunters go (Yost and Kelley 1983). The Runa of Amazonian Ecuador take many species of game at many species of fruiting trees. But 57 percent of 126 captures at fruiting trees were at *Bactris gasipaes* palms (Irvine 1987). In some cases the fruit attracts only one game species, but in other cases the attraction is more general. For example, data from Panama show that fig trees (*Ficus*) attract thirteen species of game mammals, while trees of the genus *Spondias* attract eleven species (Glanz 1982).

Because many fruits eaten by game are also eaten by humans, another reason humans visit fruiting trees is to collect fruit. In Balée's (1987) study, all of the fruits recorded as being eaten by humans were also eaten by game. With luck, a trip to collect fruit might also be a successful hunting trip.

Another method of exploiting the attraction of game to fruiting trees is to move the trees themselves to locations more accessible to hunters. Frequently these locations are old fields where, after the major harvesting of crops such as manioc and corn has ceased, trees are planted. There are many variations on this pattern, but the species that are planted (or allowed to regrow) are usually useful to humans in several ways.

For example, the Kayapó Indians of eastern Amazonia plant trees for several purposes, one of which is the attraction of game (Anderson and Posey 1985). Sixteen of these species produce fruit eaten by game, and of these sixteen all but one have fruit also eaten by humans (Posey 1984). Posey (1984) has suggested that these replanted old fields should be termed "game-farm orchards" to reflect their multiple functions.

The Bora Indians of the Peruvian Amazon also manage their old fallows, planting many wild fruit trees and building hunting platforms near some of the species to hunt the game attracted to the falling fruit (Denevan and Treacy 1987; Unruh and Flores Paitan 1987). The Runa of Ecuador plant *Bactris* palms in their fallows and actively hunt these fallows for *Dasyprocta* and *Myoprocta*, which are attracted to these palms (Irvine 1967). This practice, though infrequently documented, was probably more widespread in the past.

Agroforestry-Game Systems for the Future

Fruit and frugivory is clearly important to both game animals and humans inhabiting Neotropical forests. Traditional peoples have harvested frugivorous game species by using both wild fruiting trees and replanted trees. Some of these uses, particularly the system practiced by the Bora Indians, are preexisting examples of agroforestry systems incorporating game species. Unfortunately, investigators have not yet quantified the nature of the wildlife component of the Bora system, but these traditional methods suggest that game animals could be incorporated easily into small-scale agroforestry systems in the Neotropics.

To allow designers of agroforestry systems to evaluate certain plant species from the perspective of attracting and feeding game species, we have here gathered data from the literature that concentrates on frugivorous game animals and perennial fruit trees, although other species of plants are included. The bird, mammal, and reptile species included were chosen based on their broad importance as game animals (Redford and Robinson 1987).

Our analysis of mammals was done at the generic level, with the exception of the two species of peccaries. We collected food habit data for the opossum (*Didelphis*), the howler monkey (*Alouatta*), the spider monkey (*Ateles*), the capuchin monkey (*Cebus*), the woolly monkey (*Lagothrix*), the paca (*Agouti*), the agouti (*Dasyprocta*), the white-lipped peccary (*Tayassu pecari*), the collared peccary (*T. tajacu*), the tapir (*Tapirus*), the coati (*Nasua*), and the brocket deer (*Mazama*). The single genus of reptile is the tortoise, *Geochelone*. Because of the comparative paucity of data for birds, we collected data at the level of avian family: the tinamous (Tinamidae), the guans, curassows, and chachalacas (Cracidae), the woodquails (Phasianidae), the trumpeters (Psophiidae), the pigeons and doves (Columbidae), and the toucans (Ramphastidae). Data for fruiting plants are presented at the generic level because of the desire to provide results relevant to the entire Neotropics. Whereas many of the tree species occur over a narrow geographical range, many of the genera are widely distributed. In some cases, the results presented for a genus of tree may not apply to all of the species within that genus.

For each of the mammal, reptile, and avian taxa we compiled the genera of fruit consumed. Most of these data were gathered by biologists watching the feeding patterns of wild animals, although some come from examination of stomach contents. These data are presented in appendices 17.1 and 17.2, together with an assessment, where available, of the relationship between consumer and fruit (e.g., disperser, predator, etc.). These data, while important, were not available enough to allow generalizations. For mammals we recorded sixty-two genera of fruit-producing plants in twenty-eight families and for birds sixty-five genera in thirty-five families. However, the study of the diets of most Neotropical vertebrates is very much in its infancy. The data presented here are by no means complete and, because of the interests of the investigators, are biased toward primates. The data serve only to sketch in patterns of frugivory and to point out plant taxa that might be useful within the agroforestry context.

The fruits most frequently consumed by mammals are listed in table 17.1 and those consumed by birds in table 17.2. Five plant genera are found on both tables, *Cecropia, Ficus, Spondias, Protium,* and *Tetragastris.* Surprisingly, there is little overlap, for apparently most fruits attract either birds or mammals and not often both.

A recently completed dissertation by Bodmer (1989) provides a superb data set on food habits of the rain forest ungulates of Peru. The data provided by Bodmer were not included in our tables but are summarized here. Two species of *Mazama* showed preference for fruits of *Rheedia*, Sapotaceae, *Spondias*, *Swartzia*, Linaceae, *Iriartea*, *Euterpe*, *Maripa*, *Inga*, and *Virola*. *Tayassu tajacu* preferred fruits from the Sapotaceae, Menispermaceae, *Mauritia flexuosa*, *Jessenia*, and *Astrocar-*

Table 17.1
Fruits Most Frequently Eaten by Game Mammals

Plant Taxon	No. of Mammalian Genera (n = 12) Consuming the Fruit
Moraceae	
*Brosimum	5
*Cecropia	4
*Ficus	11
Palmae	
Astrocaryum	8
Jessenia	4
Mauritia	4
Rubiaceae	
Faramea	4
Anacardiacea	
*Spondias	8
Apocynaceae	
Lacmellea	4
Bombacaceae	
Quararibea	6
Burseraceae	
*Protium	5
*Tetragastris	6
Lecythidaceae	
Gustavia	6
Leguminosae	
Dipteryx	8
Inga	6
Meliaceae	
Trichilia	4

*Also frequently eaten by birds

Table 17.2
Fruits Most Frequently Eaten by Game Birds

Plant Taxon	No. of Avian Families (n=7) Consuming the Fruit
Moraceae	
**Cecropia*	3
Claria	2
**Ficus*	4
Pseudolmedia	2
Trophis	2
Musaceae	
Musa	2
Myristicacea	
Virola	4
Palmae	
Cryosophila	2
Geonoma	3
Sapindacea	
Talisia	2
Anacardiceae	
Metopium	3
**Spondias*	3
Boraginaceae	
Ehretia	3
Burseraceae	
**Protium*	2
**Tetragastris*	2
Flacourtiaceae	
Casearia	2
Lecythidaceae	
Lecythis	3
Melastromataceae	
Miconia	2

*Also frequently eaten by mammals

yum, while *Tayassu pecari* preferred fruits of *Iriartea,* Sapotaceae, Menispermaceae, *Astrocaryum,* and *Mauritia flexuosa.* The tapir preferred fruits of *Scheelea,* Menispermaceae, *Mauritia flexuosa,* and *Jessenia.* In general, the data presented in Appendix 17.1 are supported by those provided by Bodmer.

In order to further specify the diets of game birds and mammals we drew from the literature on hunting. Several authors, particularly Ayres and Ayres (1979), have listed species of trees to which humans go to hunt specific types of game. Table 17.3 lists twenty-five genera

Table 17.3
Game Animals and the Fruiting Trees to Which Hunters Go to Kill Them

	Mazama	*T. tajacu*	*T. pecari*	*Dasyprocta*	*Agouti*	*Tapirus*	*Cracidae*
Anacardiaceae							
Spondias lutea	1	1	1	1	1	1	1
Anacardium giganteum	1	1	1	1	1	1	1
Caryocaraceae							
Caryocar villosum	—	—	—	1	1	—	—
Myristicaceae							
Iryanthera grandis	1	—	—	1	1	—	—
Moraceae							
Bagassa guianensis	1,2	1	1	1	1	—	—
Ficus sp.	1,2	—	—	—	2	2	—
Chrysobalanceae							
Paninari sprucei	—	—	—	1	1	1	—
Rubiaceae							
Genipa americana	1	1	1	1	1	1	—
Leguminosae							
Enterolobium sp.	2	—	—	—	—	—	—
Hymenaea sp.	2	—	—	—	—	—	—

Pithecellobium sp.	—	—	—	—	—	3	—
Cassia	—	—	—	4	—	—	—
Meliaceae							
Carapa guianensis	3	—	—	—	—	—	—
Guttiferae							
Clusia sp.	—	—	—	—	—	—	3
Lauraceae							
Nectandra sp.	—	—	—	—	—	—	3
Sapotaceae							
Micropholis melinonii	—	—	—	—	—	—	3
Pouteria sp.	2	1	—	1	1	1	1
Franchetella sp.	—	—	—	3	3	—	—
Palmae							
Maximilliana maripa	1	1	—	1	1	1	—
Oenocarpus	—	—	4	4	4	—	—
Mauritia sp.	1	1,5	1,5	1	1	1	—
Jessenia batava	—	1	1	—	1	1	1
Astrocaryum sp.	—	1	1	1	1	—	—
Humiriaceae							
Humiria sp.	—	—	4	4	4	—	—
Endopleura uchi	—	1	1	1	1	1	—

1) Ayres and Ayres 1979; 2) Smith 1976; 3) Balée 1987; 4) Posey 1984; 5) Yost and Kelley 1983

of trees in thirteen families and the types of game which, according to hunters, are attracted to them. This data set is complementary to that derived from the appendices but differs from it for several reasons, including the fact that the studies were done in different geographical regions. Nine genera of trees are reputed to attract five or more species of game: *Spondias, Anacardium, Bagassa, Genipa, Pouteria, Maximilliana, Mauritia, Jessenia,* and *Endopleura.* As with tables 17.1 and 17.2, lack of information does not mean that animals are not attracted to or don't consume the fruit of the plant species.

These data point to which tree species might be included in an agroforestry scheme if game attraction is the main objective. We have presented primary information so that such a scheme can be further focused if the attraction of certain species (e.g., *Agouti*) is desired. It is preferable for the components (i.e., species) of an agroforestry system to have more than one use, so we have gathered data on use of fruits not only by game animals, but by humans as well.

Table 17.4 presents two types of data on fruit from forest trees used by humans. The first is fruit sold in markets in Iquitos, Peru; northeastern Mexico; and Belém, Brazil. The second is forest trees replanted by Indians specifically for fruit production for human consumption. The twenty-two genera listed in table 17.4 is only a small subsample of the wild fruits eaten by humans. One of the patterns evident from this table is the broad distribution of some of these tree genera and the equally broad appreciation of their fruit. For example, fruit from the genus *Pouteria* are sold in markets in Mexico and Peru and replanted by the Kayapó Indians in eastern Amazonia. This pattern, coupled with the extensive distribution of many of the species of game animals, allows broad application of many of the data presented in this paper.

Nonetheless, each agroforestry system should apply the data presented within the context of specific objectives and local conditions. We suggest that our data be used in conjunction with local botanical and zoological inventories to pinpoint species of trees that produce fruit desirable to both local human and local game animal populations. This can be done by conducting surveys that 1) evaluate which species of forest trees found in the area provide fruit desired by local human populations; 2) determine which species of game are found in the area and desired by humans; and 3) determine the diets of these game animals. Although not the focus of this paper, there may be

Table 17.4
Fruit from Forest Trees That Is Utilized by Humans

Icacinacea	
Poraqueiba	1,5
Sapotaceae	
Pouteria	1,2,3
Manilkara	2,4
Moraceae	
Pourouma	1,4
Brosimum	2
Ficus	2
Leguminosae	
Inga	1,5,4
Parkia	5
Lecythidaceae	
Bertholletia	1,3
Anacardiaceae	
Spondias	2,3
Ebenaceae	
Diospyros	2
Annonaceae	
Annona	2,5,3
Boraginaceae	
Cordia	5
Myristicaceae	
Virola	5
Caryocaraceae	
Caryocar	4,3
Myrtaceae	
Eugenia	4
Palmae	
Oenocarpus	4
Bactris	1,3
Mauritia	1
Jessenia	1
Euterpe	1,4,3
Maximiliana	3

Fruit sold in markets: 1 (in Iquitos; Padoch 1987; Padoch et al. 1987); 2 (in Mexico; Alcorn 1984); 3 (in Belém; van den Berg 1984) and planted: 4 (by the Kayapó; Posey 1984); 5 (by the Bora; Unruh and Flores Paitan 1987).

secondary uses of some of the tree species listed in the tables (e.g., timber production, building material, latex production, medicinal use) which will influence their inclusion in a specific agroforestry system. The data presented provide a starting point for choosing appropriate tree species for incorporation into local agroforestry schemes.

There may be competition between humans and game animals for fruit in a system such as the one we suggest, although there has been no indication from any investigators that this is the case. However, as Padoch (1987) has pointed out, meat is a more valuable and, when smoked, a more durable product than fresh fruit, and therefore loss of fruit to game could be viewed as an investment in the productivity of the system.

The system of incorporating game into small-scale agroforestry projects promoted here could only be applied in areas of low human population density with large tracts of relatively undisturbed habitat. These conditions are necessary because, if violated, the agroforestry-game system would attract game animals from only a small catchment basin to an area where they can be more easily harvested. Such a result would cause overharvesting and eventual extinction of the game population. However, if the catchment basin were large and not heavily disturbed, then a viable breeding population could more easily be maintained and the system could be sustainable.

ACKNOWLEDGMENTS

We would like to thank Henry Gholz, Christine Padoch, and Allyn Stearman for their comments.

LITERATURE CITED

Alcorn, J. B. 1984. Development policy, forests, and peasant farms: Reflection on Husatec-managed forests' contributions and resource conservation. *Advances in Economic Botany* 38(4):389–406.

Anderson, A. B. and D. A. Posey. 1985. Manejo de cerrado pelos índios Kayapó. *Boletin Museu Paraense. Emílio Goeldi, Botanica* 2(1):77–98.

Ayres, J. M. and C. Ayres. 1979. Aspectos da caça no alto rio Aripuanã. *Acta Amazonica* 9:287–298.

Balée, W. 1987. A etnobotánica quantitativa dos índios Tembé (Rio Gurupi, Pará). *Boletin Museu Paraense Emílio Goeldi, Botanica* 3(1):29–50.

Bodmer, R. E. 1989. Frugivory in Amazon ungulates. Ph.D. dissertation, University of Cambridge.

Boom, B. M. 1987. *Ethnobotany of the Chácobo Indians,* Beni, Bolivia. *Advances in Economic Botany* 4. New York: New York Botanical Garden.

Branan, W. V., M. C. M. Werkhoven, and R. L. Marchinton. 1985. Food habits of brocket and white-tailed deer in Suriname. *Journal of Wildlife Management*. 49:912–976.

Chapman, F. M. 1929. *My Tropical Air Castle.* New York: Appleton.

Denevan, W. D. and J. M. Treacy. 1987. Young managed fallows at Brillo Nuevo. In W. M. Denevan and C. Padoch, eds. *Swidden-Fallow Agroforesty in the Peruvian Amazon. Advances in Economic Botany* 5:8–46. New York: New York Botanical Garden.

Enders, R. V. 1930. Notes on some mammals from Barro Colorado Island, Canal Zone. *Journal of Mammalogy* 11:284–285.

Estrada, A. 1984. Resource use by howler monkeys (*Alouatta palliata*) in the rain forest of Los Tuxtlas, Mexico. *International Journal of Primatology* 5:105–131.

Estrada, A. and R. Coates-Estrada. 1986. Frugivory by howler monkeys (*Alouatta palliata*) at Los Tuxtlas, Mexico: Dispersal and fate of seeds. In A. Estrada and T. H. Flemming, eds., *Frugivores and Seed Dispersal,* 93–104. Nordecht, The Netherlands: Dr. W. Junk Publications.

Fittkau, E. J. and H. Klinge. 1973. On biomass and trophic structure of the central Amazonian rain forest ecosystem. *Biotropica* 5:2–14.

Fleming, T. H., R. Breitwisch, and G. H. Whitesides. 1987. Patterns of tropical vertebrate frugivore density. *Annual Review Ecology Systematics* 18:91–109.

Freese, C. and T. R. Oppenheimer. 1981. The capuchin monkeys, genus *Cebus.* In F. A. Coimbra-Filho and R. A. Mittermier, eds., *Ecology and Behavior of Neotropical Primates,* Vol. 1, 331–390. Rio de Janeiro, Brazil: Academia Brasileira de Ciencias.

Gautier, H., J. M. Duplantier, R. Quris, F. Freer, C. Sourd, J. P. Decoux, G. Dubost, L. Emmons, C. Erard, P. Hecketsweiler, A. Moungazi, R. Roussilhon, and J. M. Thiollay. 1985. Fruit characters as a basis of fruit choice and seed dispersal in a tropical forest vertebrate community. *Oecologia* 65:324–337.

Gholz, H. L. 1987. Introduction. In H. L. Gholz, ed., *Agroforestry: Realities, Possibilities and Potentials,* 1–6. Dordrecht, The Netherlands: Martinus Nijhoff Publishers.

Glanz, W. E. 1982. Adaptive zones of neotropical mammals: A comparison of some temperate and some tropical patterns. In M. A. Mares and H. H. Genoways, eds., Mammalian biology in South America. The Pymatuning Symposium in Ecology. Pymatuning Laboratory of Ecology. University of Pittsburgh, Pennsylvania, pp. 95–110.

Held, M. M., and M. C. M. Werkhoven. 1988. An ecological study of Cracidae in Suriname. In *Proceedings of the 2nd International Symposium on the Biology and Conservation of the Family Cracidae,* Caracas, Venezuela.(In press).

Hernández-Camacho, J. and R. W. Cooper. 1976. The non-human primates of Colombia. In R. W. Thorington, Jr., and P. G. Heltne, eds., *Neotropical Primates: Field Studies and Conservation,* 35–69. Washington, D.C.: National Academy of Sciences.

Howe, H. F. 1977. Bird activity and seed dispersal of a tropical wet forest tree. *Ecology* 58:539–550.

—— 1980. Monkey dispersal and waste of a neotropical fruit. *Ecology* 61:944–959.

—— 1982. Fruit production and animal activity in two tropical trees. In E. G. Leigh, Jr., A. S. Rand, and D. M. Windsor, eds., *The Ecology of a Tropical Forest: Seasonal Rhythms and Long Term Changes,* 189–200. Washington, D.C.: Smithsonian Institution Press.

—— 1984. Implications of seed dispersal by animals for tropical reserve management. *Biological Conservation* 30:261–281.

Howe, H. F. and G. A. V. Kerckhove. 1981. Removal of wild nutmeg (*Virola surinamensis*) crops by birds. *Ecology* 62:1093–1106.

Howe, H. F. and R. B. Primack. 1975. Differential seed dispersal by birds of the tree *Casearia nitida* (Flacourtiaceae). *Biotropica* 7(4):278–283.

Irvine, D. 1987. Resource management by the Runa Indians of the Ecuadorian Amazon. Ph.D. dissertation, Stanford University.

Izawa, K. 1975. Foods and feeding behavior of monkeys in the upper Amazon basin. *Primates* 16:295–316.

Janzen, D. H. 1981. *Ficus ovalis* seed predation by an orange-chinned parakeet (*Brotogeris jugularis*). *Auk* 98:841–844.

—— 1983. *Costa Rican Natural History.* Chicago: University of Chicago Press.

Johnson, T. B. and S. Hilty. 1976. Notes on the sickle-winged guan in Colombia. *Auk* 93:194–195.

Jordano, P. 1983. Fig-seed predation and dispersal by birds. *Biotropica* 15: 38–41.

Kantak, G. E. 1979. Observations on some fruit-eating birds in Mexico. *Auk* 96:183–186.

Kavanagh, M. and L. Dresdale. 1975. Observations on the woolly monkey (*Lagothrix lagotricha*) in Northern Colombia. *Primates* 16:285–294.

Kiltie, R. A. 1981. Distribution of palm fruits on a rain forest floor: Why white-lipped peccaries forage near objects. *Biotropica* 13:141–145.

—— 1982. Bite force as a basis for niche differentiation between rain forest peccaries (*Tayassu tajacu* and *T. pecari*). *Biotropica* 14:188–195.

Kiltie, R. A. and J. Terborgh. 1983. Observation on the behavior of rain forest peccaries in Peru: Why do white-lipped peccaries form herds. *Zeitschrift Tierpsychologie* 62:241–255.

Klein, L. L. and D. B. Klein. 1977. Feeding behavior of the Colombian spider monkey. In T. H. Clutton-Brock, ed., *Primate Ecology: Studies on Feeding Behaviour in Lemurs, Monkeys and Apes,* 153–181. London: Academic Press.

Lancaster, D. A. 1964. Life history of the Boucard Tinamou in British Honduras. Part I: Distribution and general behavior. *Condor* 66(3):165–181.

Milton, K. 1980. *The Foraging Strategy of Howler Monkeys.* New York: Columbia University Press.

Mittermier, R. A., H. DeMacedo, A. Luscombe, and J. Cassidy. 1977. Rediscovery and conservation of the Peruvian Yellow-tailed woolly monkey (*Lagothrix flavicauda*). In H. S. H. Prince Rainier de Monaco and G. H. Bourne, eds., *Primate conservation,* 95–115. New York: Academic Press.

Moskowits, D. K. 1985. The behavior and ecology of two Amazonian tortoises, *Geochelone carbonaria* and *Geochelone denticulata* in northwestern Brasil. Ph.D. dissertation, University of Chicago.

Neville, M. K. 1972. The population structure of red howler monkeys (*Alouatta seniculus*) in Trinidad and Venezuela. *Folia Primatologica* 17: 56–86.

Oppenheimer, J. R. 1982. *Cebus capucinus:* Home range, population dynamics, and interspecific relationships. In E. G. Leigh, Jr., A. S. Rand, and D. M. Windsor, eds., *The Ecology of a Tropical Forest: Seasonal Rhythms and Long Term Changes,* pp. 253–272. Washington, D.C.: Smithsonian Institution Press.

Padoch, C. 1987. The economic importance and marketing of forest and fallow products in the Iquitos region. In W. M. Denevan and C. Padoch, eds., *Swidden-Fallow Agroforestry in the Peruvian Amazon. Advances in Economic Botany* 5:74–89, New York: New York Botanical Garden.

Padoch, C., J. Chota Inuma, W. de Jong, and J. Unruh. 1987. Market-oriented agroforestry at Tamshiyacu. In W. M. Denevan and C. Padoch, eds., *Swidden-Fallow Agroforestry in the Peruvian Amazon. Advances in Economic Botany* 5:90–96. New York: New York Botanical Garden.

Posey, D. A. 1984. A preliminary report on diversified management of tropical forest by the Kayapó Indians of the Brazilian Amazon. *Advances in Economic Botany* 1:112–126.

Ramirez, M. 1988. The woolly monkeys, genus *Ateles.* In F. A. Coimbra-Filho and R. A. Mittermier, eds., *Ecology and Behavior of Neotropical Primates,* 539–575. Rio de Janeiro, Brazil: Academia Brasileira de Ciencias.

Redford, K. H. and J. G. Robinson 1987. The game of choice: Patterns of Indian and colonist hunting in the Neotropics. *American Anthropologist* 89:650–667.

Reichel-Dolmatoff, G. 1985. Tapir avoidance in the Colombian northwestern Amazon. In G. Urton, ed., *Animal Myths and Metaphors in South America,* 107–144. Salt Lake City, Utah: University of Utah Press.

Robinson, J. G. and K. H. Redford. 1986. Body size, diet, and population density of Neotropical forest mammals. *American Naturalist* 128:665–680.

Roosmalen, M. G. M.van. 1985. *Fruits of the Guianan flora.* Drukkerij Veenman BV. The Netherlands: Wageningen.

Russel, R. K. 1982. Timing of reproduction by coatis (*Nasua narica*) in relation to fluctuations in food resources. In E. G. Leigh, Jr., A. S. Rand, and D. M. Windsor, eds., *The Ecology of a Tropical Forest: Seasonal Rhythms and Long Term Changes,* 413–431. Washington, D.C.: Smithsonian Institution Press.

Sick, H. 1970. Notes on Brazilian Cracidae. *Condor* 72:106–108.

Skutch, A. F. 1983. *Birds of Tropical America.* Austin, Texas: University of Texas Press.

Smith, N. J. H. 1976. Utilization of game along Brazil's transamazon highway. *Acta Amazonica* 6:455–466.

Smythe, N. 1986. Competition and resource partitioning in the guild of Neotropical terrestrial frugivorous mammals. *Annual Review Ecology Systematics* 17:169–188.

Smythe, N., W. E. Glanz, and E. G. Leigh. 1982. Population regulation in some terrestrial frugivores. In E. G. Leigh, Jr., A. S. Rand, and D. M. Windsor, eds., *The Ecology of a Tropical Forest: Seasonal Rhythms and Long Term Changes*, 227–238. Washington, D.C.: Smithsonian Institution Press.

Sowls, L. K. 1984. *The Peccaries*. Tucson, Arizona: University of Arizona Press.

Terborgh, J. 1983. *Five Neotropical Primates*. Princeton, New Jersey: Princeton University Press.

—— 1986. Community aspects of frugivory in tropical forests. In A. Estrada and T. H. Fleming, eds., *Frugivores and Seed Dispersal*, 371–384. Dordrecht, The Netherlands: Dr. W. Junk Publishers.

Terwilliger, V. J. 1978. Natural history of Baird's tapir on Barro Colorado Island. *Biotropica* 10:211–220.

Unruh J. and S. Flores Paitan. 1987. Relative abundance of the useful component in old managed fallows at Brillo Nuevo. In W. M. Denevan and C. Padoch, eds., *Swidden-Fallow Agrofrestry in the Peruvian Amazon. Advances in Economic Botany* 5:67–73. New York: New York Botanical Garden.

van den Berg, M. E. 1984. The ethnobotany of an Amazonian market. *Advances in Economic Botany* 1:140–149.

Van Roosmalen, M. G. M. 1985. *Fruits of the Guianan Flora*. Utrecht, The Netherlands: Utrecht University. Institute of Systematic Botany.

Van Tyne, J. 1929. *The Life History of the Toucan* Ramphastos brevicarinatus. Miscellaneous Publication No. 19. Ann Arbor, Michigan: University of Michigan Museum of Zoology.

Wheelwright, N. T., W. A. Haber, K. G. Murray, and C. Guindon. 1984. Tropical fruit-eating birds and their food plants: A survey of a Costa Rican lower montane forest. *Biotropica* 16:173–192.

Yost, J. A. and P. M. Kelley. 1983. Shotguns, blowguns, and spears: The analysis of technological efficiency. In R. B. Hames and W. T. Vickers, eds., *Adaptive Responses of Native Amazonians*, 189–224. New York: Academic Press.

Utilization of Tropical Fruits by Geochelone spp. and Mammals Larger than 500 g

Numbers indicate the reference. Lower-case letters indicate frequency of utilization: r = rare; oc = occasional; m = moderate; f = frequent. Capital letters indicate the type of interaction; D = disperser; P = predator; FS = eats fruits and seeds, but no effect on seeds is documented. When the use of more than one plant species in a genus is documented, it is indicated as *spp.*; *sp.* indicates that no species was provided in the source, otherwise species names are provided.

	Geochelone	*Didelphis*	*Alouatta*	*Ateles*	*Cebus*	*Lagothrix*	*Agouti*	*Dasyprocta*	*T. pecari*	*T. tajacu*
Anacardiaceae										
Anacardium spp.	"r,1,35"	—	"r,2,4,7"	"27,35"	"35"	"28"	—	—	—	—
Spondias spp.	—	"2"	"1,2,4,7"	"D,27,35,7"	"r,13,7,12" "22,35"	—	"r,2"	"f,2"	—	"2"
Annonaceae										
Annona spp.	"1"	—	—	—	"12"	—	—	—	—	—
Duguetia spp.	"1"	—	—	"D,27"	—	—	—	—	—	—
Guatteria spp.	—	—	—	"D,27"	"12"	—	—	—	—	—
Apocynaceae										
Lacmellea panamensis	—	—	"m,2,4"	—	—	—	—	"r,2"	—	"2"
Araceae										
Maguira costanicana	—	—	"r,2,4"	—	—	—	—	—	—	—
Monstera spp.	—	—	—	"D,27"	"r,13,22"	—	—	—	—	—
Philodendron sp.	—	—	—	"27"	"27"	"22"	—	—	—	—
Bombacaceae										
Quararibea spp.	—	"2"	"r,2,4"	—	"r,13,12"	—	"r,2"	"m,2"	—	—
Boraginaceae										
Cordia spp.	—	—	—	"D,27"	"m,13,26"	—	—	—	—	—
Burseraceae										
Protium spp.	"35"	—	"35"	"r,D,10,27,35"	"f,13"	"28"	—	—	—	—
Tetragastris spp.	—	"2"	"f,2,6"	"D,27"	"r,D,11"	—	—	—	—	—

Chrysobalanaceae										
Licania spp.	"1"	—	—	"m,D,10,27," "35"	"35"	—	—	—	—	—
Convolvulaceae										
Maripa spp.	—	—	—	"P,27"	"oc,13"	—	—	—	—	—
Dilleniaceae										
Doliocarpus spp.	—	—	—	"35"	"m,13,22"	—	—	—	—	—
Euphorbiaceae										
Hyeronima spp.	—	—	"m,2,4"	"f, D,10,27"	"oc,13"	—	—	—	—	—
Guttiferae										
Clusia spp.	—	—	—	" D,27"	—	"7"	—	—	—	—
Rheedia spp.	—	—	—	"D,27"	"oc,13,12," "26"	—	—	—	—	—
Lauraceae										
Licaria sp.	—	—	—	"m,D,10,27"	—	—	—	—	—	
Nectandra spp.	—	—	"m,D,3,6"	"27"	"22"	—	—	—	—	—
Lecythidaceae										
Eschweilera spp.	—	—	—	"P,27,35"	"P,27,35"	—	—	—	—	—
Gustavia spp.	—	—	—	"27"	"f,13"	"24"	"2"	"f,2,14"	—	"2"
Leguminosae										
Dipteryx spp.	—	"2"	"m,2"	—	"m,13"	"29"	"2"	"f,FS,2,14,15"	—	"P, "
Eperua spp.	—	—	—	"P, 27"	—	—	—	—	—	—
Inga spp.	—	—	"35"	"D,27,35"	"f,13,7,12"	"S,30,31"	—	—	—	—
Melastomataceae										
Miconia argentea	—	—	—	—	"f,13"	—	—	—	—	—
Meliaceae										
Guarea spp.	—	—	—	"D,27"	—	"S,31"	—	—	—	—
Trichilia spp.	—	—	"r,2,4"	"D,27"	"26"	—	—	"m,14"	—	—
Moraceae										
Artocarpus incisa	—	—	—	—	—	—	—	—	—	—
Brosimum spp.	—	—	"m,D,2,4,6"	"f,D,10,7,35"	"12,22"	" 32"	—	—	—	—

Appendix 17.1
(continued)

	Geochelone	*Didelphis*	*Alouatta*	*Ateles*	*Cebus*	*Lagothrix*	*Agouti*	*Dasyprocta*	*T. pecari*	*T. tajacu*
Cecropia spp.	—	—	"oc,D,2,3,4," "6,7"	"D27"	"m,13,12,22"	—	—	—	—	—
Clarisia sp.	—	—	—	"35"	"12"	—	—	—	—	—
Ficus spp.	—	"2,35"	"r/f,D,2,3," "4,6,7,35"	"m,D,10,27," "35,7"	"f,12,12,13"	"29,30,31," "32"	"f,2"	"f,2"	"16,35"	"2"
Pouruma spp.	"35"	—	—	"D,27"	"7"	—	—	—	—	—
Pseudolmedia spp.	—	—	"r,D,6"	—	"12"	—	—	—	—	—
Myristicaceae										
Virola spp.	—	—	"7"	"r,D,27,x9"	"oc,13,22,26"	—	—	—	—	—
Myrtaceae										
Eugenia spp.	—	—	"r,2,4"	"D,27"	—	—	—	—	—	—
Psidium guajava	—	—	—	—	"13"	—	—	—	—	—
Oxalidaceae										
Averrhoa carambola	—	—	—	—	—	—	—	—	—	—
Palmae										
Acromia sclerocarpa	—	—	—	—	—	—	—	—	"P,17"	"16"
Attalea spp.	—	—	—	—	"35"	"7"	—	—	—	23
Astrocaryum spp.	—	"2"	—	—	"r,13,P,12," "35"	—	"2,12"	"f,FS,2,14," "15"	—	"19"
Euterpe sp.	—	—	—	"D,27"	"7"	—	—	—	—	—
Iriartea spp.	—	—	—	"D,27"	"S,12"	—	—	—	"P,18"	—
Jessenia spp.	—	—	—	"7"	—	"7"	—	—	"P,18"	"D, "
Mauritia flexuosa	"1"	—	—	—	—	"7"	—	—	"P,18"	—
Scheelea spp.	—	"2"	—	—	"f,13,P,12," "7"	—	—	"f,FS,2,14," "15"	—	"2"
Rubiaceae										
Duroia eriopila	"1"	—	—	—	"m,13"	—	—	—	—	—
Faramea spp.	—	"2"	"oc,2"	—	—	—	—	"r,2"	—	—

Genipa americana	"1"	—	—	—	—	—	—	—	—	—
Guettarda spp.	"1"	—	—	"D,27"	—	—	—	—	—	—
Sapindaceae										
Allophylus scrobiculatus	—	—	—	—	"12"	—	—	—	—	—
Cupania spp.	—	—	—	—	"m,13,26"	—	—	—	—	—
Paullinia spp.	—	—	—	"D,27"	"12"	—	—	—	—	—
Sapotaceae										
Chrysophyllum sp.	—	—	—	"f, D,10,27"	"r,13"	"FS,31"	—	—	—	—
Ecclinusa spp.	"1,35"	—	"35"	"D,27,35"	—	—	—	—	—	—
Pouteria spp.	"1"	—	—	"f, D,10,27"	—	—	—	—	—	—
Pradosia sp.	"1"	—	—	—	—	—	—	—	—	—
Sterculiaceae										
Sterculia spp.	—	—	"8"	—	"oc,13,35"	—	—	—	—	—
Theobroma cacao	—	—	—	—	"12"	—	—	—	—	—
Tiliacaeae										
Apeiba spp.	—	—	—	"D,27,35"	"f,13,12,35"	—	—	—	—	—
Violaceae										
Leonia glycycarpa	—	—	—	"D,27"	"12"	—	—	—	—	—

Numbers indicate references: 1) Moskovits 1985; 2) Glantz 1982; 3) Estrada 1984; 4) Milton 1980; 5) Branan et al.; 6) Estrada and Coates-Estrada 1986; 7) Hernandez-Camacho and Cooper 1976; 8) Neville 1972; 9) Sabatier 1983, cited by Gauthier et al. 1985; 10) Klein and Klein 1977; 11) Howe 1982; 12) Terborgh 1983; 13) Oppenheimer 1982; 14) Smythe et al. 1982; 15) Smythe 1986; 16) Sowls 1984; 17) Kiltie 1981; 18) Kiltie and Terborgh 1983; 19) Kiltie 1982; 20) Reichel-Dolmatoff 1985; 21) Terwilliger 1978; 22) Freese and Oppenheimer 1981; 23) Enders 1930; 24) Smythe 1978; 25) Enders 1935; 26) Hladik et al. 1971, cited by Freese and Oppenheimer 1981; 27) van Roosmalen 1985; 28) Kavanagh and Dresdale 1975; 29) Ramirez in press; 30) Durham 1975, cited in Ramirez, in press; 31) Izawa 1975; 32) Mittermier et al. 1977; 33) Russel 1982.

Appendix 17.2

Utilization of Tropical Fruits by Game Birds

	Tinamou	*Cracid*	*Quail*	*Trumpet*	*Pigeon*	*Parrot*	*Toucan*
Agavaceae							
Dracaena	8	—	—	—	—	—	—
Anacardiaceae							
Metopium	—	11	—	—	—	11	11
Spondias	8	16	—	—	—	—	16
Apocynaceae							
Aspidosperma	9	—	—	—	—	—	—
Couma	—	—	—	—	—	—	16
Araliaceae							
Didymopanax	—	—	—	—	—	—	D,15
Bombacaceae							
Bombacopsis	—	—	—	—	—	f,P,1	—
Bombax	—	20	—	—	—	—	—
Boraginaceae							
Ehretia	—	11	—	—	—	11	11
Burseraceae							
Protium	f,8	—	—	—	—	—	10,1,6
Tetragastris	—	*D,5*	—	—	—	—	*D,1*
Capparidaceae							
Forchhammeria	8	—	—	—	—	—	—
Celastraceae							
Maytenus	8	—	—	—	—	—	—
Combretaceae							
Terminalia	—	—	—	—	—	f,P,1	—
Connaraceae							
Cnestidium	—	—	—	—	—	—	f,3

Appendix 17.2
Continued

	Tinamou	*Cracid*	*Quail*	*Trumpet*	*Pigeon*	*Parrot*	*Toucan*
Euphorbiaceae							
Alchornea	—	—	—	—	—	—	D,6
Hura	—	—	—	—	—	P,1	—
Drypetes	8	—	—	—	—	—	—
Flacourtiaceae							
Caseania	—	—	—	—	—	P,10	f,D,7,10,1
Hippocrateaceae							
Salacia	—	—	—	—	—	—	f,3
Icacinaceae							
Calatola	—	—	—	f,4	—	—	—
Leretia	—	19	—	—	—	—	—
Lacistemataceae							
Lacistema	—	—	—	—	—	—	6
Lauraceae							
Beilschimiedia	—	—	—	—	—	—	D,1
Lecythidaceae							
Lecythis	8	8	8	—	—	—	—
Malvaceae							
Hampea	—	—	—	—	—	—	10
Mimosaceae							
Cedrelinga	—	—	—	—	—	16	—
Melastomataceae							
Blakea sp.	—	18	—	—	—	—	—
Miconia	—	6	—	—	6	—	—
Meliaceae							
Guarea	—	19	—	—	—	—	—

Swietenia	8	—	—	—	—	—	—
Trichilia	—	19	—	—	—	—	16
Moraceae							
Bagassa	—	—	—	16	—	—	—
Brosimum	f,8	—	—	—	—	—	—
Cecropia	—	6	—	—	6	—	D,10,1,6
Clarisia	—	16	—	—	—	—	16
Ficus	—	f,D,4,18	—	f,D,4,16	—	f,P,2,13,16	f,D,3,11,1,4,16
Pseudolmedia	f,8	—	—	—	—	—	D,1
Sorocea	—	—	—	—	—	—	D,1
Trophis	8	—	—	—	—	—	10
Musaceae							
Musa	—	—	6	—	—	—	3
Myristicaceae							
Virola	8	D,5,19,8,12	8	—	—	—	f,D,3,10,12,1,6
Myrtaceae							
Psidium	—	6	—	—	—	—	—
Nyctaginaceae							
Neea	—	—	—	—	—	—	11
Palmae							
Bactris	—	19	—	—	—	—	—
Cryosophioa	8	—	—	—	—	—	—
Euterpe	—	—	—	—	—	—	16
Geonoma	8	8	8	—	—	—	—
Oenocarpus	—	—	—	—	—	—	16
Socratea							
Sabal	8	—	—	—	—	—	D,1
Astrocaryum	—	—	—	—	—	—	f,3
Iriartea	—	—	—	—	—	—	f,3
Podocarpaceae							
Podocarpus	—	18,19	—	—	—	—	—

Appendix 17.2
Continued

	Tinamou	*Cracid*	*Quail*	*Trumpet*	*Pigeon*	*Parrot*	*Toucan*
Rubiaceae							
Faramea	—	—	—	—	—	—	D,1
Hamelia	—	6	—	—	—	—	—
Sapindaceae							
Allophylus	—	—	—	f,4	—	—	—
Cupania	—	—	—	—	—	—	f,3
Dipterodendron	—	—	—	—	—	—	6
Talisia	—	—	—	—	—	11	11
Sapotaceae							
Manilkara	8	—	—	—	—	—	—
Pouteria	8	—	—	—	—	—	—
Tillaceae							
Apeiba	—	—	—	—	—	16	—
Ulmaceae							
Celtis	—	—	—	—	—	*f*,4	—
Violaceae							
Rinorea	8	—	—	—	—	—	—

The numbers represent citations in which game birds were observed eating fruit. Capital letters were used to designate the type of interaction recorded (D = seed disperser, P = seed predator). An "f" indicates fruits that were frequently eaten or were preferred food items. Numbers indicate references: 1) Janzen 1983; 2) Janzen 1981; 3) Van Tyne 1929; 4) Terborgh 1986; 5) Howe 1982; 6) Skutch 1983; 7) Howe and Primack 1975; 8) Sick 1970; 9) Lancaster 1964; 10) Howe 1977; 11) Kantak 1979; 12) Howe and Kerckhove 1981; 13) Jordano 1983; 14) Wheelwright et al. 1984; 15) Chapman 1929; 16) Van Roosmalen 1985; 17) Howe 1980; 18) Johnson and Hilty 1976; 19) Held and Werkhoven 1988. In order, the bird families listed are Tinamidae, Cracidae, Phasianidae, Psophiidae, Columbidae, Psittacidae, and Ramphastidae.

EIGHTEEN

Common Property Resources in the Neotropics: Theory, Management Progress, and an Action Agenda

PETER H. MAY

Forests in much of the American tropics contain economically useful plant species whose multiple market and subsistence products complement agricultural production and off-farm income of peasant households (Hecht, Anderson, and May 1988). Land uses employed by such households, when pressure on fallow cycles is not severe, tend to enable forests to regenerate. Conversion to monocrops such as sugarcane, rice, or pasture is more permanent and damaging to the tropical forest environment. Changes in property regimes resulting from permanent land use conversion and technological innovation in the cases of *babaçu* palms and natural rubber in Brazil have a number of social equity effects. The initial rights over trees and land contribute to the ultimate distribution of rewards from agricultural intensification and industrial innovation.

The economics literature on agrarian transformation largely suggests that technical innovation in response to changing factor or market prices or their proportions will lead to a more efficient agriculture: more food, fiber, and fuel requiring less human toil to produce and at a lower cost to consumers (Hayami and Ruttan 1971). Farmers displaced from relatively less efficient forms of production will be absorbed by other sectors, whose growth is stimulated by the added surplus derived from a more efficient agriculture (Johnston and Kilby 1975).

The aim of rural development has principally been to alter existing agricultural systems or unexploited natural environments to stimulate output growth. Infrastructure improvements and selective subsidies facilitate producers' access to new frontiers and enable them to reorganize the production process by adopting technical and institutional innovations. The environmental and social costs of moving to a more "efficient" state may, however, exceed the benefits derived from the alteration. Norgaard (1984) described as "coevolutionary" those development processes that are sensitive to the biological constraints imposed by local environmental conditions but at the same time enable local inhabitants to transform that environment for their needs.

A focus on the efficiency or even coevolutionary potential of a given development path is necessary but not sufficient. To these criteria must be added an analysis of the equity effects of changes in control over resources (Dasgupta 1982). Rapid capitalist development has displaced a powerless peasantry at a rate inconsistent with the capacity of the nonagricultural sectors to absorb them. The question of fairness is critical to any objective assessment of the returns from innovation. The issue becomes not just "How much?" but also "For whom?"

Processes of change in tropical land use often deprive indigenous and colonist populations of native plant species from which they derive a wealth of goods (wood, fibers, medicines, rubber, oils, and dyes) and unpriced services (e.g., watershed and soil protection, nutrient recycling). In many cases, these resources have been managed for generations under common property regimes for sustained yield. Deforestation upsets the delicate balance that characterizes these management systems.

Extractive resources in the tropics are not, however, threatened only by land conversion. They are also sometimes subject to degradation resulting from overexploitation by their traditional users due to expansion in market demand or population growth (Repetto and Holmes 1983). Economists have tended to blame resource depletion on an insufficient measure of individual control over resource use exercised through property rights (Furubotn and Pejovich 1972). Privatization is the usual prescription given by neoclassical economists for resource degradation problems that arise from the tendency of individual profit seekers to exploit open-access resources at a level beyond their sustainable equilibrium.

Placing blame for the malaise of common-property resource degradation on burgeoning human populations (Hardin 1968) is often erroneous. In Amazonia, for example, there is no basis for arguing that resource degradation is caused by population pressure on scarce land resources. The problem is that the resources at the disposal of the vast majority tend to be restricted due to a highly skewed distribution of access to property rights (Hecht 1985). In contraposition to the neoclassical economists' insistence on privatization, a forceful swing toward enclosure—barring entry to resource use by concentrating property rights in the hands of a few—has often increased rather than reduced the rate of predatory resource exploitation (Runge 1984). Property rights delimitation may also result in pressures on limited remaining resources traditionally managed in common, thus hastening their degradation. Such results have been defined as a "tragedy of the non-commons" (May 1986).

The theoretical and historical roots of the non-commons tragedy, discussed here, are a revealing prelude to its unfolding in the cases of *babaçu* and natural rubber in Brazil. Countering this tragic trend, Amazon social movements have proposed institutional shifts in property rights systems from which funding agencies seeking equitable solutions to Neotropical resource problems may draw lessons for their work.

Extractivism, Exploitation, and Innovation

Extractive activities are associated with severe human exploitation. The social relations through which extractive products are gathered and make their way into markets have tended to be ones of debt servitude or intermediary manipulation. In many cases, gathering and rudimentary processing are done by people possessing control over neither the resource nor the placement of goods in the market. These people therefore have little bargaining power with which to affect the share of the product's value that accrues to them; are often forced to live under terms of dependency to landowners, concessionaires, or merchants; and are the first to suffer from the periodic booms and busts endemic to extractive industries. Where demand is intense and the physical resource base is accessible as well as limited, extraction may result in rapid degradation of the plant resource. By limiting supply, degradation will further push prices upward and thus increase pressure on the resource. Extinction is a frequent result of this cycle.

A process of market evolution leading to degradation of extractive resources has been proposed by Homma (1981) and by Repetto and Holmes (1983) for the case of subsistence resources in general. A description of this process, following the tenets of static equilibrium analysis, is presented in figure 18.1, and adaptation of Homma's (1981) theory. Initially, demand (D_0) for the extractive product is presumed to be dormant, while supply (S_0) is high. Since the resource is conceived as an open-access "commons," anyone can obtain access to the resource to extract the product. Although an open-access condition is posited to exist, the costs of extraction and marketing are severe due to absence of infrastructure or marketing channels. Hence, the supply curve hugs the Y-axis up to some point at which the minimum entry-level marginal costs would be redeemed by the market price. Since neither producers nor consumers are satisfied with the price conditions available in this first stage, there is no extraction; the resource remains untouched.

In the second stage (figure 18.2), infrastructure development in the area of an extractive resource enhances the product's marketability and reduces its extraction cost. Supply (S_1) is able to satisfy the initial meager demand (Q_1). However, the resource is underexploited at this demand level. More frequently, demand leads supply in stimulating entrepreneurs to develop the market infrastructure necessary to initiate trade in the extractive product. This order of events occurred in the case of Amazon natural rubber. In response to successful innovation in the production of tires and other rubber products, expansive domestic and international markets shifted the demand curve to the right (D_1). The better prices offered, in turn, stimulated the organization of more intensive rubber exploitation systems.

Due to the resource's fixed and limited character, the cycle will eventually hit up against a supply limitation. Supply at some point (in this case S^*), becomes perfectly inelastic to price, and the curve becomes vertical. An ecological-economic equilibrium is attained at the point where Q_2 is harvested; at demand levels beyond this point, high prices will induce producers to overexploit the resource, eventually causing its diminished productivity if not total extinction. The supply curve shifts back and upward to S_2, characterizing conditions

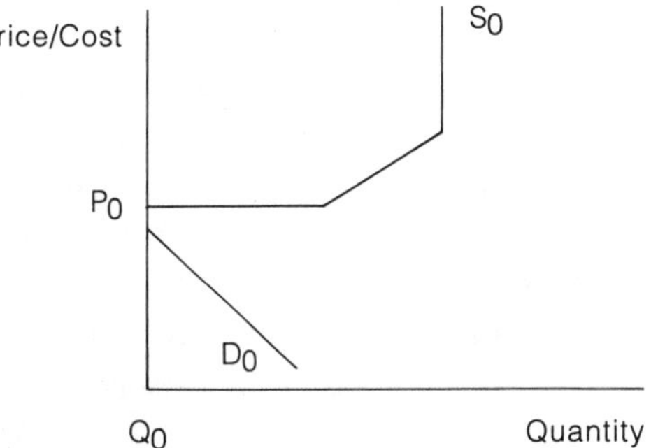

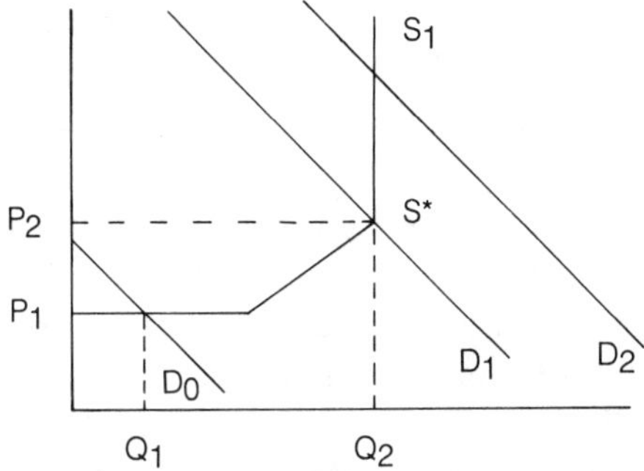

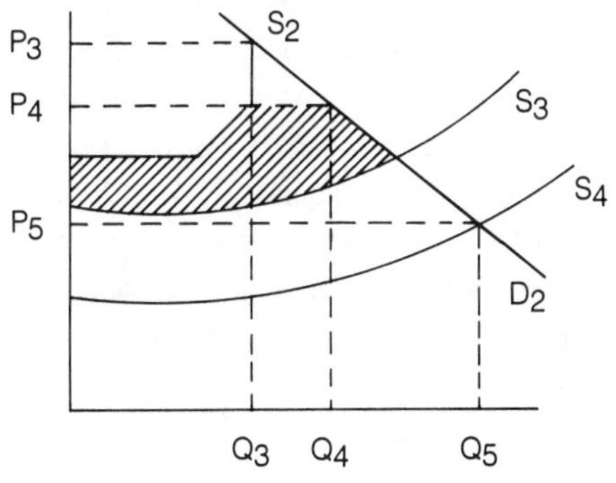

Figure 18.1 (top) Model of the extractive produce market.

Figure 18.2 (middle) Demand-induced resource exploitation leads to degradation

Figure 18.3 (bottom) Demand-induced resource exploitation leads to degradation

Technical innovation substitutes extractive production. Adapted from Homma (1981) and May (1986).

of resource exhaustion with a resultant increase in exploitation costs. But, because demand is unsatiated, pressure on the resource continues. Such pressures, under open-access conditions, lead to the traditional "tragedy of the commons" (Hardin 1968:1243).

Institutional options to restrain the rate of exploitation of the extractive resource may be pursued at this point. The specific direction taken by decision makers (whether representatives of a central state or, alternatively, community or tribal leaders) to alter property institutions will determine not only the sustainability of the resource, but also the partitioning of benefits from its exploitation. Some societies have been able to define property rights so as to ensure broad enjoyment of surplus, while others have parceled out the rewards of resource control to a circumscribed elite. The extent to which rights to productive resources are concentrated affects not only the distribution of rents from resource exploitation, but also the measure of welfare enjoyed by the broader society.

In the cases of natural rubber and *babaçu* palms, the initial usufruct rights of peasants over the trees are in the process of being restricted. At the same time, deforestation and land-use conversion to pastures are not only diminishing the productive base of the extractive system but also reducing access to land for shifting cultivation. If land is plentiful enough elsewhere, however, the peasant family may be able to relocate to a different area, where extractive resources are still abundant and cattle still scarce. What happens, however, on reaching the frontier, is that the peasant family is forced onto land of lesser productivity for shifting cultivation, placing greater pressure on the resources that remain. Similar to the pressures brought about through market forces, subsistence needs will increasingly result in resource degradation. This process results in the same shift of the supply curve upward and to the left as in S_2 in figure 18.2. This process has been termed a "non-commons tragedy" (May 1986), resulting from the privatization of natural resources.

Given technical feasibility, several things might occur to make up for scarcity of the extractive product caused by increasing pressure on the resource. One option is discovery of synthetic substitutes. Another is genetic enhancement of the native species and its development under rationalized plantation conditions as a crop. In either case, the initial investment to make the necessary transition is fueled by the potential for innovation rents derived from the high prices caused by unsatisfied consumer demand in the last stage. In figure 18.3, these rents are equivalent to the area between S_2 and the supply curve for the synthetic industry or plantation system (S_3). Since, at its initial stage, agroindustrial innovation is likely to be limited to one or a few firms, moreover, the amount supplied (Q_4) will be limited to that necessary to derive monopoly profits. As more firms enter the field, however, marginal production costs for the sector as a whole will become homogenous (S_4), and an equilibrium price and quantity will be attained (P_5 and Q_5).

Where extractive products are recognized as perennial crops suitable for expanded production in plantations or synthetic substitutes are found, however, rationalization of production methods undercuts employment and wages in extraction. By reducing the costs of production, the rationalized system eventually eliminates the extractive system or reduces its share to an insignificant segment of the overall market. Since prices have been reduced over time by rationalization, extractive industries often require price floors or subsidies on production costs to maintain competitiveness. Those who control extractive resources are able to bargain for such policies because they typically constitute cohesive elites able to manipulate political action on their behalf. But such policies only temporarily stave off the decline in extractive production.

As with land-use conversion, the benefits of agroindustrial innovation are partitioned in accordance with claims over resources. The extraordinary gains obtained through control over innovative processes or improved crops nurture the infant industry while enabling it to undercut the extractive system. Those who control proprietary rights to the innovative process are made better off at the expense of the original producers. Since production costs are lower, demand may be more fully satisfied, and the condition of society as a whole is improved. Yet, unless some claim is placed on innovation rents to compensate those displaced, the distributional consequences of the shift are negative for extractive producers.

Producers of many extractive goods rely for a considerable proportion of their incomes upon their sale or trade. It is difficult for them to quickly adjust their labor or land allocation among other activities in areas where extractive resources predominate. The shift from collection of native plant resources to annual crop cultivation or livestock husbandry is drastic for people who may have little experience with the crops or management practices appropriate to the area. Many are forced out of rural areas entirely by the shift in technology.

Relying on the essential adaptability of humans to their environment (Norgaard 1984) is one answer to adjustments required by those dependent on extractive resources. Producers will eventually adapt to shifts in markets for extractive products. Such adaptability, however, does not guarantee responsiveness to exogenous conditions that may result in resources being depleted or excluded through property rights delimitation.

The Property Rights Controversy

No "right" exists without social institutions to defend it. Each society creates unique institutions and rules to define and defend property. In fact, the concept of property itself lacks any meaning outside a social context of relations between people. Property rights have been defined as "sanctioned behavioral relations among men that arise from the existence of things and pertain to their use" (Furubotn and Pejovich 1972:1139).

One branch of economics, that of "institutional economics," proposes that the particular institutions devised to articulate and protect property rights in a given society have arisen from conflict between self-interested individuals over rights to use and benefit from resources. These institutions have evolved so as to create a context for resolving such conflicts, providing "security for some claims and aspirations" (Randall 1978:3, citing; Commons 1934). At the core of the conflict-resolution process are transactions that involve "alienation and acquisition of the rights of property and liberty created by society" (Randall 1978:3). Working rules for resolving conflicts through transactions are the result of "deliberate collective action." Such rules include the rights themselves, the duties that such rights impose on others, and the limits and responsibilities of the right holder as defined by "liabilities" imposed by the society through the rule of law.

Rights over resources having "open-access" characteristics have been particularly thorny issues in economics and law. Such resources are those in which benefits or costs are not restricted to individuals because the resources themselves are unbounded or the benefits or costs flowing from them are indivisible and hence not easily traced as regards their effects or source. Typical examples of such resources are air, fisheries, and tropical forests. The effects of one such resource user's action on others—users or nonusers—are termed "externalities."

The "New Resource Economics" (NRE) in neoclassical microeconomic theory (Runge 1984) has proposed an approach to resolve open-access resource problems. The NRE school assumes that rational economic actors will sort out their conflicts through market exchange. However, this approach avoids questions of equity and fairness in either the initial distribution of rights or wealth or that which arises from alterations in resource utilization.

The NRE paradigm suggests that most, if not all, problems associated with open-access resources can be resolved through bargaining. All that needs to be done is to define who is liable for damages accruing from use of the resource. If the party causing the externality is liable, he will be required to compensate the injured party with a sum at least equal to the cost resulting from his action. On the other hand, rather than make compensatory payments, and provided technology is available, he might "internalize" the externality if the cost of doing so is less than the minimum payment necessary to offset damages.

If liability does not fall on the acting party, however, the injured parties to the externality must pay its producer to compensate his cost if they desire its abatement. This approach, first formulated by Coase (1960), requires the condition that there be no costs in making transactions between acting and injured parties, since the costs of organizing to make such claims are often prohibitive. Only where the liability is placed fully upon the producer of the externality will there be a result in which no one is made worse off.

However, it is rare that liability will be placed on the producers of externalities in an open-access resource situation, either because they are difficult to identify or because administrators "contribute, foster, aid and indeed legitimate the special claims of small but highly organized groups to differential access to tangible resources which are extracted from the commons" (Crowe 1969:1106). Those who sustain the injury must pay the price. But those affected are typically many, and the transactions involved in obtaining contributions to a pool are so costly as to exceed the benefits to be obtained. Alternatively, affected parties might organize to seek a change in the (liability) rules of the game. Yet the organizational solutions to externality problems are themselves public goods, which are prone to free ridership and noncontributive behavior (Olson 1965).

Because of these problems inherent in market solutions, popular opinion and political action have tended to favor a government standard-setting role rather than rely on privatization and the magic of the marketplace in the case of environmental externalities. In less-developed nations, however, privatization of property rights over resources has been perceived by governments as a more appropriate means to avert open-access resource problems. As shown in the case studies below, however, such an approach often promotes severely inequitable consequences and may end up being more costly to the state than efforts to adopt common-property management methods.

The Tragedy of the Non-Commons

The ownership of land can confer the power to withhold land from production (Taylor 1978). Such a power to constrain land utilization can lead to increased pressure on remaining land resources that were traditionally managed for sustained yield as common property. Where resources have been privatized so as to curtail benefits obtained through common management, and those excluded are denied compensation due to lack of either bargaining power or of legal legitimation of property rights, there exists a tragedy of the non-commons (May 1986).

The enclosure movement that took place from the seventeenth to the nineteenth centuries in England provides an example of a non-commons tragedy. Overgrazing was not the principal stimulus for landowners to enclose the common fields, as Hardin (1968) has suggested. Regulation of stocking rates on the common fields was generally practiced through requirement of a home feed base during non-grazing periods and at night (Tate 1967). Rather, the increased profitability to the English manorial lord of sheep production for wool at the onset of technological development in the textile industry led to the enclosure of the commons. Such enclosure resulted in severe rural immiseration, and forced many people to migrate to cities, where they created what Marx called the "reserve army of the unemployed" whose downward pressure on wages helped stimulate the Industrial Revolution. During the same period, in France,

> as forest lands became increasingly profitable as sources of timber for sale vis-a-vis their traditional role as sources of livestock forage, firewood for home consumption, and building material for the peasant vil-

> lage, the feudal lords changed from administrators and protectors to profit-seeking entrepreneurs. . . . The result was a weakening of the village system and dispossession of the peasantry. The peasant was transformed from a co-equal owner on the commons with secure tenure to a landless worker on the feudal estate. This is the true "tragedy of the commons." (Ciriacy-Wantrup and Bishop 1975:720)

Similar enclosures are underway in many developing countries today. Those who formerly managed resources in common are rapidly being excluded from access to their benefits. To combat inequities arising from resource exclusion, some demonstration of the viability of collective management may be required. Resources treated as "no-one's property," and hence theoretically subject to overexploitation, are often in fact regulated by common-property institutions so as to avert degradation. Demonstrative cases of this type include those of traditional range management (Orlove 1975; Orlove 1977; Netting 1976; Gilles and Jamtgaard 1981), as well as a variety of systems used for managing fisheries, wildlife, and tree tenure documented in BOSTID (1986) and Fortmann and Bruce (1988).

It may not be enough to support efficient collective management of resources, however. As a legal lever against the state's expropriation of land to benefit timber corporations in Indonesia, Dove (1983) suggests that individuals must establish a signature of ownership to ensure control over forest fallows formerly regulated through common property institutions. Thus attempts to preserve traditional management systems may need to give way to efforts to prove efficient and continuous occupation of forest lands to forestall their preemptive enclosure by powerful interests. How to achieve this without necessitating damaging and irrevocable deforestation is one of the greatest challenges that face those concerned with tropical resource development today. The following case studies shed light on some of the options that have emerged.

The Case of Natural Rubber

The classic case of forest product extractivism in Brazil is that of rubber tapping in the Amazon (Santos 1980). Usufruct rights to land containing rubber trees (*Hevea brasiliensis*) was conceded to entrepreneurs who imported workers from the dry Northeast. These workers were allotted tracts of land along navigable streams, where they made periodic rounds of sparsely distributed trees (often averaging fewer than one rubber tree per hectare), which they tapped for the valuable latex. Once gathered in liquid form, the latex was smoked and solidified into large balls for trade with concessionaires (*seringalistas*).

Rubber producers devised land management and property rights practices that suited the sparse natural occurrence of *Hevea*. Because rubber trails (*estradas*) of different households often intersected, it made little sense to define holdings by property boundaries. Instead, property rights were defined as exploitation rights to specific trees or *estradas*, and rights over other resources were expressed through common property regimes.

The *seringalistas'* agents plied their territories in riverboats, trading commodities needed by the tappers for rubber and other extractive products. By extending credit for commodities at high interest and taking advantage of the virtual monopsony entailed by their control of the rubber stands, *seringalistas* kept the rubber tappers in virtual debt slavery. Because Brazil was, until the first decade of the twentieth century, the only major exporter of latex, a product in ascendent demand after the development of the vulcanization process, the rubber barons reaped immense profits from this trade.

Yet the *seringalistas'* splendor was short-lived. *Hevea* seedlings smuggled out of Brazil to Kew Gardens and thence to Malaysia came into production in the years immediately preceding World War I. In a matter of months, rubber prices dropped precipitously in the international market, devastating the fortunes of the Amazonian gentry. Latex production continued, however, for a growing domestic market, with prices maintained by the federal government at a level that enabled extractive production to continue. Despite efforts to promote plantation production of *Hevea* in Brazil, the trees soon fell prey to a leaf blight (South American leaf blight) that destroyed them. This fungus (*Microcyclus ullei*) did not affect sparsely distributed rubber trees in species-diverse forests and was nonexistent in Southeast Asia.

In the 1950s, Brazil substituted imports of foreign rubber products with domestically produced goods by allowing foreign subsidiaries of multinational tire manufacturers to establish operations. The growing domestic automotive market boosted demand for rubber to a point where, today, Brazil imports more than 70 percent of all natural rubber it uses and produces even more synthetically. The difference between the domestic support price and foreign market prices for natural rubber imports subsidizes both the extractive industry and increasingly successful attempts to plant rubber in "escape" zones of South American leaf blights (Dean 1987).

Production relations in Brazilian extractive rubber production to this day are little altered. In a growing number of cases, however, gatherers' dependence on concessionaires has been replaced by ties to merchants, as *seringalistas* have sold off their vast holdings to southern land speculators. The latter have invested in the clear-cutting of forests to establish pastures and have expelled rubber tappers from the native forest on which they depend for their livelihood.

Since the early 1970s, a social movement has arisen in the western Amazon to contest this pattern of land occupation. Through nonviolent *empates* (stalemates), rubber tappers have impeded land clearing operations in areas possessing economically important tree species. In 1985, representatives of a number of rural workers' unions and producer associations met in Brasília to found the National Council of Rubber Tappers and demand the creation of "extractive reserves" that would ligitimate producers' property rights over the forest resource on which they depend.

Such reserves would be common-property forest management units, recognizing traditional patterns of occupation conceded under long-term agreement by state governments to forest-dwelling communities that would designate rights and responsibilities of individuals to the resources within them. This alternative form of property has been increasingly recognized by state and national governments as an appropriate mechanism for conflict resolution over extractive resource access and use.

Relative autonomy obtained through legitimation of usufruct rights to the resource still may leave producers highly vulnerable to the vagaries of the market and to monopsony exercised by those who control trade. These problems have been clearly identified by the extractivist movement in the western Amazon and has led to the search for alternative marketing channels, cooperative purchase of industrial commodities, and product diversification as measures to surmount them.

The Babaçu *Palm Case*

In the Mid-North region of Brazil, composed of portions of the states of Maranhão, Piauí, and Goiás, historical patterns of land occupation led to dominance of the *babaçu* palm in the secondary forest

(Anderson and May 1985). For many years, these palms were managed within low intensity shifting cultivation systems by forest dwellers who made use of the palm's multiple products and revered it as a "tree of life." Its leaves furnished their shelter, its husk their fuel; they and their animals derived nourishment from its oil kernel, starchy outer husk, and palm heart, while a substantial regional vegetable oil industry arose to make use of *babaçu* kernels. It is estimated that as many as 420,000 rural households depend on *babaçu* for part of their incomes (Mattar 1979).

Babaçu products are particularly valuable to the subsistence economy because they are obtained in the period between peak labor demands in annual crop production and are produced primarily by women and children, whose labor is not critical to agricultural or off-farm activities. For these reasons, subsistence farmers protect the palms, retaining them within agricultural fields, though they cut other species or use them for fences.

In what ways does *babaçu* constitute a common property resource? Land at the Amazon frontier in Maranhão was initially subject to open access. Property rights over palms and other extractive plant resources in the *babaçu* zone were established informally by peasant production groups who occupied the "free lands" at the frontier (Alencar 1983; Luna 1985; Tendler 1980; Almeida 1986). There, shifting cultivators exercise usufruct rights to fallow-cycle products of the plots from which they harvest annual crops. This gives them an incentive when clearing the plot to retain *babaçu* palms and other useful plant species that provide them with essential goods and services during the subsequent fallow. Under some tenures, each household is considered to retain harvest exclusivity over *babaçu* palms in the immediate surroundings of its dwelling (Soares 1981). Presence of dense stands of *babaçu* is one criterion for the location of rural settlements and farm sites (Moran 1981).

On traditional private estates in the *babaçu* zone, peasants who reside on a given property are granted exclusive usufruct rights to the *babaçu* palms that grow there, on the condition that they sell extracted kernels through the landowner. Charcoal produced from *babaçu* fruit husks and the numerous other subsistence products derived from the palms are part of their usufruct rights held in common with other residents. Those who are granted these rights are thus motivated to exclude others from trespassing to collect fruits or other goods from the property, hence sustaining resource productivity by reducing predatory exploitation.

Since the early 1970s, expansion in the state road network and generous rural credit terms to large-scale investors stimulated land-use change in Maranhão. With infrastructure development and regionally targeted subsidies, the palm forests, and those sustained by them, are being subjected to a non-commons tragedy. Deforestation of palms to make way for sugarcane and improved pasture and enclosure of lands that were formerly available for shifting cultivation have placed pressure on resources that remain accessible, forcing peasants to cut down the very palms on which they rely for sustenance (Santos 1984). The conversion of large areas of the *babaçu* zone to pasture also implies a reduction in labor used in agriculture and has resulted in the expulsion of peasants from traditional estates, commonly managed lands, and unclaimed frontier territories.

Perhaps the most significant factor in the decision to clearcut *babaçu* stands is the ranchers' desire to minimize their interaction with the rural labor force. Despite expulsion, poor families that depend on *babaçu* continue to enter converted properties, breaking fruit to extract kernels and using the husks to make charcoal. Landowners then cut the palms simply to discourage trespassing, alleging that the husks left on the ground wound their cattle or that charcoal production could cause wildfires (Anderson and May 1985).

The answer provided by those proposing a technical solution to the *babaçu* problem is to develop industrial technology that will allow a more thorough utilization of *babaçu* byproducts. This will purportedly enable the industry to pay enough for the fruit that landowners will no longer cut the palms. But this solution also generates a non-commons tragedy. By selling whole fruit rather than the kernels produced by peasants in local cottage industries, the landowner will be able to appropriate a larger share of the price. He will need peasants only for the harvesting of fruit, a seasonal activity; there will be no need to secure a permanent resident labor force. Fruit harvesting requires vastly fewer workers than the current cottage industry, so the number of workers employed would be much smaller. This innovation thus has the potential to cause considerable labor displacement.

This outcome is the result of either palm destruction or technical innovation in the *babaçu* processing industry. Both result in peasants being excluded from property rights they formerly held. The chances that those displaced will be absorbed by other sectors of the regional economy are poor, given the highly capital-intensive pattern of rural and urban investment. The only alternative is emigration to already overcrowded cities or to the rapidly closing Amazonian frontier.

The New Resource Economics formulation of transactions between producers and victims of externalities suggest that gainers will compensate losers. There are several considerations that undermine the validity of this reasoning in the tragedy of the non-commons. First is the obvious fact, patent in the *babaçu* case, that bargaining power is unequal. Given an inequitable prior distribution of property rights, landowners are under no pressure to compensate peasants' losses from *babaçu* eradication. Peasants lack the political cohesion of small groups necessary to effectively mount an organized movement that will force a change in the rules of the game.

With increasing spread of pastures and reduction of land available for agricultural use, poor families have become more dependent on *babaçu* for their survival. Although the state government in Maranhão has approved a law protecting *babaçu* stands from indiscriminate deforestation, this law, up until now, has never been applied. Small farmers, acting through their rural workers' unions, are cited in this law as "official notifiers of infractions related to the cutting of babaçu palms," and high fines are to be meted out against infractors. This at least partially confers legitimation of small farmers' property rights over the palms, but it does not furnish them any power to ensure its enforcement. In consequence, as occurs with other common property resources in developing nations, *babaçu* palms do not have any effective protection.

What options exist for policy intervention to confront the problems of rural unemployment and reduction in the supply of basic foods provoked by the process of resource privatization in the Brazilian *babaçu* case? Technological alternatives that envision *babaçu* processing at the farm level rather than in large factories could reduce transport costs and, under restricted conditions, result in an increase rather than a decline in rural unemployment. These conditions exist in areas where both the current rate of *babaçu* extraction is low and there is little potential for agropastoral development. In other areas, however, either access to resources has already been preempted by privatization or current exploitation levels imply that employment impacts of industrialization would be severe. Such circumstances compel institutional and not solely technological solutions, so as to resolve conflicts resulting from restriction in resource access.

It is necessary to consider how to regularize usufruct rights to extractive resources that remain fenced within others' lands, or under ill-defined public property regimes. One solution would be to propose long-term usufruct contracts between proprietors and peasants, specifying responsibilities for protection and management of resources and assuring the landowner of a secure return. Another variant of this option, described above, is the creation of extractive reserves. In this approach, public or untitled lands under squatter occupancy that contain significant economically valuable and renewable forest resources are demarcated and placed under long-term stewardship by extractive producers. The extractive reserve concept provides a framework for sustained utilization and protection of forest resources through definition of property rights to the benefit of local producer communities (Allegretti and Schwartzman n.d.; Schwartzman in press).

The coordination of contractual negotiations between landowners, small farmers, rural workers, and the state to establish usufruct rights to extractive resources—which up to now has been left to highly imperfect markets—requires political definition to avoid the aggravation of the problems presented here.

Role of Funding Organizations

The moment is ripe for selective intervention to promote institutional modifications and support refinement of common-property management practices in use by Neotropical forest dwellers. Such actions should be geographically concentrated and involve support to grassroots and intermediary organizations, government agencies, and training and research institutions that agree to collaborate toward common ends. Some of the issues that should be the focus of financial support include:

1. Research on the legal and customary definitions of common-property management regimes and assistance to forest dwellers in devising codes specifying rights and responsibilities to resources on which they depend, including rates of exploitation of extractive products, land use practices, human carrying capacity, and tenure over flora and fauna.

2. Education of forest dwellers about their rights under existing property and labor laws, as well as efforts to define alternative property arrangements within national and local legal frameworks.
3. Defense of native peoples and other forest dwellers against pressures that threaten to cause non-commons problems through efforts to legitimate customary property and usufruct rights over extractive resources.
4. Analysis of market trends and strategic entry points for a range of forest products, as well as alternative marketing channels, processing options and transport systems to reduce dependence on intermediary chains and increase returns to the producer.
5. Long-term support for research on methods to enrich natural and degraded forests, once these are securely under the control of local communities, with species whose products are beneficial to subsistence economies and will have high market value over the predictable future.

ACKNOWLEDGMENTS

Financial support for the dissertation on which this paper is based (May 1986) was provided by the New York Botanical Garden's Institute of Economic Botany (under a USAID grant), the National Science Foundation, the Inter-American Foundation, and Cornell University. Segments of this paper were published in Susana B. Hecht and James Nations, eds., *The Social Dynamics of Deforestation: Processes and Alternatives,* Cornell University Press, 1988. I am grateful to the many *babaçu* and rubber extractivists whose patient responses to inquiries provide the substantive basis for my conclusions. For any errors in my interpretation, I alone remain responsible.

LITERATURE CITED

Alencar, T. H. de. 1983. Terra de Promissão: Luta pela subsistência de um povoado na frente de expansão do Sudoeste do Maranhão. Masters thesis, Museu Nacional de Antropologia, Rio de Janeiro, Brazil.

Allegretti, M. and S. Schwartzman. Extractive reserves: A sustainable development alternative for Amazonia. Report to World Wildlife Fund, U.S., Washington, D.C.

Almeida, A. W. de. 1986. Estrutura fundiâria e expansão camponesa. In J. M. Gonçalves ed., *Carajás: Desafio Político, Ecologia, e Desenvolvimento.* São Paulo, Brazil: Brasiliense/CNPq.

Anderson, A. 1983. The biology of *Orbignya martiana* (Palmae): A tropical dry forest dominant in Brazil. Ph.D. dissertation, University of Florida, Gainesville.

Anderson, A. and P. May. 1985. A palmeira de muitas vidas. *Ciência Hoje* 3(19):58–64.

BOSTID. 1986. *Proceedings of the Conference on Common Property Resource Management.* Washington, D.C.: National Academy Press.

Ciriacy-Wantrup, S. V. and R. C. Bishop. 1975. "Common property" as a concept in natural resources policy. *Natural Resources Journal* 15:713–727.

Coase, R. 1960. The problem of social cost. *Journal of Law and Economics* 3: 1–44.

Commons, J. 1934. *Institutional Economics.* New York: Macmillan.

Crowe, B. L. 1969. Tragedy of the commons revisited. *Science* 166:1103–1107.

Dasgupta, P. 1982. *The Control of Resources.* Oxford: Basil Blackwell.

Dean, W. 1987. *Brazil and the Struggle for Rubber.* Cambridge: Cambridge University Press.

Dove, M. 1983. Swidden agriculture and the political economy of ignorance. *Agroforestry Systems* 1:85–99.

Fortmann, L. and J. Bruce, eds. 1988. *Whose Trees? Proprietary Dimensions of Forestry.* Boulder, Colo.: Westview Press.

Furubotn, E. and S. Pejovich. 1972. Property rights and economic theory: A survey of recent literature. *Journal of Economic Literature* 10:1137–1162.

Gilles, J. and K. Jamtgaard. 1981. Overgrazing in pastoral areas: The commons reconsidered. *Sociologia Ruralis* 21:129–141.

Hardin, G. 1968. The tragedy of the commons. *Science* 162:1243–1248.

Hayami, Y. and V. Ruttan. 1971. *Agriculture Development.* Baltimore: Johns Hopkins University Press.

Hecht, S. B. 1985. Environment, development and politics: Capital accumulation and the livestock sector in Amazonia. *World Development* 13:663–684.

Hecht, S. B., A. B. Anderson, and P. May. 1988. The subsidy from nature: Shifting cultivation, successional palm forests and rural development. *Human Organization* 47(1):25–35.

Homma, A. K. 1981. Uma tentativa de interpretação teórica do extrativismo Amazônico. Simpósio sobre Sistemas de Produção em Consórcio para Exploração Permanente dos Solos da Amazônia. EMBRAPA-CPATU/GTZ, Belém, Brazil.

Johnston, B. F. and P. Kilby. 1975. *Agriculture and Structural Transformation: Economic Strategies in Late-Developing Countries.* New York: Oxford University Press.

Luna, R. C. 1985. *A terra liberta: Um estudo da luta dos posseiros pela terra no vale do Pindaré/MA.* São Luís, Brazil: Universidade Federal do Maranhão.

Mattar, H. 1979. Industrialization of the babaçu palm nut: The need for an ecodevelopment approach. Conference on Ecodevelopment and Ecofarming. Seminar Center of the German Foundation for International Development, Berlin.

May, P. 1986. *A Modern Tragedy of the Non-Commons: Agro-Industrial Change and Equity in Brazil's Babaçu Palm Zone.* Latin American Program Dissertation Series. Ithaca, New York: Cornell University.

Moran, E. F. 1981. *Developing the Amazon.* Bloomington: Indiana University Press.

Netting, R. M. C. 1976. What alpine peasants have in common: Observations on communal tenure in a Swiss village. *Human Ecology* 4:135–146.

Norgaard, R. B. 1984. Coevolutionary agricultural development. *Economic Development and Cultural Change* 32:525–546.

Olson, M. 1965. The logic of collective action; public goods and the theory of groups. Cambridge, Mass.: Harvard University Press.

Orlove, B. 1975. Surimana: Decaimiento de una zona, decadencia de un pueblo. *Antropología Andina* 1–2:75–110.

—— 1977. *Alpacas, Sheep and Men.* New York: Academic Press.

Randall, A. 1972. Market solutions to externality problems: Theory and practice. *American Journal of Agricultural Economics* 54:175–183.

—— 1978. Property institutions and economic behavior. *Journal of Economic Issues* 12:1–21.

Repetto, R. and T. Holmes 1983. The role of population in resource depletion in developing countries. *Population and Development Review* 9:609–632.

Runge, C. 1984. The fallacy of privatization. *Journal of Contemporary Studies* 7:3–17.

Santos, M. P. 1984. Palmeiras em Chama: Os trabalhadores rurais e as empresas agroindustriais do Cerrado Maranhense. CARITAS Brasileira, São Luis, Brazil. Mimeograph.

Santos, R. 1980. *História Econõmica da Amazõnia (1800–1920).* São Paulo, Brazil: T. A. Queiroz.

Schwartzman, S. J. In press. S. Hecht and J. Nations, eds., *The Social Dynamics of Deforestation: Processes and Alternatives.* Ithaca, New York: Cornell University Press.

Soares, J. 1981. *Campesinato: Ideologia e Política.* Rio de Janeiro, Brazil: Vozes.

Tate, W. 1967. *The English Village Community and the Enclosure Movements.* Cambridge: Cambridge University Press.

Taylor, K. 1978. *Sugar and the Underdevelopment of Northeastern Brazil, 1500–1970.* Gainesville: University of Florida Press.

Tendler, J. 1980. Shifting agriculture, land grabbing, and peasant organization on Brazil's northeast frontier. Consultant Report to the World Bank, Washington, D.C. Mimeograph.

NINETEEN

Valuing Land Uses in Amazonia: Colonist Agriculture, Cattle, and Petty Extraction in Comparative Perspective

SUSANNA B. HECHT

The Dynamics of Destruction

The destructive patterns of deforestation are well advanced in the Western Amazon. In 1980, less than 8,000 square kilometers of Rondônia's forests had fallen. Acre's forests were still largely intact. By the end of the decade, some 60,000 square kilometers, or 17 percent, of the state of Rondônia had been cleared. In Acre—more distant, with fewer roads and better political organization of forest inhabitants—some 4 percent (more than 600,000 hectares) had been incinerated (FUNTAC 1990). In spite of the increasing international censure, the rates of deforestation in the Amazon basin more than doubled during the last decade (Myers 1989).

The potential hydrological and climatic impacts, the incalculable costs of species extinction, and the often irreversible degradation of renewable resources of soils and timber forests are cause for serious concern. The social costs of this process include explosive urbanization, massive regional migration with high rates of colonist turnover, a great deal of violence, and marginalization of forest peoples and indigenous populations. While the debates that surround both the environmental and social consequences of deforestation remain contentious, there is agreement that strategies for the protection and management of tropical forests are urgently needed. Until very recently, ideas about conservation for forests involved parks or reserves—set asides—where human habitation was excluded. This conception was altered by parks managed by native peoples such as the Kuna and Cuyabeno reserves. The extension of this idea to include forest peoples was the keystone for elaborating the idea of extractive reserves, first articulated by rubber tappers of Acre as a means of protecting their forests and thus their livelihood from deforestation.

In the Brazilian Amazon, deforestation has been the outcome of several dynamics. These include forest clearing for agriculture and pastures, infrastructure development, government policy that included a lucrative array of financial incentives that made investment quite attractive for people of means during the 1970s to the mid-1980s, and the general allure of a resource frontier, especially when the gold rush began and the mahogany forests were made accessible. But deforestation is also driven by the unique role of land in certain kinds of developing economies. Land maintains its value in inflationary economies, is a secure investment when economic volatility enormously increases risk in manufacturing or other sectors of the economy, and rises in value as infrastructure extends into new areas. In the Amazonian case, these generated an extraordinary speculative boom in the value of land (Mahar 1989). Securing land claims can assure access to both above- and below-ground resources, as well as social resources such as cheap credits, incentives, tax breaks, etc. The cumulative pressures of all of these resulted in rises in land values for reasons that often had very little to do with actual production and management of land resources. In Amazonia, where overlapping forms of land rights make competing ownership claims common, possession is nine-tenths of the law, and the law is *O lei do mais forte*—the law of the jungle. To secure land in Amazonia is to clear it. Under the circumstances, the larger the area in possession, the greater the effective returns, whether from land speculation, from landed resources of minerals and precious woods, subsidized credits, or indemnification. Thus rapid clearing is occurring in Amazonia as a means of controlling land and consolidating access to the economic benefits of landed property.

Small farmers—lured by colonization programs and the relatively low price of lands compared to other regions, or expelled from other rural zones by changes in access to land, holding fragmentation, and changes in farming structures that made their holding sizes uncompetitive—also found themselves hacking down forest. Land was cleared to establish claim, and to attempt to wrest a living in the face of the region's poor soils, endemic malaria, weak infrastructure, institutional indifference, and risible producer prices. Not surprisingly, land was often abandoned as the Jeffersonian rural ideal collided with Amazonian reality.

The Issue of Sustainability

Three Models of Amazonian Occupation

The forest destruction now at hand might have been tolerable if the replacement land uses—agriculture and pasture—were sustainable. As it stands now, agriculture and pasture are but short moments of production in a larger process of degradation.

Cattle

Livestock raising, the activity ultimately responsible for most of the deforestation in the Brazilian Amazon, is characterized by low production parameters (one animal per hectare) and a pasture life that is generally less than twelve years for most soil types in Amazonia (Hecht 1985; Fearnside 1980; Buschbacher 1986; Serrão and Toledo 1990). Poor grasses, poor management and an array of financial "rents" that could be captured, whether land was well managed or not, resulted in a situation in which now more than 50 percent of areas converted to pasture is abandoned (Hecht 1985; Toledo and Serrao 1990).

Livestock activities have been associated with several nonproductive but highly profitable dynamics that have little to do with animal production per se. Livestock activities were often the vehicle for capturing the more than $1 billion of highly subsidized credit and fiscal incentives applied to the sector (Browder 1988). In addition, tax holidays, the capital gains associated with rising land prices, the value of land in an inflationary economy, and the real potential for finding mineral or timber resources on a given piece of land have in part fueled this land use in spite of the fact that its rates of return in terms of animal production are often negative (Hecht, Anderson, and May 1988b). The expansion of livestock has been associated with extreme land concentration and a great deal of rural violence (Asselin 1982;

Schmink 1982; Martins 1984; 1990; Wagner 1990). Because livestock generates minimal employment, and most of this of a temporary nature, its labor market characteristics are very undesirable for regional development (Mahar 1989). Permanent labor requirements average less than one person per 1,000 hectares, while temporary work in deforestation and brush clearing—ephemeral activities—dominate the longer-term labor demands. The combination of large holdings, violence, and the limited employment associated with this type of land use has resulted in extraordinary rates of urbanization as recent migrants and local small farmers have had minimal access to land and to jobs.

Small Farmers

The agriculture of small farmers is also beset with serious problems. A survey of settler attrition (Hecht, forthcoming) in the entire Amazon basin indicates an average rate of more than 50 percent (see table 19.1), and even by Amazonian standards, the turnover rates in the western Amazon are very high. This reflects a diversity of factors ranging from agronomic problems of soils and pests, to institutional constraints such as the costs of credit, to land tenure difficulties. National and international market pressures and the problems of unequal exchange can also be significant. Recent analyses suggest also that the nature of the labor markets and the availability of small-scale extraction opportunities is central in determining success or failure among the poorer strata of colonists (Moran 1990). The instability of colonist settlement regardless of its final cause translates as an expanding, but unstable agricultural deforestation frontier and accelerated rural to urban migration (Baxx 1990; Moran 1990; Hall 1989).

Petty Extraction

The widespread occurrence of nontimber resource extraction has been overlooked by regional analysts and development planners until quite recently. The popular misconception of extractive activities as a moribund vestige of a semifeudal past described in planning documents and discussions of the regional economy have tended to overlook several features of extractive economies that could modify the way these systems should be viewed in both economic and social terms.

Table 19.1
Attrition Rates in Amazonian Settlement

Country	Attrition	Population in Wage Labor	Extraction	Speculation
Brazil				
Rondonia	23%/yr	49.7%	yes	yes
	63%	—	—	—
TransAm.	38%	75–100%	yes	yes
	13%/yr	71%	—	—
	67%	n.d.	yes	yes
Acre	27%	n.d.	yes	yes
Bolivia				
Alto Beni	50%	n.d.	yes	n.d.
Nuevo Mundo	68%	65%	n.d.	yes
Santa Cruz	32%	35%	yes	yes
Yacapani	47%	66%	n.d.	n.d.
San Julian	—	—	—	—
Peru				
Uchiza/Nuevo Horizonte	80%	—	—	—
Peru Oriental	30%	—	—	—
Alta Huallaga	40%	—	—	—
Tingo Maria	80%	—	—	—
Jenaro Herrera	57%	100%	yes	yes
Colombia				
Guaviare	15%	68%	yes	yes
	28%	40%	—	—
Caqueta	32%	24%	yes	yes

The products of petty extraction often have significant markets in the local economies (Padoch et al. 1985; Anderson et al. 1987) and are generators of substantial amounts of foreign exchange (Prance 1989). Production for market is important, but the role of extraction in the household economy is crucial for providing subsistence or cash supplements that cannot be generated through agricultural or wage activities. Extractive activities are tightly but flexibly integrated into market and credit systems. Petty extraction is rarely a solely subsistence system and has and continues to generate substantial surpluses that have maintained elites for centuries. Rubber, for example, generated more than $3 trillion since 1923, and the taxes on this rubber alone amounted to some $500 billion (Cultural Survival 1989). These revenues were not invested in the region, except in forms of elite conspicuous consumption, and certainly not in the welfare of extractors or in enhanced production technologies. Extractive activities provide a significant subsidy to rural domestic welfare through goods such as fuel, fodder, fiber, and medicinals. The importance of this "subsidy from nature" is the particular importance of access to these resources for poorer households (Hecht, Anderson and May 1988b; see May, this volume).

What is now critical, in light of the high rates of tropical forest destruction, is that this activity appears to be relatively ecologically sustainable and generates income with far less environmental damage than the other alternatives on offer. Table 19.2 shows the total deforestation in extractive reserves in the state of Acre. Slightly more than 1 percent of the land area is cleared, and this is for mostly domestic agricultural production. Archival reports, travel accounts, and early photographs clearly indicate that these *seringais,* or rubber estates, have been occupied for more than one hundred years (Falção 1906; Akers 1912; Church 1904). The compatibility of extractive activities, human populations, and forest ecosystems argues for using elements

Table 19.2
Areas of Human Disturbance in Extractive Reserves

			Area of Disturbance	
Reserve	County	Area	Absolute (ha)	% of Total Area
Cachoeira (Chico Mendes)	Xapuri	24,973.00	782.71	3.13
Macaua	Sena Madureira	103,000.00	370.00	0.36
Santa Quiteria	Assis Brasil	4,054.00	167.00	4.12
	Brasileia	40,151.00	794.00	1.98
	SUBTOTAL	44,205.00	961.00	2.17
San Luiz do Remanso	Rio Branco	6,395.00	111.00	1.74
	Xapuri	33,175.00	973.00	3.93
	SUBTOTAL	39,570.00	1,084.00	3.74
(Total for the state)		211,748.00	3,197.71	1.51

Source: FUNTAC 1990.

of the extractive economies as potential development "platforms," rather than continuing a research emphasis on land uses that require wholesale ecosystem conversion.

The Good, the Bad, and the Ugly: Comparing Forests with Other Land Uses

Extractive activities provide a means of "valuing" a tropical forest in ways that translate into economic frameworks more easily than trying to assess the economic benefits of ecosystem services or the potential worth of an unknown species. Government planning documents have often compared extractive activities—usually rubber extraction—to its competing land uses on a per hectare basis that takes into account only one moment in time. Thus the average or highest return per hectare at a given moment in a parcel's economic life in an unsustainable land use is compared with an activity, extraction, which produces a lower per hectare value but is able to be maintained at that level of production more or less indefinitely. By avoiding the dimension of time, the critical question of sustainability is circumvented.

Another problem in evaluating extractive activities is that the statistical information is sparse and highly aggregated. It is usually based on tax information rather than on production data. Since extractive production is quite decentralized and marketing channels can elude government vigilance, the volume of these products and their regional economic impact can be seriously underestimated. The usual statistical sources on extraction, such as Instituto Brasileiro de Geografia e Estatisticas (IBGE) data, provide little insight into the dynamics of production.

There have been some recent efforts to quantify the value of extractive products and the forests that produce them, notably the studies of Peters, Gentry, and Mendelsohn (1989); May (1986); Anderson (1990); LaFleur (1989); Padoch et al. (1985); and Padoch and de Jong (1989). These studies have all used quite different methods to evaluate the returns from a standing forest. While an analysis of each of these efforts is beyond the purview of this paper, all show that extraction can be profitable for rural households.

The studies by Peters, Gentry, and Mendelsohn and by LaFleur specifically compare the returns from extraction with agriculture and cattle ranching. The methods used in these studies vary from this analysis. Peters, Gentry, and Mendelsohn (1989) calculated the potential return to forest extraction by analyzing the botanical composition of the forest and then pricing the potential production of its renewable products. LaFleur (1989) developed his simulation from data derived from IBGE statistics and from Hecht and Schwartzman (in prep.). Peters' results showed a highly positive return to extraction compared with other uses on a per hectare basis. LaFleur used land-use area "modules" that were then reduced to persons per hectare and were then structured into family units of six people holding a different number of hectares per household. His simulation showed that extraction was less lucrative than agriculture or livestock, but that its case could be argued on the grounds of sustainability.

The analysis of the three main forms of land use in Amazonia that follows is elaborated differently than either Peters's or LaFleur's.

Methods

The model developed for this study analyzes the costs and returns for the main land-use alternatives. The model focuses on household incomes, productivity of land use over time (i.e., its sustainability), and the costs of forest destruction as costs of land recuperation and returns foregone from forest products. The costs of land recuperation are used as the indicator of the value of the forest. This values the forest as a form of fertilizer and ignores the value of its numerous ecological services and long-term economic potential.

The study does not analyze the costs and benefits through the usual techniques of discounting net present value because a subsequent paper will deal in more detail with the costs and benefits under the current configurations of the Amazonian land markets. In this paper, a model of colonist short-cycle agriculture, livestock, and petty extraction in the Western Amazon is elaborated where only the aggregated gross returns are compared. The costs of clearing, planting, and maintaining farms is not calculated here. Thus a forest, farm, and pasture—the basic productive apparatus—are *given* at the beginning of the run. This biases the model in favor of cleared land uses because it assumes no development or management costs. As mentioned in the previous section, study after study has shown that pasture and agriculture are often unprofitable because of high costs relative to returns and ultimately unsustainable. This paper is an exercise that compares the various land uses under the best possible conditions and calculates *only* environmental costs and ignores production costs. The model deliberately casts forest land uses that replace forests in the best possible light to show how extraction compares to production returns of pasture and farming under the most advantageous circumstances. These costs and returns are placed in the context of deforestation in Acre in order to provide some sense of the regional costs and benefits.

Land use and deforestation in Acre in given in table 19.3. Satellite imagery from LANDSAT TM-5 (FUNTAC 1990) shows that about 633,131 hectares were cleared, with more than half the disturbed area

Table 19.3
Land Use and Disturbance in Acre

Disturbance[1]	Area (ha)	% of Total Area
Total territory	15,348,890	
Disturbed territory	3,006,100	
Deforested territory	526,900	
Land Use[2]	**Area (ha)**	**% of Total Area**
Area in Agriculture	691,751	3
Permanent Crops	16,403	3
Annual Crops	52,771	10
Area in Pasture	457,725	87
Total Converted Area in Production	526,899	
Deforested or Disturbed Areas in Acre[3]	**Area (ha)**	**% of Total**
Urban	9,442	1.40
Second growth	60,565	9.57
Small clearings	72,230	11.41
Colonization projects	87,607	13.84
Pasture	349,456	55.19
Agriculture	45,378	7.17
Waterways	9,431	1.33

1. J. Malingreau and C. Tucker 1988. 2. Census Agropecuario 1985. 3. FUNTAC 1990.

in pasture, 32 percent in agriculture, about 10 percent in second growth, and small areas in cities. Another study by Malingreau and Tucker (1988) based on NOAA 9 satellites indicated that about half a million hectares were deforested in Acre by 1985, and another 3 million were badly disturbed. The disturbed areas constitute zones that are primarily abandoned pasture. This set of remote-sensing images corresponded well with the agricultural census data. About 13 percent of the deforested area is in agriculture, of which 10 percent is in annual cropping systems. Perennial cropping, in spite of its ecological aptness in Amazonia, has in fact declined in Acre by some 28 percent since the 1980 census. The remaining area is in pasture. The Malingreau and Tucker data suggest that roughly 25 percent of Acre's forest cover has been modified or deforested, a sharp contrast to the 4.15 percent recorded by FUNTAC. For the sake of comparison, the implications of both levels of deforestation have been analyzed, but I am inclined to favor the FUNTAC estimate.

The Costs and Returns from Different Land Uses: Determining the Value of Production The details on the value of production for each of the three land uses are derived from farm-level research carried out by the author, agronomic research from EMBRAPA and EMATER, the FUNTAC/UCLA/EDF/IEA project on household income formation in extractive reserves, from World Bank and IDB data from regional development projects in Rondônia and Acre, recent field data on prices, agricultural census data, Statistical Abstracts for Acre, and the academic literature for the western Amazon. These data provide general ranges of the returns per hectare of various land-use activities. The model is informed by field situations in the western Amazon. While tremendous variations exist in on-ground management, the results reflect the general patterns that many researchers have observed.

Annual Cropping Systems The annual cropping systems practiced by colonists are variable and complex. Studies such as those of Moran (1990), Milliken (1988), Coy (1987), and the FAO/World Bank (1987) have helped clarify some of the ranges in yields that can be expected.

A standard INCRA model of 100 hectares is used as the landholding size of which only 50 hectares may be cleared as per the Codigo Florestal. The clearing behavior patterns are based on data presented by Fearnside (1984) and Browder (1988a,b). The yield declines on a swidden plot typically generate a pattern clearing of new lands to maintain a relatively high flow of cash from annual crops and because the returns to labor are higher because yields are higher in the early part of the production cycle. Typically, the pattern of relay cropping begins with 2 hectares of rice and 1 hectare each of beans and corn. After the first harvest, manioc is planted and eventually turned to pasture. The final phase is a fallow. While planting perennial crops is an option, under the prevailing conditions of often problematic land title, and production problems and riskiness of markets for these perennial crops, as well as their high labor, input, and cash demands at the establishment stage, the tree crop agriculture under most colonist conditions is quite difficult (FAO/WB 1987; Milliken 1988; Gill 1987; Coy 1987; Moran 1990). The model of expanding clearing for annual crops while planting pasture, renting it, and fallowing degraded pasture is the least labor-demanding of the potential colonist uses and also one of the most widespread patterns.

Parameters for crop productivities and prices given in table 19.4 are based on FAO (1987) evaluations of Polonoreste, field studies by Coy (1987), and Browder (1988). These numbers generate an optimistic income level, since several studies indicate that the per capita income in settlement zones is lower than the income predicted from this model.

Livestock The livestock data are based on technical data derived from EMBRAPA, EMATER, and others. These represent the standard production parameters for livestock on soils similar to those found in Acre in other areas of the Amazon. The data are presented in table 19.5.

Table 19.4

Crop Productivities and Per Hectare Value of Production

Ranges	Yields	Return/ha (in US$)
Rice	1,400 kg/ha	213/ha
Beans	500 kg/ha	139/ha
Corn	1,200 kg/ha	141/ha
Manioc	12,000 kg/ha	370/ha
Rental Pasture		
Good Quality	—	10/ha
Low Quality	—	5/ha

Sources: FAO 1987; FGV 1987; FGV 1988; Coy 1987.

Table 19.5
Livestock Parameters

	Regular	Optimistic
Area in pasture	150 ha	150 ha
Animal support	.5 animal units/ha	1.00
Number of animals	80	150
% takeoff	7%	14%
Number of animals culled	5.6/year	16.8/year
Average animal weight	350 kg	350 kg
Total kg/year	1,960	5,880
Return	US $1,960	US $5,880
Total return after 10 years	US $19,600	US $58,800
Return after 15 years	US $19,600	US $58,800
Average past life	7–10 years	7–10 years

Table 19.6
Extractive Household Incomes[1]

	Cash Returns Extractive Activities (US$)		Total Cash from Extraction (US$)	Value of Agriculture (US$)	Total Returns (US$)
	Rubber	Nuts			
SLR[2]	424	396	820	683	1,503
C[3]	480	851	1,331	648	1,979

Source: FUNTAC/UCLA/EDF/TEA field project data.

1. This number includes products that are both consumed and marketed. In SLR, about 9% is marketed. At C, some 16% of agricultural produce market circuits. This value does not include livestock assets. 2. São Luis de Remanso, the first official extractive reserve. 3. Cachoeira, the second major extractive reserve, home to Chico Mendes. Land conflicts associated with this reserve eventually cost Mendes his life. These two reserves are used as examples of "average" and "better" extraction.

Extraction The data on extraction are derived from a study of household income formation carried out in 1989 by FUNTAC/UCLA/EDF/IEA (table 19.6). The details of this production system are elaborated in greater detail in Schwartzman and Allegretti (forthcoming), Allegretti (1990), and in a forthcoming FUNTAC document. These data present the results of two surveys in the first extractive reserves created, Sao Luis de Remanso (SLR) and Cachoeira. They include the cash returns to extractive activities of rubber and Brazil nuts and the value of the agricultural products. The value of these products is determined by their cost in Rio Branco, not the middleman's price. Since purchased food is quite expensive (as much as 30 to 50 percent higher) when purchased in the *seringal,* the value to the household in this model is relatively low. It is worth noting that the value of agriculture exceeds that of rubber for both SLR and Cachoeira and is greater than nuts in SLR. It constitutes fully 45 percent of the income (real and shadow priced) at SLR, and about 33 percent in Cachoeira. Women and children are largely charged with the day-to-day management of the agriculture. While discussions of reserves have focused on forest products, the importance of agriculture to households cannot be underestimated, and any interventions in cropping systems must place women squarely at the center of the activity.

Recuperation Costs The ecological costs of species loss, climate change, transformation of hydrological regimes, and loss of ecosystem services that attend forest destruction is incalculable. To ascertain the ecological costs, two indicators of the value of a standing forest were used: the loss of income from extractive products and the cost of recuperating a degraded area back to a pasture. This represents the least expensive means ($260 per hectare) of getting degraded land back into at least marginally productive use. These are based on EMBRAPA data developed by Kitamura, Filho, and Serrão (1985). Establishment of a rubber plantation costs close to $1,000 according to World Bank figures for Rondônia. Henry Knowles, in charge of forest regeneration at the Trombetas mine, holds that mixed forest regeneration costs roughly $8,000 (Knowles 1988).

Comparing Land Uses This model uses 300 hectare area as the analytic unit. This corresponds roughly to the size of a rubber tapper *colocação* (area of usufruct). The standard colonization lot is 100 hectares, so the colonist model includes three households. The average ranch in Acre includes some 650 hectares, although more than 52 percent of herds is on holdings greater than 1,000 hectares. For the purposes of analysis here, two laborers are maintained on the cattle operation. An optimistic and average scenario for each land use is presented.

To quickly reiterate, only the value of production and ecological costs, are analyzed here. Production costs are not included. This means that this model is the most optimistic presentation of the value of land uses that replace forest because it assumes that production has no cost.

Table 19.7 summarizes the gross value generated in the various land uses using a ten-year and a fifteen-year project run. At the end of the production cycle, deforested land is then recuperated using the methods for recovering degraded pasture—fallow and one year's fertilizer. These costs, plus the value of extractive products, are subtracted from the total gross returns.

Household incomes are highest for improved livestock, with good agriculture and good extraction roughly comparable, followed by average extraction. Average agriculture and average ranching earn an annual US $1,200 and $1,350 respectively. These numbers are most characteristic of ranching and farming for much of the western Amazon.

The value of production for the best-case agriculture and livestock are roughly comparable in the ten-year run for cumulative returns for area given to production. Average farming is followed by best-case extraction, then average extraction; with average livestock generating the lowest revenues over ten years. In terms of average *household* income from these land uses, a colonist household with a ten-year revenue earns about $20,000—the same as the more lucrative case of extraction. Household returns for average agriculture over ten years is some $12,000, $3,000 less than average extraction and $1,700 less than average livestock.

In terms of aggregated values the value per hectare over ten years in terms of area cleared for production (about 150 hectares), better agriculture ranks highest, good livestock is second, average agriculture is third, with both extractions following. Average livestock ranks last. The average annual returns per cleared hectare follows the same pattern with the range in value descending from $40 to about $9.10, not a very substantial sum given the devastation engendered by clearing. Gross value per hectare over the entire holding follows the same pattern but with extraction closing the gap on agriculture, while livestock's return is decidedly worse.

Household returns after fifteen years show good livestock generating the greatest returns, followed by extraction of both types. Good agriculture, at $21,931, provides an income comparable to poor ex-

Table 19.7
Comparative Gross Value of Production in Acre on 300-Hectare Holding

	Colonist (3 households)		Livestock		Extraction	
	Good	Avg.	Good	Avg.	Good	Avg.
Household Income/yr[1] (US$)	2,005	1,200	5,880	1,370	2,500	1,520
Total Income 10 yr[1] (U$)	60,153	36,200	58,800	13,720	19,790	15,030
Total Income 15 yr (US$)	65,793	39,120	58,800	13,720	29,685	22,950
Producing Area[2] (150 ha)	401/40	240/40	392/39	91/9	131/16	153/15
Total Holding (300 ha)	200/20	120/12	196/19	45/4.5	65/6.5	50/5
Producing Area (150 ha)	438/29	260/17	352/39	91/6	195/16	153/10
Total Holding (300 ha)	219/14	130/8.6	176/11	45/3	98/6.5	76/5
Recuperation Costs (US$)	92,960	93,960	117,450	117,450	—	—
Return B-C	−28,217	−54,810	−58,800	−103,730	+29,685	+22,950

[1]Mean return/ha aggregated over 10 years and average annual return/ha over 10 yrs.
[2]Mean return/ha aggregated over 15 years and average annual return/ha 15 yrs.

traction, while average agriculture and poor livestock, at $13,000 and $13,720, are roughly comparable. The household income from the land for the latter activities are less than $1,000 per year.

The aggregated returns per hectare after fifteen years show that the returns for cleared and productive land is highest in agriculture at $438, followed by ranching, average agriculture, both extractive systems, and average livestock. The average yearly returns per hectare after fifteen years follow the same pattern, ranging from a high of $29 to the modest $6 in pasture. If the entire holding is considered, that is, cleared or producing lands and conservation lands together, the average annual value of returns per hectare over fifteen years drops to about $14 per hectare to a low of $3 per hectare. These are not very compelling returns on the production per hectare over time, and one could ask whether the destruction of forest is really worth the modest absolute benefits above the admittedly modest incomes and per hectare returns in extraction.

If some indicator is used to value the loss of forest, the situation looks more dismal. In this model (as in the real world) land uses other than petty extraction are not sustainable and go out of production. If they are to enter into another cycle of production, they require either a long fallow or fertilization plus fallow. Thus if we place the value of a forest equal to the one-year cost of applying enough fertilizer to the land to get it back to a poor short-term pasture, or $261 per hectare for three years, we derive the following results: extraction, which has no recuperation costs associated with it, and which maintains a continuous stream of benefits over the entire period of analysis has a positive value of $29,685 and $22,950, depending on the type and area of extraction. Best agriculture produced value over the project life minus recuperation costs result in *negative* values of more than $51,657 to $78,000. Even if fertilizer were used only for one year, regular agriculture would still generate slightly negative returns. Livestock activities are catastrophic, generating negative values of some $58,000 to $103,730 in overall costs.

The per hectare value of areas in agricultural production, and the handsome economic returns to livestock in the short run, are purchased at very high environmental costs, which were calculated using the most economically conservative low-input- and low-labor-demanding recuperation strategies. These costs do not reflect in any way potential environmental damage associated with germ plasm loss or more general hydrological and climatological problems that may be associated with large-scale clearing. It is, of course, a somewhat academic exercise to imagine what the costs of recuperating large areas might mean in real financial terms. In the developed world the efforts to reduce pollution, recuperate land and waters, and maintain regulatory agencies at state and federal levels involves about 3 percent of the GDP of Europe and the EEC and their multitrillion-dollar economies. In Brazil, the state and private enterprises view environmental costs as pure externalities. But if, for the sake of argument, the costs of the Malingreau and Tucker deforestation level is analyzed, recuperation costs of returning 3 million hectares of degraded land in Acre to some level of production would involve $781 million in direct recuperation. If the FUNTAC number is used (about 600,000 hectares) the bill for recuperation would weigh in at $165,247,191 dollars. This represents probably the lowest imaginable recuperative costs, yet the tax revenues for the entire state of Acre from rural and urban sectors was only $4,054,037 in 1987.

This exercise of comparing costs has some advantages over other efforts. First, the model is based on a great deal of regional western Amazonia field data and uses household income generation as the point of departure. Peters, Gentry, and Mendelsohn's model was based on potential returns given species productivity and distribution. LaFleur's model is based on IBGE (Brazilian Statistical Institute) data that is too generalized. Both studies provide valuable insight into the relative merits of the competing land uses in terms of revenues, but the model presented here is the only one that incorporates the costs of land degradation.

Extraction is a land use that is not very lucrative for its inhabitants, but its returns from productive activities based on actual use of land resources is better than the alternatives. The problem, however, is that if land markets are such that land as a commodity is of greater interest than land as an input to production, there is little incentive to work for sustained activities. In this light, extractive reserves, which effectively remove forest lands from land markets, provide the tenurial basis through which sustainable livelihoods for people and forests can be maintained. But virtually all analysts understand that extraction as it is currently elaborated is a land use of great economic hardship and one of exploitation through rents, through markets, and, in many cases, through continued debt and debt peonage.

This study focuses on the economy of rubber, but rural households throughout the Amazon engage in forms of forest extraction for subsistence and cash. While the rubber tappers' movement has been very effective at making their plight heard, most extractors have no voice, and as the process of enclosure proceeds, they find themselves increasingly part of a landless strata with few alternatives outside of urban or frontier migration (May 1986; Baxx 1990). But even if land claims are recognized, this alone cannot assure a decent living.

Strategies to improve the lot of extractors are complex, because it is important to remember that extractors derive their livelihoods from an array of activities including hunting, fishing, agriculture, animals, artisanal activities, and petty commerce among the many ways they have to assure subsistence and cash. Enhancing and maintaining these activities are also part of ameliorating conditions for these rural populations.

While a larger rethinking of regional development approaches in the Brazilian Amazon might well be warranted, in the short and medium term even incremental improvements such as those suggested below would represent an enormous effort and financial commitment, although one that can be justified in social and environmental terms. The marginal increase in economic value is so slight compared to the "cost" of the forest, that significant efforts to enhance extractor livelihoods and density are likely to pay off more than the research and investment strategies in land uses that replace forests.

Improving the extractor household situation will involve

1. Maintaining tenurial structures that permit continued access to forest resources;
2. Gathering basic data on productivity and ecology of major products used in extraction and information on the potential of new products;
3. Analysis of the interventions in forests and forest-based agriculture that could improve production and income while enhancing sustainability;
4. Analysis of marketing dynamics, particularly the roles of small-scale commerce (often carried out by extractors themselves) and forms of transportation;
5. Analysis of small-scale, flexible product-processing plants and their regional potential;
6. Careful analysis of the social relations characteristic in differing

kinds of extraction, including the gender division of labor, that prevail;

7. Analysis and comparison of each of the occupation models in terms of the regional economic linkages and employment creation; and
8. Elaboration of community property systems that recognize and compensate intellectual property rights through novel trademarking agreements.

LITERATURE CITED

Akers, C. 1912. *The Rubber Industry of Brazil and the Orient.* London: Methuen.

Allegretti, M. 1990. Extractive reserves: An alternative for reconciling development and environmental conservation in Amazonia. In A. Anderson, ed., *Alternatives To Deforestation,* 252–264. New York: Columbia University Press.

Anderson, A. 1990. Extraction and forest management by rural inhabitants in the Amazon estuary. In A. Anderson, ed., *Alternatives to Deforestation,* 3–23. New York: Columbia University Press.

Anderson, A., A. Gély, J. Strudwick, G. Sobel, and M. Pinto. 1987. Um sistema agroflorestal na varzea do Estuario Amazonico. *Acta Amazonica* 15(1–2):195–224.

Asselin, V. 1982. *Grilagem: Corrupção e violencia em Terras do Carajas.* Petrópolis, Brazil: Vozes.

Baxx, K. 1990. Shanty Town, the final stage of rural development? In D. Goodman and A. Hall, eds., *The Future of Amazonia.* New York: St. Martin's Press.

Browder, J. 1988. Colonists in Rondônia. Paper presented at the Latin American Studies Association Meeting. New Orleans.

—— 1988. Public policy and deforestation in the Brazilian Amazon. In R. Repetto and M. Gillis, eds., *Public Policies and the Misuse of Forest Resources,* 247–298. Cambridge: Cambridge University Press.

Buschbaker, R. 1986. Tropical deforestation and pasture development. *Bioscience* 36(1):22–28.

Church, G. E. 1904. The Acre region and the Caoutchouc region of southwest Amazonia. *Geographical Journal* 23(5):596–613.

Coy, M. 1987. Frente pioneira e programa Polonoroeste. In G. Kohlhepp, ed., *Homem e Natureza na Amazonia,* – Tübingen, Germany: ADLAF.

Cultural Survival. 1989. *Marketing Non-Timber Tropical Forest Products: Prospects and Promise.* A Workshop Report. Cambridge, Mass.: Cultural Survival.

Falcão, A. 1906. *Album do Rio Acre.* Belém, Brazil: Falcao.

Fearnside, P. 1980. The effects of cattle pasture on soil fertility in the Brazilian

Amazon: Consequences for beef sustainability. *Tropical Ecology* 21(1):125–137.

—— 1984. Land clearing behavior in small farmer settlement schemes in the Brazilian Amazon and its relation to human carrying capacity. In A. Chadwick and S. Sutton, eds., *Tropical Rainforest: The Leeds Symposium*. Leeds, England: Leeds Philosophical and Literary Society.

Food and Agriculture Organization (FAO) and the World Bank (WB). 1987. *Projetos agropecuarios Polonoroeste: Examen tecnico*. Rome: FAO/World Bank.

Fundação de Tecnologia do Estado do Acre (FUNTAC). 1990. *Monitoramento da cobertura do Estado do Acre*. Rio Branco, Brazil: FUNTAC.

Gill, L. 1987. *Peasants, Entrepreneurs and Social Change*. Boulder, Co.: Westview Press.

Hall, A. 1989. *Developing Amazonia*. Manchester, England: Manchester University Press.

Hecht, S. B. 1983. Cattle ranching in the eastern Amazon: Environmental and social implications. In E. Moran, ed., *The Dilemma of Amazonian Development*, 155–188. Boulder, Co.: Westview Press.

—— 1985. Environment, development and politics: Capital accumulation and the livestock sector in eastern Amazonia. *World Development* 13(6):663–684.

—— Forthcoming. Rethinking colonist attrition. *World Development*.

Hecht, S. B., R. Norgaard, and G. Possio. 1988a. The economics of cattle ranching in the Eastern Amazonia. *Interciencia* 13(5):174–188.

Hecht, S. B., A. B. Anderson, and P. May. 1988b. The subsidy from nature: Shifting cultivation, successional palm forests, and rural development. *Human Organization* 47(1):25–35.

Hecht, S. B. and S. Schwartzman. The good, the bad and the ugly. In prep.

Hiraoka, M. 1985. Cash cropping, wage labor and urban migration in the Peruvian Amazon. *Studies in Third World Societies* 32:199–243.

Instituto Brasileiro de Geografia e Estatisticas (IBGE). 1986. *Censo agropecuario*. Rio de Janeiro, Brazil: IBGE.

Italiano, E., E. Morães, M. Dias Filho, A. Serrão. 1982. *Fertilização de Pastagem de Capim Colonião em Degradação*. Comunicado Tecnico No. 31. Manaus, Brazil: EMBRAPA.

Kitamura, P., M. Dias Filho and E. Serrão. 1982. *Analise economico de algumas alternativas de manejo de pastagens cultivadas*. Belém, Brazil: EMBRAPA.

Knowles, H. 1988. Paper presented at Conference on Alternatives to Deforestation, February 4–8, Belém, Brazil.

La Fleur, J. 1989. Efficiency and extraction. Paper presented at the Symposium on Extractive Economies in Tropical Forests of the National Wildlife Federation. Washington, D.C.

Mahar, D. J. 1979. *Frontier Development Policy in Brazil: A Study of Amazonia*. New York: Praeger Publishers.

—— 1989. *Government Policies and Deforestation in Brazil's Amazon Region*. Washington, D.C.: World Bank.

Malingreau, J. and C. Tucker. In press. Large scale deforestation in the southeastern Amazon Basin. *Ambio*.

Martins, J. de Souza. 1984. *A Militarização da questao agraria no Brazil*. Petrópolis, Brazil: Vozes.

—— 1990. The political impasses of rural social movements in Amazonia. In D. Goodland and A. Hall, eds., *The Future of Amazonia*. New York: St. Martin's Press.

May, P. 1986. The tragedy of the non-commons. Ph.D. dissertation, Cornell University, Ithaca, N.Y.

Milliken, B. 1988. Dialectics of devastation. Master's thesis, University of California, Berkeley.

Moran, E. 1982. *Developing the Amazon*. Bloomington, Ind.: University of Indiana Press.

—— 1990. Private and Public Colonization Schemes in Amazonia. In D. Goodman and A. Hall, eds., *The Future of Amazonia*. New York: St. Martin's Press.

Myers, N. 1989. *Deforestation Rates in Tropical Forests and their Climatic Implications*. London: Friends of the Earth.

Padoch, C., and de Jong. 1989. Production and profit in agroforestry: An example from the Peruvian Amazon. In J. Browder, ed., *Fragile Lands of Latin America*. Boulder, Co.: Westview Press.

Padoch, C., C. Inoma, W. de Jong, and J. Unruh. 1985. Amazonian agroforestry: A market oriented system in Peru. *Agroforestry Systems* 3(1):47–59.

Peters, C., A. Gentry, and R. Mendelsohn. 1989. Valuation of an Amazon rainforest. *Nature* 339:656–657.

Prance, G. 1989. Economic prospects from tropical ethnobotany. In J. Browder, ed., *Fragile Lands of Latin America*. Boulder, Co.: Westview Press.

Salinas, J. 1987. *Recuperación con pasturas de areas degradadas: Avances en la investigación*. Cali, Colombia: CIAT.

Schmink, M. 1982. Land conflicts in Amazonia. *American Ethnologist* 9(2):341–357.

Schwartzman, S. and M. Allegretti. Extractive reserves: a sustainable alternative for Amazonia. In S. Hecht and J. Nations, eds., *The Social Dynamics of Deforestation: Processes and Alternatives*. Ithaca, New York: Cornell University Press.

Serrão A. and A. Homma. 1982. Recuperação e melhoramento de pastagem en area de floresta. Document 17. EMBRAPA/CPATU.

Serrão, A., and J. Toledo. 1990. The search for sustainability in Amazonian pastures. In A. Anderson , ed., *Alternatives to Deforestation*, 195–214. New York: Columbia University Press.

Wagner, A. 1990. The state and land conflicts in Amazonia. In D. Goodman and A. Hall, eds., *The Future of Amazonia*. New York: St. Martin's Press.

TWENTY

Buying in the Forests: A New Program To Market Sustainably Collected Tropical Forest Products Protects Forests and Forest Residents

JASON CLAY

During the last years there has been considerable interest throughout the industrialized West in the destruction of tropical forests, particularly rain forests as articles in *Cultural Survival Quarterly, The Ecologist, The Economist, Scientific American, Time* and other publications attest. This interest has resulted from a renewed concern about the greenhouse effect and global warming, as well as about their relationship to the massive deforestation of Brazilian rain forests, where 1,800 fires of 1 square kilometer were reported on a single day during the summer of 1988.

As a consequence of this increased awareness and personal concern, various strategies to halt the destruction of tropical forests are being suggested by a number of states, international bodies, and nongovernmental organizations, including some from within Brazil (CEDI, *Manchete*). Few such strategies have been implemented, however. Most focus on stopping the destruction by highlighting the negative aspects of deforestation for the region or, more often, the world, and by criticizing the policies of the governments that allow or even encourage the destruction of forests.

While highlighting the consequences surrounding deforestation is certainly important, such approaches are unduly negative, often antagonistic, and at times border on environmental imperialism. To date, little attention has been paid to the poor, whose desperate conditions lead them to cut the forests (either for themselves or in the employ of others) or to the countries whose indebtedness is one of the major justifications for get-rich-quick schemes in the forests. The role of Western consumers in deforestation is often studiously avoided by neglecting to carefully examine demand in Europe and North America for such commodities as tropical hardwoods, meat, bananas, oranges, palm oil, petroleum, aluminum foil, and steel.

Even less attention has been given to the identification and elaboration of positive alternatives that could protect the remaining tropical forests or reclaim those areas already degraded by the "development" schemes being spawned throughout the world's rain forests, most recently in the Amazon, during the last two to three decades.

At this time a number of rain forest communities—indigenous peoples, extractivists such as Brazil's rubber tappers, and long-established peasant communities—are attempting to maintain their ways of life and expand their sources of income by intensifying, extending, or otherwise modifying their sustainable extractive activities in forest areas (for a review of their traditional resource management systems, see Clay 1988). Botanists, anthropologists, and environmentalists have long argued that healthy forests are capable of generating more income and employment than the same areas when cleared and put into pasture or agricultural crops. Recent studies indicate that in both the mid and long term, sustainably harvested secondary forest products generate even more income per acre of land than the sale of tropical timber (Peters, Gentry, and Mendelsohn 1989; Peters and Hammond 1989). While high levels of income from extractive activities in forests are theoretically possible, few theoreticians have worked with actual communities or producer groups to insure that income from sustainable activities amounts to more than a small fraction of what is possible or, more importantly, what will be necessary to conserve the resource base in the face of both internal and external pressures on the land (Repetto 1988; Reid and Miller 1989).

One major bottleneck in the expansion of sustainable income-generating activities in tropical forests is that information about forest products is largely unavailable to potential buyers (for exceptions, see Cavalcante 1988; Le Conte 1922; FAO 1985; Pesce 1985), so consequently, there is a lack of markets. This situation, in turn, generates few incentives for those who would like to expand their extractive activities. Yet the theoretical premise for development of sustainably extracted resources in tropical forests still holds: through the expansion of local, national, and international markets for secondary forest products, a sound economic as well as ecologic argument can be made for harvesting from existing forests rather than clearing them for alternative economic uses.

The Marketing Solution

Cultural Survival, a Harvard University based nonprofit organization that has been working with indigenous peoples since 1972, has decided that the best way to protect the future viability of forest communities and to ensure the future of the forests that they depend upon is to expand the market for sustainably harvested forest commodities (Booth 1989; Christianson 1989; McCabe 1989; Reuter News Service 1989; Petit 1989; Seibert 1989). It is time to stop talking about how valuable such products might be and begin proving it in the marketplace (Clay 1989a; Clay 1991a; Clay 1991b; CGBD 1989). By helping communities organize to expand production and diversify income, we will also help them strengthen the community-level organizations that are essential for the defense of their lands, forests, and ways of life.

Cultural Survival came to this conclusion reluctantly, however. For the last decade most of the organization's activities (and budget) focused on land rights, local organization, and resource management projects. Recently, we have seen that without income, indigenous peoples and other long-term forest residents have little chance of defending themselves, their way of life, or the environment that they and future generations depend upon. By the same token, helping groups to produce products for markets that do not exist is a misguided activity at best and, at worst, can actually discourage the type of community organization and development that cultural survival and conserving resources depend upon. Developing markets, then, can be an essential step in the vertical integration of any human rights work with such groups.

Cultural Survival waited for others, particularly the larger environmental organizations that talk so much about the value of forests, to develop the markets that forest residents so desperately need. In fact, we avoided market-oriented activities, thinking others with more experience would surely become involved. But for all their talk, none has done so.

Cultural Survival also was concerned that market activities would divert too much of our small staff's attention to activities that are not central to the goals of the organization. Over the last few years, however, Cultural Survival has funded sustainable resource management projects with indigenous peoples that now generate commodities that could benefit from stronger prices and expanded markets. We have also created markets for our publications, which generate nearly 60 percent of the organization's total budget, allowing us to be more independent of grants from governments, multilateral agencies, or foundations. It is this same ability to generate funds without depending on outside sources of income (or the values and priorities that are often attached to such grants) that we encourage among the indigenous communities that we support. Appropriate, replicable development in such communities rarely results from significant international aid. To put it another way, lack of grant money is not usually the problem.

We decided to begin trading in forest commodities as a nonprofit entity because helping remote forest dwellers enter national and international economic and political systems on their own terms is part of our charter as a nonprofit organization. We believe that we can serve a useful purpose not only by helping indigenous peoples to enter the market more on their own terms or to expand the market in which they participate, but also by helping to reshape old markets (such as the Brazil nut trade) or to create new ones (e.g., for fruits, oils, or flours) so that more benefits accrue to producers. Traditionally, nearly all the value of rain forest products has been added after they leave the hands of the forest residents.

During the 1989 Brazil nut harvest, for example, remote nut collectors in western Brazil were paid only three to four cents per pound for unshelled nuts, while shelled whole nuts in New York were fetching between $1.70 and $2.20 per pound. What must be done is to eliminate some of the intermediaries en route between forest residents to manufacturers. This can be accomplished by strengthening indigenous cooperatives and organizations so that they can purchase their members' produce. Producer groups add value by transporting produce further through the market and, in some instances, by processing the raw products themselves. This will allow the groups to garner a larger percentage of the final price. In the case of Brazil nuts, producers will receive from three to four times the current price for their unprocessed nuts by merely transporting them to the factory gate.

During the first five to ten years of the project we will not be able to buy significant quantities of shelled, vacuum-packed, and nitrogen-flushed nuts directly from the producers. Instead, we will buy them directly from the factory to which they sell. We will, however, be able to net about 10 percent of the New York wholesale price, which we will return to the producers for projects that they design and run. During the first six months of trading, for example, Cultural Survival generated enough income to finance the construction of the first harvester-owned nut-shelling factory in the Brazilian Amazon. This factory has been located in the town of Xapuri, Acre, the home of slain rubber-tapper leader Chico Mendes. The factory should produce 30 to 50 metric tons of processed nuts each year and, during the off season, will be available for drying fruits, herbs, and medicinal plants for export. The first factory is seen as a prototype. Through it, we will be able to judge the costs and benefits of such factories in the forest (see La Fleur 1989 for a preliminary analysis). If it proves to be worthwhile, a number of bilateral agencies have expressed interest in financing similar factories in other parts of Brazil. One other advantage the factory offers is employment of the urban poor, thus cementing a political alliance between active forest extractors and the urban poor.

The same skills used to organize producer cooperatives also allow groups to establish consumer cooperatives, where members can be charged only a 30 percent markup for goods rather than the 1,000 percent charged by many merchants, who, as a result, never actually pay cash for produce to their permanent indebted clients. Such activities will not only generate income, purchasing power, skills, and employment but also increase the political clout of members.

Stronger grassroots organizations and a growing international market specifically for sustainably collected products coming from indigenous communities allows elimination of many intermediaries that traditionally added most to the price of rain forest commodities in New York or London. By eliminating some of these intermediaries, our trading can convert this margin into funding for projects with indigenous organizations.

From this margin Cultural Survival can guarantee producers a higher and more stable price (in dollar equivalents) for their commodities, generate income for appropriate and sustainable development projects in tropical forest areas, and provide United States and European manufacturers with a product that is comparably priced to those of alternative sources. In addition, some manufacturers have agreed to designate a percentage of their profits for rain forest conservation, land rights, and resource management projects.

During the first few years, when output is still small, we will be able to choose to supply only those companies that agree to pay higher prices to producers and/or give a percentage of their profits for projects with forest dwellers. But even when we begin to market rain forest products through more traditional companies, significant benefits will accrue to the producing groups through the creation of expanded markets alone.

Another benefit of marketing rain forest products is that it will generate employment not only for forest residents, but also in the villages and cities in the rain forest areas. Few realize that the majority of people who live in rain forest areas such as the Amazon live in cities. The local employment generated through transport, trade, processing, and manufacturing will help these people realize that their futures depend upon the future of the forests. Just as the economic and political position of forest residents will be enhanced through an expanding market for their products, so too will be the position of urban residents. And, importantly, the possibility of a political alliance between urban and rural residents ultimately could prove essential in generating legislation and policies to protect tropical forests.

Perhaps the most important role of Cultural Survival's efforts to market rain forest products, however, is at least to demonstrate the potential to other producers, exporters, international companies, and governments of making significant money through the maintenance of tropical forests and the marketing of sustainably collected products. (Other reports support this projection, see, for example, Little 1990; *The Economist* 1989a; *The Economist* 1989b; Freitag 1989; Larson 1989; Schwadel 1989; Tye 1989). There are already indications that bilateral and multilateral development agencies will begin to redirect at least some of their development efforts in tropical forest areas. If other nonprofit organizations, particularly environmental or indigenous rights groups, become involved in similar efforts, an even greater number of groups will benefit.

Marketing—The First Year

During the first year, investigations have focused on the rain forest products already for sale in the Amazon that might have new or expanded markets in their country of origin, in the United States, or in

Europe. Most people realize that tropical forest products rarely make their way into United States or European markets. Few realize, however, that each country with tropical forests has large internal markets for materials which easily could be, but usually are not, supplied by their own forests. In Brazil, for example, most people in São Paulo and Rio have never heard of, much less eaten, the most common fruits from the Amazon. This is a logical place to increase the marketing of such items. Expanded markets inside the countries of origin could help to create national support for the protection of fragile, local resources, and for fragile indigenous and environmental organizations as well. Expanded markets outside the country of origin, at least initially, however, will in all likelihood make most rain forest products more culturally acceptable within the producing country and, as a result of the overall expansion of markets and production, render the products more valuable and readily available as well.

During the first year we have focused most of our attention on the Brazilian Amazon, but we have begun to collect samples of products from other Amazonian countries, Southeast Asia, Central America, and Africa. Initially we work with groups that we have supported in the past, gradually incorporating other groups that we have not worked with before. Our initial research efforts focus on products that are already being collected or produced in quantities sufficient for export from the region. Product samples are then sent to interested manufacturers in order for them to experiment with the new products. No orders are placed with local groups (or expectations raised) until actual orders are in hand. Once we receive a shortlist of items that interest companies, and a general idea of quantities desired, then we approach indigenous groups to see if such quantities are available and ecologists and botanists to see if (or how) such quantities can be harvested sustainably.

We have collected a wide variety of product samples, (more than three hundred in the first eight months) including nuts, fruits, oils, flours, fragrances, tubers, spices, fiber, and medicinal plants. With fruits, for example, the likely new markets are products that could be flavored rather than made entirely with an imported rain forest fruit (e.g., ice cream, yogurt, milk, candy, soda, snacks, wine coolers, soaps, shampoos, creams, or various natural food products). We are currently in touch with more than thirty companies in these categories that have already received forest product samples and are conducting tests. Fruits that are being tested initially are *abrico, açaí, bacaba, bacuri,* cacao (chocolate), cashew, *cupuacu,* custard apple, *graviola,* guava, jackfruit, mango, passion fruit, and *patoa.* We are also experimenting with expanded markets for items (e.g., coffee, cacao, vanilla, and allspice) produced in agroforestry production systems—where native hardwoods are interplanted with other tree crops—on reclaimed deforested or degraded lands.

The Market Demand

Cultural Survival has been approached by progressive manufacturers in the United States and Europe who wish to purchase "politically correct" rain forest products. Their desire is to locate products that they can incorporate into or use as substitutes in current products, or around which they can create new product lines. These manufacturers share a common concern with stimulating extractive production through direct purchases; including public education material on their packaging about the product ingredients, the rain forests, and the people who live in them and collect the food items; donating a portion of their profits back to forest-dwelling groups for sustainable resource management projects; and providing consumers with appealing, competitively priced products. Of course, they also want to make a profit. To date, such companies have expressed interest in making a profit rather than a killing.

It is clear that such companies could offer the most advantageous entry into the market for forest dwellers. Because of their concern about forests, they are more patient with the development of controls on product quality and quantity. They are also more sensitive to the cash flow problems inherent in the project during its initial stages. Eventually, however, if the marketing of tropical forest products is to expand significantly, mainstream companies who have no direct political interest in tropical forest issues will probably have to become involved. Even so, in the short term demand for rain forest products far exceeds supply. One company wants to purchase more Brazil nuts than the rubber tappers in Acre currently collect, much less shell and process. Another company has placed an order for copaiba oil that exceeds Brazil's total exports of this oil in 1985 (the year for which we have the most recent, official, published data).

For the most part, however, initial efforts have been aimed at determining which traditional rain forest products have the greatest market potential and how many different groups are producing them that could pool their commodities in order to collectively meet increased demand. Because the commodities are unknown, and because there is a certain amount of risk involved in bringing out new products, manufacturers (even those politically disposed toward the project) will enter the market slowly. For this reason, we will need to import a large number of samples to provide potential buyers with the widest range from which to choose. From the manufacturers' point of view, if it doesn't smell good or taste good they will be able to develop only limited markets. Other important considerations are competitive prices, guaranteed supplied, and quality control.

Some companies have already agreed to our suggestion of a marketing strategy for rain forest products that incorporates the seasonality of the rain forest into the appeal of the product. Thus the limited two-month supply of a fruit could be part of the sales pitch for marketing strategy rather than a shortcoming. Ben and Jerry's Homemade has expressed interest in an advertising strategy that would allow them to identify enough rain forest fruit flavors so that they can have a new ice cream every four to six weeks throughout the summer if necessary. The seasonality of forest fruits would be part of an informational blurb on the ice cream containers.

With the current all-time high level of interest in tropical forests, many companies are convinced that using rain forest products in existing product lines could be a wise marketing strategy. Because marketing trends shift relatively rapidly, it is important to move quickly to expand the use of rain forest–generated materials in new markets. If we can capture a small corner of the market, due in part, at least, to the current interest in tropical forests, we can create a demand/taste for forest products that will not disappear when the latest rain forest fad fades. There are many indications, however, that "green" consumerism is here to stay.

Identifying the Pitfalls

During the first year, information will be collected about current production and price levels throughout the year of various food and nonfood items that are collected sustainably. The marketing project will focus on materials that are not cultivated, so that an expansion of markets would lead to increased value for standing forests. It is quite likely, however, that existing forests will be "upgraded," e.g., more valuable species will be planted in greater density within forests over time, reflecting increased demand for a given product. An issue that will be researched within the context of this project is the extent to which upgrading can take place without irrevocably affecting genetic diversity. Since the use of a trademark ensures that the product is produced in a sustainable resource management system, we will be monitoring this issue closely in the areas where we purchase products. It is possible, however, that the demand generated through this project could contribute to destructive upgrading or even plantation agriculture in other areas, particularly in the case of a few valuable commodities. If this happens, Cultural Survival will have to expose to potential consumers those who carry such products. And, finally, it is also likely that degraded forests, or even areas converted to pasture or agricultural use, might be replanted with more valuable tree species as a result of expanding markets.

This raises two other important issues about the philosophy that undergirds the project. First, we are willing to purchase products from degraded or deforested areas that have been replanted to tree crops including, preferably, more than two different species. It is our belief that assisting in the replanting of such areas through market incentives far outweighs the benefits of trading only in products from more pristine rain forests. We would not, however, trade in commodities that are grown in areas deforested specifically for that purpose.

Second, Cultural Survival will not trade in tropical timber, at least initially, even from indigenous or long-established peasant communities living in forest areas. We do not believe that enough is known about tropical forest regeneration to make guaranteed claims that hardwoods, for example, are sustainably harvested. We do not know of a single example of tropical forests being logged sustainably, even though there are numerous claims for this feat. As more research is undertaken and more information is generated and a more defensible position becomes generally accepted, we might, of course, reconsider this policy.

Another concern with adding value to rain forest products in the forest itself is that local processing of commodities not result in environmental degradation (e.g., forest destruction through the undue use of fuel [wood] or electricity generated by hydroelectric dams built in rain forests).

We will also monitor the effect that increased marketing of an item has on the nutrition and health of local rural and urban populations. Clearly, all research must be undertaken on such issues as long as production for external markets continues, but no single organization can coordinate, much less undertake, all the necessary research. There is considerable room for collaboration on such research topics with other individuals and organizations, both NGOs and academic. However, the rain forest residents whose future depends on the accuracy of the findings must be incorporated into the research teams as quickly as possible.

Research undertaken during the first year has indicated where large holes in our knowledge exist and where additional information is still needed, even for items already traded in large quantities. In subsequent years, as other marketable commodities are identified, we will identify additional research needs. Much of the necessary information is already known to local producers, researchers, and food processors, but it needs to be brought together and examined in light of possible market expansion. Our own project and research staff will need to be expanded, even though we will not be able to undertake most of the research ourselves. It will have to come from academics and organizations who have been working on these issues for years and who want their information to be used by a wider audience. We have, for example, long drawn on a network of some 2,500 social scientists and environmentalists, who will also be called upon more systematically to help us with this research or to have their postgraduate students undertake research that will ultimately prove vital to the sustainable use of the world's rain forests.

Finally, another aspect of the project that will be closely monitored is its overall impact on producer groups. While we have never encountered any indigenous peoples that are completely outside the market economy (or even any who *want* to be outside it), we realize that such a project could radically change values and priorities, causing considerable chaos and conflict. Still, if people are allowed the space and the time to make the choices that affect their lives, their decisions will have a better chance of success. This project at least allows people to enter the market more on their own terms than ever before. Ultimately, they will argue among themselves and decide what can be changed and what can't. We have found that groups change all the time and if they have the power to choose, they are willing to live with their decisions. It must be kept in mind, however, that very few people in the world want to live exactly as their ancestors did.

How the Markets Work—Two Examples

How the development of markets for rain forest products works is perhaps best understood by a quick review of our efforts to date with three companies, Community Products and Ben and Jerry's Homemade of the United States, and The Body Shop of England. Ben Cohen, founder of Community Products and cofounder of Ben and Jerry's, after expressing enthusiasm for the project and performing initial tests on samples, ordered Brazil nuts and cashew nuts for Rainforest Crunch, a nut brittle Community Products is manufacturing and selling both as a candy and as a flavoring that Ben and Jerry's can mix into vanilla ice cream. Since Brazil nuts are collected in the forests, any market expansion would increase demand for them and thus be a compelling economic reason to leave the trees (and the forests) standing.

Community Products began to manufacture the nut brittle in September. By October, Ben and Jerry's began to sell the ice cream in 2.5 gallon containers. It was manufactured in pint containers in late February. The candy was an instant success, back-ordered weeks before Christmas. Community Products gives 40 percent of all profits from the sale of the brittle for rain forest conservation programs, notably sustainable resource management projects. In addition, high school and college groups sold the candy (along with raw Brazil and cashew nuts) during International Rainforest Week and door-to-door before the holidays to earn money for projects in Brazil. Again 40 percent of the retail price was earmarked for projects in Brazil. About $25,000 was raised in this manner. One donor agreed to match all funds raised for projects in Brazil through such efforts.

It is not our intent, or Ben and Jerry's wish, to subsidize through project funding organizations whose produce is being marketed already but rather to fund projects with other groups to help them enter the global market on more advantageous terms. For this reason, profits earmarked for projects will be continuously directed either at new groups or at new product marketing or processing systems.

Ben and Jerry's has also asked Cultural Survival to buy rain forest fruits for a new sorbet that they would like to market in 1990 to benefit

Indians and/or rubber tappers. They are testing seven fruits now but will need some 50,000 pounds of canned or dried fruit to supply the demand for one summer. (Fruit ice creams are in high demand in the United States only during the summer months.) It is doubtful that such quantities can be treated before 1991.

While Community Products and Ben and Jerry's have agreed to purchase significant quantities of produce, they will not advance the money for their purchases. Consequently, large sums of money will be tied up for a few months, the amount of time required to transport the products by boat from the producers to the final destination in Vermont. Thus cash flow is one of the initial bottlenecks for the project.

Ben and Jerry's have agreed, however, to exhibit sample food items imported by us or manufactured by other small, committed manufacturers in their regular marketing displays at national food shows throughout the year and to introduce us to other progressive companies that might want to use some sustainably harvested forest materials in their own products.

The Body Shops, a company with more than 450 stores in thirty-three countries, produces cosmetics, lotions, and sundry other items for the body; it is starting a new rain forest line of cosmetics (lotions, soaps, shampoos, massage oils, and scrubbers) and promoting the issue with its customers. The Body Shops is approaching the project in a slightly different way, however. The company believes in paying Third World producers First World prices for their products. Thus producers would receive substantially higher prices initially, but no support for the projects. With higher incomes and guaranteed markets, groups are expected to identify and fund their own development projects.

The Body Shops has received nearly three hundred potential products and is presently experimenting with thirty flours, oils, nuts, fibers, pigments, fragrances, soaps, and fruits. Unlike Ben and Jerry's, The Body Shops has agreed to advance the money needed to purchase products for them at the source. It has also agreed to accept whatever quantities we are able to obtain initially, to help us regularize and increase marketable produce in one or several areas as producer organizations increase in number, size, and organizational capacity. They are committed to a two- to three-year period of regularizing product quantity and quality.

In general, during the first few years, we will look for products that are already being produced by groups that could benefit from having a more direct link to a foreign manufacturer and that already have an international market that can be expanded. This approach will allow us to create a substantial volume in a short period of time that will in turn generate sufficient income to enable us to begin work with less well organized producers and products that do not currently have an international market.

We will also be focusing mostly on products that come directly from undisturbed forests. However, we realize that colonists who have already moved into some areas and cut trees must be organized to use these areas sustainably and productively, or they will move on to clear new areas. For this reason, we will also work with peasant farmers and indigenous peoples who are establishing systems of agroforestry on degraded lands. We will not, however, work with groups who cut forest in order to claim that the areas are degraded and then receive project assistance. We in no way wish to create a demand for products that would lead to the clearing of forests to create plantations of more valuable trees.

By the end of 1990, at least twelve companies will be selling rain forest product lines. Many more, however, will be testing samples and determining if it is possible to produce a commodity given the current availability and quality of forest products. It will take time to begin to expand the marketing of other nonfood items, as well as to identify products from other countries.

In the final analysis, creating markets for rain forest products requires addressing the issues of ecological sustainability, producer interest, and "green" consumer demand in the West simultaneously. If this marketing approach begins to show signs of success in the Amazon, there is no reason we can't take the show on the road.

LITERATURE CITED

Arthur, D. Little. 1990. Packaging in Today's Environmental Climate. ADL Center for Produce Development and Center for Environmental Assurance. Reference 90492–01, Cambridge, Mass.

Booth, W. 1989. Study offers hope for rain forests: Balance of development and conservation proposed. *Washington Post,* (June 29): A1 and A22.

Cavalcante, P. B. 1988. *Frutas comestiveis da Amazonia.* Belém, Brazil: Museu Paraense Emílio Goeldi.

Centro Ecunêmico de Documentação e Informação (CEDI). 1989. Amazonia. *Tempo e Presence* vol. 11, no. 244–245.

Christianson, J. 1989. Getting the scoop on the rain forest. *St. Louis Post Dispatch* (July 11).

Clay, J. W. 1988. *Indigenous Peoples and Tropical Forests: Models of Land Use and Management from Latin America.* Cambridge, Mass.: Cultural Survival.

—— 1989a. *Rainforest Marketing Project Proposal.* Cambridge, Mass.: Cultural Survival.

—— 1989b. Radios in the rain forest: Instead of being destroyed by outsiders' technologies, some tribal groups are using them to maintain their cultures. *Technology Review* 92(7):52–57.

—— 1990. Guest column: A rain forest emporium. *Garden* 14(1):2–7.

—— 1991a. Building and supplying markets for non-wood tropical forest products. Paper presented at February meeting of American Association for the Advancement of Sciences.

—— 1991b. Some general principles and strategies for developing markets in North America and Europe for non-timber forest products. A talk presented to "Alternativas Economicas para as reservas Extractivistas" Feb. 1991.

Consultative Group on Biological Diversity (CGBD). 1989. *Marketing Non-Timber Tropical Forest Products: Prospects and Promise.* Workshop Report. Cambridge, Mass.: Cultural Survival.

Dean, W. 1987. *Brazil and the Struggle for Rubber.* New York: Cambridge University Press.

The Ecologist. 1989. Amazonia: The future in the balance *The Ecologist* 19(6):223–227.

The Economist. 1989a. Rain forest products: Growing profits. *The Economist* (September 9–15).

—— 1989b. The month Amazonia burns. *The Economist* (September 9–15):000–000.

—— 1989c. The perils of greening business. *The Economist* (October 14).

Food and Agriculture Organization (FAO). 1985. *Food and Fruit-Bearing Forest Species: Three Examples from Latin America.* Rome: United Nations.

Food and Drug Administration (FDA). n.d. *Requirements of Laws and Regulations Enforced by the US Food and Drug Administration.* Health and Human Services Publication No. (FDA) 85-1115. Washington, D.C.: Government Printing Office.

Freitag, M. 1989. Luring "green" consumers. *New York Times* (August 6).

Le Conte, P. 1992. *L'Amazone Bresilienne.* Paris: Augustun Challand.

Le Fleur, J. R. 1989. *Projector do apoio ao desenvolvimento da Cooperativa Agroestraivista de Xapuri Ltda.* Xapuri, Acre, and Recife, Pernambuco. July.

Larson, E. 1989. Strange fruits: How Frieda's Finest established a brand name in an industry that had never had one. *Inc.* (November).

Mahar, D. J. 1989. *Government Policies and Deforestation in Brazil's Amazon Region.* World Bank, World Wildlife Fund, Washington, D.C.: and The Conservation Foundation.

Manchete. 1989. Amazonia. *O Desafio do Futuro.* Special Edition. September.

McCabe, M. 1989. Marketing the rain forests: Renewable products vs. short-term profits. *San Francisco Chronicle* (Oct. 1, 1989), p. 21.

Panayotou, T. n.d. Maximizing the new present value of tropical forest: The role of nontimber forest products. Typescript.

Parfit, M. 1989. Whose hands will shape the future of the Amazon's green mansions? *Smithsonian* (November).

Pesce, C. [1941] 1985. *Oil Palms and Other Seeds of the Amazon.* Translated and edited by Dennis V. Johnson. Algonac, Mich.: Reference Publications.

Peters, C. M., A. H. Gentry, and R. O. Mendelsohn. 1989. Valuation of an Amazonian rainforest. *Nature* 339 (29 June):655–656.

Peters, C. M. and E. J. Hammond. 1989. Fruits from the flooded forests of Peruvian Amazonia: Yield estimates for natural populations of three promising species. *Advances in Economic Botany* 8:159–176.

Petit, C. 1989. Foreigners hope to save the rain forest by marketing it. *San Francisco Chronicle* (September 28) A1.

Reid, W. V. and K. R. Miller. 1989. *Keeping Options Alive: The Scientific Basis for Conserving Biodiversity.* Washington, D.C.: World Resources Institute.

Repetto, R. 1988. *The Forest for the Trees? Government Policies and the Misuse of Forest Resources.* Washington, D.C.: World Resources Institute.

Reuter News Service. 1989. Saving the Amazon the natural way: Using rainforest products to prevent their destruction. *Ottawa Citizen* (October 29), C6.

Schultes, R. E. 1979. The Amazonia as a source of new economic plants. *Economic Botany* 33(3):259–266.

Schwadel, F. 1989. Retailers latch on to the environment. *Wall Street Journal* (November 13).

Scientific American. 1989. Managing Planet Earth. Special Issue. September.

Seibert, S., M. Margolis, and D. Foote. 1989. Hug a tree, kiss an herb: Saving Brazil's forests. *Newsweek* (May 1) p. 50.

Time. 1989. Torching the Amazon: Can the rain forest be saved? *Time* (September 18).

Tye, L. 1989. Environmentalists map ways for consumers to help save the forests. *Boston Globe* (April 10).

Urra, J. and R. Fuller. 1989. Nondestructive investment possibilities in the Amazon rainforest. Typescript.

TWENTY-ONE

Neotropical Moist Forests: Priorities for the Next Two Decades

ROBERT J. A. GOODLAND

> The poverty of the poor—and the greed of the rich—are the greatest polluters. . . . Indira Gandhi

The Problem: Loss of Neotropical Moist Forest

The loss of biodiversity or the extinction of species is both effect and evidence of unsustainable resource use. Biodiversity is mainly being lost because tropical forests are being lost. These forests support half the world's five to thirty million species on seven percent of the world's land area. Latin America alone may support fifteen million species, possibly one third of the world's total. But these richest of the planet's terrestrial ecosystems are dwindling at the rate of 74,000 acres per day. The main causes of deforestation, cattle ranching and the shifting cultivation of forested lands by poor people deprived of other alternatives, have been identified. And the perilous consequences of the loss of biodiversity have been widely discussed recently (Ehrlich and Ehrlich 1981).

Today's extinction rates are unprecedented; the problem is urgent. Our global society has ten to thirty years left to conserve some tropical rain forest habitats. In the Philippines, Haiti, El Salvador, Sri Lanka, parts of India, and Bangladesh it is already largely too late; a major proportion of these countries' biodiversity is extinct. In addition, most extant species may be "living dead," i.e., no longer breeding or with nonviable populations (Janzen 1986). If *our* generation

The World Bank does not accept responsibility for the views expressed herein, which are those of the author and should not be attributed to the World Bank or its affiliated organizations.

does not conserve tropical forest species, they won't be there for the next.

United States–based readers should not feel that tropical deforestation is not their problem; it is. The United States' inexorable demand for tropical commodities such as rubber, beef, oil palm, cocoa, and tropical timber and veneer exerts irresistible pressure on tropical forests. The timber market alone brings in $1 billion annually.

If conversion of tropical forests created sustainable jobs or permanent benefits, it would be difficult to criticize, but it does not. Tropical forest is being destroyed for the slightest and most ephemeral gains. One of the most damaging land uses is the creation of pasture for a few years' production of minutely cheaper beef hamburgers, a major cause of forest loss in Central America.

Biologist Edward O. Wilson described the gravity of the problem cogently:

> The worst thing that can happen—will happen—in the 1980s is not energy depletion, economic collapse, limited nuclear war, or conquest by a totalitarian government. As terrible as these catastrophes would be for us, they can be repaired in a few generations. The one process ongoing in the 1980s that will take millions of years to correct is the loss of genetic and species diversity by the destruction of natural habitats. This is the folly our descendants are least likely to forgive us. (Wilson 1988)

The question for us is: What are we—not our children—going to do about all this?

What Are Not Solutions to the Loss of Neotropical Moist Forests

Solutions proposed to stem the loss of Neotropical moist forests (NMF) often include research into how NMF can be used, how to improve shifting cultivation, how shifting cultivators can progress beyond subsistence, how to correct or circumvent the poor nutrient status of many tropical soils and waters, and how to achieve sustained-yield forestry.

I have become skeptical of such approaches for two reasons. First, even if all the minutiae of NMF ecosystem structure, nutrient cycling, and linkages were elucidated and quantified, the number of people who could be supported permanently on these lands would be so small as to be irrelevant and not even interesting for politicians. Although this has been known for years and was emphasized by D. H. Janzen as long ago as 1973 in his classic *Tropical Agroecosystems,* NMF is still regarded as an underexploited resource, wasted until it is exploited. Irresistible economic incentives are still awarded by governments to exploit NMF, which then vanishes in a few years.

Second, time has run out for research. If samples of NMF are not conserved right now, there will be no raw material left to research in the near future. Exceptions to this devil's advocacy, such as the study of restoration or rehabilitation of degraded or former NMF, are discussed later in this paper. But the main assertion remains: lack of knowledge of NMF ecosystems is not the reason they are not being used sustainably.

Solutions to the Loss of NMF

Solutions to NMF loss can be grouped into four categories: 1) population management; 2) economic incentives; 3) education/political pressure; and 4) site-specific alternatives.

Population Management

Environmentalists must reinforce population management efforts; population control is the sine qua non of sustainable development for most or all NMF countries. Environmentalists first helped to identify the issue (e.g., Ehrlich's *Population Bomb* 1968) but since then have assumed that demographers and those who specialize in population questions have been pushing their agenda as much as possible. Given religious and other sensitivities, and the right-to-life and antiabortion movements, many concerned with the environment may believe that progress has been as fast as should have been expected. I no longer hold this view. Environmentalists need to renew coalitions with population activists and emphasize that family planning must be part of environment/natural resource conservation efforts. Sustainability can be achieved only when high population-growth rates fall.

In this regard, the liberation of women is essential to environmental sustainability, as well as being a key equity issue. The president of the World Bank recently emphasized that

> women do two thirds of the world's work. Their work produces 60 to 80 percent of Africa's food, 40 percent of Latin America's. Yet, they earn only one tenth of the world's income and own less than one percent of the world's property. (Conable 1987a; Conable 1987b)

Women constitute an unacceptably large proportion of the poorest of the poor and of landless laborers; in addition, they endure the burden of the double day: both job and housekeeping. Resolution of this situation lies in girls' education and in improvements in agriculture (particularly reuniting nutrition with agriculture), health and in agrarian reform (Dankelman and Davidson 1988; Goodland 1991). Bearing fewer children must be made a reasonable choice for women; at the moment, for many women it is not a choice.

Fostering a coalition between groups promoting environmental, women's, and population-management issues would greatly help to advance each of these causes and help ensure sustainability of resource use and the future of forests as well.

Economic Incentives

Incentives promoting environmental abuse and the wanton destruction of forests should be removed, while incentives promoting sustainability should be created. Abusive incentives are fairly well known (Binswanger 1989); they include tax holidays for Amazonian cattle ranchers, subsidized loans for cattle producers in Central America, and access to credit only for landowners who have "improved" (i.e., deforested) their holdings. Land speculation, often promoted by economic incentives, is at the bottom of much NMF destruction. Environmental economists are needed urgently to identify and distinguish incentives that contribute to and those that militate against sustainable use of forested lands and then to correct the prevailing system.

Creation of incentives that promote sustainable resource use is a fairly new field in which much work is needed. Integration of externalities into cost-benefit calculations is a necessary step in all project planning. The costs of tropical forest destruction—erosion, decreased fertility and dry season water, increased floods, sedimentation, species extinction, etc.—can be included in economic accounting much more than is commonly done at present.

Incentives to preserve forests such as Edward Goldsmith's rent scheme, recently endorsed by the United Nations Commission on Sustainable Development (Brundtland 1987), also need to be promoted. Goldsmith suggests that wealthy countries pay sovereign NMF owners an annual rent to conserve their forests. The rent would cease as soon as the NMF owner cuts the forest.

Two related economic mechanisms also need to be implemented widely: "debt-for-nature" swaps and "carbon dioxide sink" forests (growing trees sequester CO_2). Swaps only marginally reduce the countries' indebtedness, but they significantly boost their conservation status. The creation of carbon dioxide sink forests, funded by large-scale CO_2 emitters, also could be part of a major resource transfer from industrialized countries to areas where warm, adequately wet climates and abundant land (e.g., deforested NMF lands) are favorable for the rapid growth of carbon dioxide-consuming trees. While this scheme would not conserve biological diversity, it could relieve pressures on intact natural forest and create sustainable employment for people who are now, or who might otherwise become, shifting cultivators.

To promote these and other incentives to conservation and sustainable resource use, ecologists must learn to talk with economists and decision makers. Emphasizing the close linkages between economic and environmental concerns, World Bank Vice President Shahid Husain (1987) pointed out that "many direct governmental subsidies are unsound both in environmental and in economic terms. They add to a country's fiscal burden, encourage the wasteful use of scarce resources and frequently benefit the large landowners." For instance, Brazil's two major programs to subsidize beef cattle production in the Amazon, combined with the destruction of forest resources those programs encourage, represent a total social cost of no less than $7.2 billion.

Education/Political Pressure

The gamut of environmental interest groups, from consciousness raising, education, nature lovers, and lobbying groups through high-powered political pressure groups, has an enormous potential to promote sustainable natural resource management and conserve NMF.

Responding to pressure from environmentalists, Coca-Cola's Minute Maid orange juice subsidiary donated 40,000 acres of Belizean tropical forest for conservation, plus $50,000 toward its management to the Massachusetts Audubon Society (see Burley, this volume). Coca-Cola now seems to be the only major multinational to conserve a sample of the natural ecosystem from which it derives profit. Can other corporations that have destroyed NMF in pursuing agriculture (rubber, chocolate, pineapple, palm oil, etc.), mining, or ranching in the tropics be similarly convinced to conserve some area of forest? The success of Earth First! and the Rainforest Action Network in encouraging Burger King to eschew tropical rain forest beef attests to the possibilities.

Groups exist to make use of whatever talents and interests dedicated environmentalists may have. If your forte is golfing with chief executive officers of multinationals, then join a well-entrenched, establishment organization. Activists courageous enough to interpose their bodies between trees and chainsaws or bulldozers or between harpoons and whales will join Greenpeace, Chipko, Sea Shepherds, or Earth First! Writers, politicians, rabble-rousers, economists, media-seducers, number-crunchers, fund-raisers, and bumper sticker–stickers—all are needed. Everyone has to find his or her own tribe.

Eventually, and sooner rather than later, the nationals of tropical forest nations must manage their own resources. One highly effective action is to support conservation and environmental organizations in such nations. This includes linking advocates of jungle dwellers with advocates of tropical forest conservation and environmental conservation in general.

Site-Specific Alternatives

In most cases where NMF is being destroyed, feasible alternatives exist for the people doing the cutting. Sometimes the alternative will be found outside the NMF region. For example, landless Amazonian peasants and shifting cultivators would be better off resettled in the vast *cerrado* savannas of central Brazil. Often the alternative will be found in the NMF region but involves innovative planning. For instance, Amerindians are often the best guardians of NMF, and their incorporation in forest reserves and parks as rangers or the equivalent can often be particularly advantageous to both the native population and the park's administration and visitors (Amaru IV 1980). When unpressured by invading colonists or business interests, native peoples frequently manage resources sustainably; they also have a vast knowledge of those resources. Their incorporation in scientific and nature tourism schemes can give the forest dwellers sustainable, cash-producing employment while conserving the NMF. In addition, if rangers are paid commensurately with how many species the ecotourist sees, even if only charismatic megafauna, the resident ranger community has a strong incentive to conserve.

Plantation forests may offer an alternative in areas where logging is a major cause of deforestation. Plantations relieve pressure to cut intact natural forests and are several times more productive of commercial timber and fuel wood. Plantations create more employment than can be readily generated in natural forest and are economically profitable on their own (Goodland 1990). When the incentive of carbon dioxide sink forests is added, then tree plantations should become even more lucrative. These types of projects have already begun.[1]

Research and Studies

I emphasize again that much research and many studies already exist unread and unimplemented. The highest priority of those concerned should be the implementation of what is known and its translation into action as outlined above. As a secondary priority, several trial and pilot study topics are particularly important:

1. Restoration/rehabilitation of degraded areas;
2. Improvement of existing sustainable resource use patterns for people already in tropical forests;
3. Configuration and size of conservation units;
4. Inventory of all biota; and
5. Identification of linkages within and between ecosystems.

Restoration and Rehabilitation of Degraded Areas

Ecologist Dan Janzen's work in the Costa Rican province of Guanacaste shows that abandoned cattle pastures and ranches can be restored to an approximation of tropical forest, in this particular case a semideciduous one (pers. comm.). Janzen's work could be applied elsewhere with enormous benefit. Uhl's (1986) work in Pará, Brazil, shows how the forest edge can be encouraged to expand into heavily degraded pasture. More work along these lines is important.

Preference should be given to using degraded areas for development projects and tree plantations. How this should be done also deserves further research, although the basics are already known.

Improve Existing Sustainable Practices

Most Amerindians live sustainably in NMF already; very few other people do. Extractive reserves such as those in the Brazilian state of Acre are located in forests rich in naturally occurring rubber and Brazil nut trees, which are tapped and harvested. Residents of these reserves augment the incomes they receive from rubber and nuts by collecting still other forest products, including fish. Extractive reserves merit research: their carrying capacity should be estimated, as should their enrichment potential. Raising indigenous biota (e.g., caimans, turtles, palm hearts, agoutis, pacas, orchids, or iguanas) in these managed forests may also prove profitable. Much can be learned from Amerindians in this regard.

In emphasizing and improving the riches that can be extracted from the forest, care must be taken to avoid attracting more people from outside the NMF areas. Research should aim at improving living standards of people already inside NMF and accommodating their population increase.

Conservation Units

Where to place conservation units in order to conserve the most biodiversity needs investigation. The optimal size and shape or general configuration of effective reserves require applied research. For a further discussion of the subject, see Appendix D of Ledec and Goodland (1988).

Biotic Inventory

Although we must intensify the struggle to prevent species extinction, realistically, much loss is unavoidable. Therefore major emphasis should be placed on taxonomic inventory of tropical forest species and their collection and conservation, dead or alive (preferably the latter), before they disappear. Peter Raven (1986) of the Missouri Botanical Garden is a leader in stepping up tropical taxonomy; he emphasizes the need to train more taxonomists, as well as to increase funding of collections and their curation.

Identification of Linkages

Research into linkages between NMF and other biotypes is often closely related to issues raised in investigating placement of conservation units. Janzen (1986) found that biota in seasonal forest closely depends on often distant moister forest for survival. Such linkages are likely to be common, and their elucidation is a prerequisite for sustainable conservation. Seasonal migration corridors between dry and moist ecosystems need also to be identified if conservation efforts are to succeed.

Six Implications of Sustainability

A primary goal of sustainable development is to achieve a reasonable and equitably distributed level of economic well-being that can be perpetuated continually for many human generations. Sustainability implies a transition away from economic growth based on depletion of nonrenewable resource stocks (e.g., oil) and toward progress (i.e., improvement in the quality of life) based more on renewable resources over the long run.

Sustainability has six major implications:

First, well-being depends on three categories of value:

1. Economic efficiency;
2. Equitable distribution of economic resources; and
3. Noneconomic, including religious and spiritual values, dignity, esteem, esthetics, and liberty (Shihata 1988; Taylor 1986).

Development efforts should seek optimal compromises among these three values, since some trade-offs are inevitable.

Second, although it is impossible to predict with precision the likely interests of future generations, it is prudent to assume their need for natural resources (soil, air, water, forests, fisheries, biodiversity, energy, minerals) will not be markedly less than ours. Therefore, sustainable development implies using renewable resources in a manner that does not degrade them. It implies harvesting such resources on a sustained-yield basis rather than "mining" them to near extinction. Whales, tropical forests, and coral reefs are examples of renewable resources that are often mined rather than harvested sustainably. High discount rates (commonly 10 percent and more) discourage long-term investments and promote short-term investments.[2] For such reasons it is difficult, for example, to invest in forestry in developing countries.

Third, sustainable development implies using nonrenewable (exhaustible) mineral reserves in a manner that does not unnecessarily preclude easy access to them by future generations. It will be easier in the future to use today's scrap metal if it is recycled rather than dumped in a dispersed manner.

Fourth, sustainable development also implies depleting nonrenewable energy resources at a rate slow enough to ensure a high probability of an orderly societal transition to renewable energy sources (including solar, biomass, wind, and hydroelectric energy) when nonrenewable energy becomes substantially more costly. Goldemberg et al. (1988) show that the transition to renewable sources is practical, affordable, and feasible now, without any technological breakthroughs.

Fifth, agricultural sustainability implies the permanent maintenance of biological productivity on the site, with the (nonsubsidized) costs of imported inputs (e.g., diesel, biocides) and nutrients (e.g., fertilizer) not exceeding the commercial value of the site's production. Sustainability needs resilience, the ability to return to an initial state; resistance to, or the ability to cope with, change (such as the addition or reduction of species or weather cycles); and persistence, which means that the agroecosystem succeeds for a long time relative to the life spans of its components (Conway 1985). Agricultural sustainability can be facilitated by decreasing current emphasis on the export of primary commodities, increasing processing and value added (e.g., export furniture and tires rather than logs or crude rubber), diversification of exports, efficient import substitution, shortening supply lines, and some measure of food self-sufficiency akin to the well accepted goal of energy self-sufficiency. This is important for developing countries because the index of "real" (purchasing power) prices of nonfuel commodities was lower in 1986 than at any time since 1870, when comparable data first became available.

Sixth, not all projects can be sustainable; all nonrenewable resource extraction such as oil drilling and iron mining, for example, is by definition nonsustainable. However, even in such endeavors sustainability can be attained by linking them with complementary projects that are sustainable. A portion of the receipts from the nonrenewable extraction can be invested annually in a renewable asset. The growth rate of the renewable asset would then provide a permanent source of income that could be adjusted to equal that part of the receipts of the latter that are consumed (El Serafy 1989; Daly 1991).

Economist Herman Daly (1991) further recommended that if projects are to be sustainable, then it is inappropriate to calculate the benefits of such projects or alternatives by comparing them with an unsustainable option. That is, it is inappropriate to use a discount rate that reflects rates of return on mainly unsustainable uses of capital. For example, if a managed forest can yield 4 percent and is judged uneconomic in comparison with a discount rate of 6 percent for a project that is unsustainable (perhaps even the clear-felling of the same forest), then the decision boils down to sustainable versus unsustainable use. If our policy is sustainable development, then we choose the sustainable course; the fact that it has a negative present value at an unsustainable discount rate is irrelevant. The discount rate must reflect the return on alternative sustainable uses of capital if sustainability includes efficient allocation. The efficiency allocation rule (maximize present value) cannot be allowed to subvert the sustainability of development by application of an unsustainable discount rate.

Urgent Actions for the Near Future

These sustainability considerations are not new. As individuals, the more concerned and thoughtful already use mass transit, drive smaller cars, conserve energy, recycle waste, provide for their descendants, and behave thriftily. As a society, though, we allow ourselves and our globe to be operated on a system of "empty world" or frontier economics created when the world was essentially empty of people, when resources were scarcely tapped, when air and water were clean, and when land and forests were abundant.

That idyllic era has long since vanished, but the economic orthodoxy remains to be modernized to reflect today's world, which is full (or overfull). Possibly the single most influential change toward sustainability is to revamp neoclassical economics to accommodate environmental sustainability, including ethical and nonanthropocentric values. (Daly 1984; Daly 1985; Daly 1987; Daly 1989; Warford 1986; Soderbaum 1986; Randall 1981; Goodland and Ledec 1987; Jansson 1984).

As a society, it is both important and urgent that we start to approach sustainability now. Such an approach will be painful and arduous. However, the costs of inaction are intolerable, and the costs of postponing action soar daily. Many feel that for some values, in some places, it may be too late. Indeed, large parts of our planet already have exceeded their carrying capacity.

NOTES

1. In October 1988 one modest coal-fired utility, Applied Energy Services of Arlington, Virginia, internalized voluntarily and for the first time what is arguably the world's most pervasive negative externality: carbon dioxide emissions. AES estimated how much carbon dioxide (387,000 tons) would be emitted by its new 180 megawatt Uncasville, Connecticut, coal utility over its estimated forty-year life. Then AES calculated how many trees should be planted to absorb that amount of carbon dioxide, above those likely to be used for fuelwood and other needs. AES, USAID, CARE, and the Peace Corps then agreed with the government of Guatemala to plant about 52 million trees over a ten-year period; 385 square miles of woodlot, agroforestry, and fire-protection zones eventually would result. (Two square miles of forest absorb one megawatt's worth of carbon emissions; 2 million hectares of new forest will absorb about 1 gigaton per year of carbon; 1,000 megawatts therefore need 2,500 square miles of forest.) This suggests that greenhouse gas emissions could be used, voluntarily or by taxation, to subsidize the sink function of the environment for their absorption. Carbon dioxide sink forests cannot solve the greenhouse problem, but they can certainly buy time for its solution.

2. An example may help illustrate: If a whaler can make a 15 percent per annum profit by exterminating whales over ten years, then invest the proceeds in a different sector, what economic incentive is there to make a 10 percent per annum profit by harvesting whales? In fact, the unquantified value of whale survival (expressed in political pressures) is high enough to outweigh the effects of discounting—otherwise all governments, bar two or three, would not have agreed to limit whaling.

Appendix 21.1

The Hamburger Connection

If cattle gain 50 kilograms per hectare per year and are slaughtered after four years (but let's assume eight years); and half their weight is nonmeat (skin, bones, etc.), then each cow produces 200 kilograms or 1,600 hamburgers. If one hectare of tropical moist forest weighs 800 tons, then 800 tons = 0.5 tons of forest per hamburger. At 10,000 square meters per hectare – 1,600 = 6.25 square meters of forest per hamburger. After ten years, this means $3 per hectare per year. In summary, one hamburger translates into 0.5 ton of forest. The cumulative effect of such hamburger consumption is equivalent to millions of years of evolution and to thousands of species: is it worth it?

Conversion of all of Amazonia (4 million square kilometers) to cattle pasture would produce (1,600 hamburgers per hectare for 8 years) 6.4 × 10″ hamburgers, enough for three burgers per day for the world's (five billion) population for 43 days (6.4 × 10″ per 5 billion = 128 burgers per person 13 = 43 days). Would the irreversible loss of the Amazon forest for a month of hamburger for the world's population be a worthwhile trade-off?

	Total Cost $ Billions	Cost per Hamburger
SUDAM Tax Credit Programs	1.2	0.135
Central Bank Rural Credits	1.5	0.083
Timber Destruction	4.5	0.524
Total Social Cost	7.2	0.742

For every quarter-pound unit of Amazon beef that costs Brazilian ranchers $0.26 to produce, Brazil absorbs $0.742 in social costs. Stated differently, one metric ton of Amazon beef (before processing) embodies about $6,500 in social costs (of which subsidies alone represent $1,900). In stark contrast, between 1971 and 1983, Brazil paid an average import price for foreign produced and processed beef of $1,086 per metric ton. In other words, the Brazilian government could have acquired nearly two metric tons of foreign beef for the same amount of capital it spent subsidizing the production of one metric ton of Amazon beef (Browder 1988).

Source: C. Uhl 1986, C. Uhl and G. Parker 1986, and J. Browder 1988.)

LITERATURE CITED

Amaru IV. 1980. *The Once and Future Resource Manager.* Washington, D.C.: World Wildlife Fund.

Binswanger, H. P. 1989. Fiscal and legal incentives with environmental effects on the Brazilian Amazon. World Bank discussion paper ARU69. Washington, D.C.

Browder, J. O. 1989. The social costs of rain forest destruction: The hamburger debate. Blacksburg, Va.: Virginia Polytechnic Inst.

Brown, B. J., et al. 1987. Global sustainability: Toward definition. *Environmental Management* 11(6):713–719.

Brundtland, G. H. 1987. *Our Common Future.* Oxford, England: United Nations World Commission on Environment and Development. Oxford University Press.

Caufield, C. 1985. *In the Rainforest.* New York: Alfred A. Knopf.

Conable, B. 1987a. *Address to the World Resource Institute.* World Bank. Washington, D.C.

—— 1987b. Giving the environment its due at the World Bank. *Environmental Protection Agency Journal* (Sept.).

Conway, G. R. 1985. Agroecosystem analysis. *Agricultural Administration* 20:31–55.

Daly, H. E. 1984. Alternative strategies for integrating economics and ecology. In A. M. Jansson, ed., *Integration of Economy and Ecology,* 19–30. Stockholm: Wallenberg Foundation.

—— 1985. The circular flow of exchange value and the linear throughput of matter-energy: A case of misplaced concreteness. *Review of Social Economy* 42(3):279–297.

—— 1987. *The Steady-State Economy: Alternative to Growthmania.* Population-Environment Balance. Washington, D.C.

—— 1987. Filters against folly: The impossible, the undesirable and the uneconomic. *Environmental Economics* 1(1):5–9.

—— 1991. Sustainable development: From concept to theory towards operational principles. *Population and Development Review* 78(1):31–50.

Dankelman, I. and J. Davidson. 1988. *Women and Environment in the Third World.* London: Earthscan.

Davidson, J. 1985. *Economic Use of Tropical Moist Forest.* International Union for the Conservation of Nature, Commission on Ecology Paper 9. Gland, Switzerland.

Ehrlich, P. R. 1968. *The Population Bomb.* New York: Ballantine Books.

Ehrlich, P. R., and A. H. Ehrlich. 1981. *"Population Bomb": Extinction: the Causes and Consequences of the Disappearance of Species.* New York: Random House.

El Serafy, S. 1989. The proper calculation of income from depletable natural resources. In Y. Ahmad, and S. El Serafy, eds., *Environmental Accounting.* World Bank. Washington, D.C.

Fearnside, P. M. 1988. Deforestation and international economic development projects in Brazilian's Amazonia. *Conservation Biology* 32(2):18–29.

Fearnside, P. M. and G. de L. Ferreira, 1984. Roads in Rondônia: Highway construction and the farce of unprotected reserves in Brazil's Amazon forest. *Environmental Conservation* 11(4):297–360.

Friends of the Earth. 1987. *A Hard Wood Story: Europe's Involvement in the Tropical Timber Trade.* London: Friends of the Earth.

Friends of the Earth with National Association of Retail Furnishers, 1987. *The Good Wood Guide.* London: Friends of the Earth.

Foreman, D. and B. Haywood. 1987. *Ecodefense: A Field Guide to Monkeywrenching.* Tucson, Ariz.: Ned Ludd Books.

Fornos, W. 1987. *Gaining People, Losing Ground: A Blueprint for Stabilizing World Population.* The Population Institute. Washington, D.C.

Ginsberg, M. 1953. *The Idea of Progress: A Revaluation.* London: Methuen.

Goldemberg, J., T. B. Johansson, A. K. Reddy, and R. H. Williams. 1988. *Energy for a Sustainable World.* New Delhi, India: Wiley.

Goodland, R. J. A. 1988. A major new opportunity to finance biodiversity preservation. In E. O. Wilson, ed., *Biodiversity.* Washington, D.C.: Smithsonian Institution Press.

—— 1990. Tropical moist forest management: The urgency of transition to sustainability. *Environmental Conservation* 17(4): 303–318.

—— 1991. Tropical deforestation. Washington, D.C., World Bank, Environment Dept., Tech. Paper 43.

Goodland, R. and J. Bookman. 1977. Can Amazonia survive its highways? *The Ecologist* 7(10):376–380.

Goodland, R. and H. S. Irwin. 1975. *Amazon Jungle: Green Hell to Red Desert? An Ecological Discussion of the Environmental Impact of the Highway Development Program in the Amazon Basin*. New York: Elsevier Scientific.

Goodland, R. and G. Ledec. 1987. Neoclassical economics and principles of sustainable development. In R. Costanza and H. E. Daly, eds., *Ecological Economics*, 19–46. Amsterdam, The Netherlands: Elsevier.

Grainger, A. 1987. The future of the tropical moist forest. Resources for the Future. Washington, D.C. Typescript.

Gradwohl, J. and R. Greenberg. 1988. *Saving the Tropical Forests*. London: Earthscan.

Hansen, S. 1988. *Debt-Equity Swaps*. World Bank. Washington, D.C.

Hardin G. and J. Baden, eds. 1977. *Managing the Commons*. Freeman: San Francisco.

Hargrove, E. C., ed. 1986. *Religion and the Environmental Crisis*. Athens, Ga.: University of Georgia Press.

Hecht, S. 1985. *Dynamics of Deforestation in the Amazon*. World Resources Institute. Washington, D.C.

Hoban, T. M. and R. O. Brooks. 1987. *Green Justice: The Environment and the Courts*. Boulder, Colo.: Westview.

Husain, S. 1987. "On the Record", Washington, D.C. The World Bank.

Janzen, D. H. 1973. Tropical agroecosystems. *Science* 182:1212–1219.

—— 1986. The future of tropical ecology. *Annual Review of Ecology and Systematics*. 17:304–324.

Jansson, A. M., ed. 1984. *Integration of Economy and Ecology: An Outlook for the Eighties*. Stockholm: Wallenberg Foundation.

Kirchner, J. W., G. Ledec, R. Goodland, and J. M. Drake. 1985. Carrying capacity population growth, and sustainable development. In D. Mahar, ed., Rapid population growth and human carrying capacity. World Bank Tech Paper 690. Washington, D.C.

Lal, R. 1986. Conversion of tropical rainforest: Agroeconomic potential and ecological consequences. *Advances in Agronomy* 39:173–264.

Ledec, G. 1985. The political economy of tropical deforestation. In H. J. Leonard, ed., *Divesting Nature's Capital: The Political Economy of Environmental Abuse in the Third World*, 179–226. New York: Holmes & Meier.

Ledec, G. and R. Goodland. 1985. Environmental implications of the carrying capacity concept. In D. Maher, ed., *Rapid Population Growth and Human Carrying Capacity: Two perspectives*. World Bank Staff Working Paper 960. Washington, D.C.

Ledec, G. and R. Goodland. 1988. Environmental perspectives on tropical land settlements. In D. A. Schumann and W. L. Partridge, eds., *The Human Ecology of Tropical Land Settlement*, Boulder, Colo.: Westview.

Ledec, G. and R. Goodland. 1988b. *Wildlands: Their Protection and Management in Economic Development*. World Bank. Washington, D.C.

Midgley, M. 1983. *Animals and Why They Matter.* Athens, Ga., University of Georgia Press.

Norton, B. G., ed. 1986. *The Preservation of Species: The Value of Biological Diversity.* Princeton, N.J.: Princeton University Press.

Pearce, D. W. 1985. Sustainable futures: Economics and the environment. Inaugural lecture, University College, London, December 5.

Pillet, G. and T. Murote, eds. 1987. *Environmental Economics: The Analysis of a Major Interface*. Geneva: R. Leingruber.

Pollock, C. 1987. *Mining Urban Wastes: The Potential for Recycling*. Worldwatch Institute Paper 76. Washington, D.C.

Randall, A. 1981. *Resource Economics: An Economic Approach to Natural Resource and Environmental Policy.* Columbus, Ohio: Grid.

—— 1986. Human preferences, economics, and the preservation of species. In B. G. Norton, ed., *The Preservation of Species,* 79–109. Princeton, N.J.: Princeton University Press.

Raven, P. H. 1986. We're killing our world. (American Association for the Advancement of Science Keynote) Missouri Botanical Garden. St. Louis, Missouri. Typescript.

Redclift, M. 1984. *Development and the Environmental Crisis.* London: Methuen. 149 p.

Regan, T. 1983. *The Case for Animal Rights*. Berkeley, Calif.: University of California Press.

—— 1987. *The Struggle for Animal Rights*. Clarks Summit, Pa.: International Society for Animal Rights.

Repetto, R. 1987. Population, resources, environment: An uncertain future. *Population Bulletin* 42(2):1–44.

Repetto, R. and M. Gillis, eds., 1988. *Public Policy and the Misuse of Forest Resources*. Cambridge, England: Cambridge University Press.

Rich, B. 1985. Multinational development banks: Their role in destroying the global environment. *The Ecologist* 15(1–2):21–38.

—— 1986. Multilateral development banks and the environment. *Ecological Law Quarterly* 12(4):681–745.

Sagoff, M. 1984. Ethics and economics in environmental law. In T. Regan, ed., *Earthbound: New Introductory Essays in Environmental Ethics,* 147–178. New York: Random House.

Schatan, J. 1987. *World Debt: Who Is Going to Pay?* London: Zed Books.

Shihata, I. F. I. 1988. The World Bank and human rights: An analysis of the legal issues and the record of achievements. *The Denver Journal* (June 3).

Singer, P. 1975. *Animal Liberation: A New Ethics for our Treatment of Animals*. New York: Avon Books.

Singer, P., ed. 1985. *In Defense of Animals*. Oxford, England: Blackwell.

Sivard, R. 1987. *World Military and Social Expenditures: 1986*. World Priorities. Washington, D.C.

Soderbaum, P. 1986. Economics, ethics and environmental problems. *Journal of Interdisciplinary Economics* 1:139–153.

Taylor, P. W. 1986. *Respect for Nature: A Theory of Environmental Ethics*. Princeton, N.J.: Princeton University Press.

Tokar, B. 1987. *The Green Alternative: Creating an Ecological Future*. San Pedro, Calif.: R. & E. Miles.

Uhl, C. 1986. Our steak in the jungle. *BioScience* 36:642.

Uhl, C. and G. Parker. 1986. Is a quarter-pound hamburger worth a half-ton of rainforest? *InterCiencia* 11(5):210.

U.S. Congress. House. 1987. *A Bill to Protect the World's Remaining Tropical Forest*. HR3010. Office of Technology Assessment. 1987. *Technologies to Maintain Biological Diversity*. Washington, D.C.: U.S. Congress.

Warford, J. J. 1986. *Natural Resource Management and Economic Development*. World Bank. Washington, D.C.

Wilson, D. 1984. *Pressure: The A to Z of Campaigning*. London: Heinemann Books.

World Bank. 1987. *Environment, Growth, Development*. World Bank, Development Committee. Washington, D.C.

World Bank. 1982. *Tribal Peoples and Economic Development: Human Ecologic Considerations*. World Bank Report No. 28. Washington, D.C.

Worldwatch Institute. 1987. *On the Brink of Extinction: Conserving Diversity of Life*. Worldwatch Institute Paper 78. Washington, D.C.

World Resources Institute. 1985. *Tropical Forest Action Plan: A Call to Action*. 2 vols. Washington, D.C.: World Resources Institute with the Food and Agriculture Organization, World Bank, and United Nations Development Program.

Index

Numbers appearing in *italics* indicate illustrations.

Buriti (Mauritia flexuosa), 177
Burkill, I. H., 309
Burley, F. William, 204–5, 421
Burseraceae, *64*, 321, 326, *338*, *339*, 350, 355
Buschbacher, R., 175, 381
Bushmaster (*Lachesis*), 74

Cabecar people, studies, 269
Cabecita (Perebea), 66
Caboclos (Amazonian peasants), 32–33, 182; *Caboclos/ribereños*, 131–33; *Caboclos/ribereños* studies, 135–50
Cacao (*Theobroma* sp.), 63, 71, 187, 407; Combu Island, 188, *192–93*; income from, 189, *189*; Ucayali floodplain cultivation, 166
Cacao del monte: *Theobroma obovatum*, 323; *Theobroma subincanum*, 331
Cachoeira extractive reserve, 391
Caesearia, 339
Café de hoja ancha (*Vismia* spp.), 328
Café del monte (*Vismia* spp.), 328
Caiman crocodilus (caiman, *lagarto*), 142; hunting of, 115–16, 124
Caimito (*Chrysophyllum cainito*), 63, *67*, 71
Caimito blanco (*Pouteria durlandii*), 331
Caimito tushma (*Sloanea* sp.), 327
Cairina moschata (Muscovy duck), *68*, 72–73
Caladium anthurium (*anturio*), *64*
Calardra palmarum (larvae), *93*
Calathea spp. (*bijao, chinula*), *66*, 72, *76*; leaves, 70
Calathea wallisii (*achira blanca, achira morada*), 329
Calatola, 356
Calyptrocarya glomenrulata (*colcha macho*), 327
Calzavara, B. B. G., 180, 187
Camino, A., 101
Camisea River, Machiguenga settlement, 86
Camitillo amarillo (*Solanea* sp.), 321
Camnosperma, 76
Camnosperma panamensis (*sajo*), *64*
Camona (*Iriartea*), 315, 323
Camonilla (*Iriartella*), 315, 323
Camote (*Ipomoea batatas*), *64*
Campa people, 307
Campbell, C., 291, 293
Campbell, D. G., 42, 44
Campesinos, Beni Biosphere Reserve, diet of, 118
Campo cerrado (scrub-grassland), 30
Caña brava (Gynerium), 65
Cana cana (*Costus* sp.), 324, 332
Cannaceae, *64*
Canna sp. (*achira*), *64*, 72
Caobo (*Platymiscium*), *65*
Capisenango (*Potalia amara*), 322, 329
Capital: access to, and agricultural choices, 169; short-term accumulation, 278
Capitalist development, and peasantry, 360
Capparidaceae, 355
Capsicum spp. (*aji*, chili peppers), 38, *67*, 71
Capuchin monkeys (*Cebus*), 88, 337
Capybaras, 142; hides of, 114; hunting of, *95*; managed populations, 256; in Yuquí habitat, 122
Carapa, 76
Carapa guianensis (*tangare*), *66*, 187, *341*
Carbohydrates, in diet, 102
Carbon dioxide emissions, 427*n*1
Carbon dioxide sink forests, 420, 422, 427*n*1
CARE, 261, 266
Careiro, land use changes, 136
Caricaceae, *64*
Carica papaya (papaya), *64*, 71; in Machiguenga diet, *95*
Caricillo morado (*Lasiacis sorghoidea*), 328
Cariseco (*Billia colombiana*), *65*
Carneiro, R. L., 28, 40, 42
Carretera Marginal (Perimeter Road), Peru, 307
Carvajal, G. de, 182
Caryocar, *343*
Caryocaraceae, 321, 326; eaten by game animals, *340*; eaten by humans, *343*
Caryocar glabrum (*almendra colorada*), 321, 326
Caryocar villosum, *340*
Cascadilla (*Pentagonia macrophylla*), 331
Cascadilla amarilla (*Pentagonia gigantifolia*), 323, 331
Cascarilla (*Cephaelis*), *67*